HVAC Pump
Handbook

Heating, Ventilating, and Air Conditioning

HVAC Pump Handbook

James B. Rishel, P.E.
Systecon Inc.
West Chester, Ohio

McGraw-Hill

New York   San Francisco   Washington, D.C.   Auckland   Bogotá
Caracas   Lisbon   London   Madrid   Mexico City   Milan
Montreal   New Delhi   San Juan   Singapore
Sydney   Tokyo   Toronto

Library of Congress Cataloging-in-Publication Data

Rishel, James B.
  HVAC pump handbook / James B. Rishel.
    p.   cm.
  Includes index.
  ISBN 0-07-053033-5 (alk. paper)
    1. Hydronics—Equipment and supplies—Handbooks, manuals,etc.
  2. Pumping machinery—Handbooks, manuals, etc.   I. Title.
  TH7478.R57   1996
  697—dc20   96-13590
                                                        CIP

McGraw-Hill

A Division of The **McGraw-Hill** Companies

1 2 3 4 5 6 7 8 9 0   DOC/DOC   9 0 1 0 9 8 7 6

ISBN 0-07-053033-5

*The sponsoring editor for this book was Robert Esposito, the editing
supervisor was Paul R. Sobel, and the production supervisor was
Donald F. Schmidt. It was set in Century Schoolbook by Ron Painter of
McGraw-Hill's Professional Book Group composition unit.*

Printed and bound by R. R. Donnelley & Sons Company.

McGraw-Hill books are available at special quantity discounts to use
as premiums and sales promotions, or for use in corporate training programs. For more information, please write to the Director of Special
Sales, McGraw-Hill, 11 West 19th Street, New York, NY 10011. Or
contact your local bookstore.

 This book is printed on recycled, acid-free paper containing a minimum of 50% recycled de-inked fiber.

This book is dedicated to my wife Alice for her patience and to all of the employees of Systecon Inc. for their knowledge and assistance

Contents

Preface

The purpose of this book is to provide a handbook on the application of pumps to heating, ventilating, and air conditioning systems, hereafter known as HVAC. It will include detailed descriptions of pump installations for all of the major HVAC systems such as hot, chilled, cooling tower, and condenser water as well as those for boilers and energy storage systems for ice and hot or chilled water.

Disclaimer: This book offers no final answers on how to design a specific HVAC water system. It has brought together technical data and, hopefully, has provided answers to particular pumping applications in this industry.

The format for this book has been developed to provide a working handbook. There may appear to be an excessive amount of cross referencing and many variations of the same formula. The reason for these inclusions is to provide rapid access to the desired subject. The water system designer who uses this book should be able to reach a pumping subject quickly without having to hunt through several chapters. Care has been taken to insure that tables, figures, equations, and pages all have different designations to reduce any confusion.

Almost all of the technical information required for applying pumps to these systems is included in this book. It is hoped that it can become a single source of information for the HVAC pumping system designer. With the advent of electronic, on-line data services for the HVAC industry, much additional information will continue to be available to the water system designer.

This handbook is being written at a time of great changes in our methods of communicating technical information. This technological revolution is probably the greatest since the invention of the printing press. Also, digital electronics is just now bringing its tremendous potential to the way we design these water systems, select equipment for them, and control the flow of water in them. Recognizing the electronic revolution that we are in, an effort has been made to point the reader toward new methods of information transmission that will become commonplace in the near future.

Another significant event in the HVAC field is the realization of the great capability of the variable speed pump in saving energy and improving the performance of water systems. So far, most variable speed pumps in the HVAC industry have been applied to hot and chilled water distribution systems. Now they are being installed on other systems such as cooling tower and condenser water applications. The ongoing increase in cost and unavailability of electrical energy, along with the continued reduction in cost of variable speed drives, will result in most HVAC pumps being variable speed during the twenty-first century.

Two great facts have thrust themselves forward as this book was prepared. These are:

1. There is so much inexactness in the data used to design water systems and their pumps. For example:
   a. What do we mean when we use the word "water"? Do we mean distilled or pure water? Or do we mean water furnished by the local water company? All of the data furnished in this book makes no reference as to what the water is when properties such as its specific gravity or viscosity are defined. It is presumed that the scientific data included pertains to pure water, but that is not what is coursing through most of the HVAC systems.
   b. Pipe and fitting friction is at best an inexact science. Hydraulic Institute estimates that the variation in the roughness factor, ϵ, can be as much as -5 to $+10\%$ for steel pipe, and the listed losses for steel and cast iron fittings can vary from ± 10 to $\pm 50\%$. We, at this writing, have absolutely no information for the friction loss through a reducing tee or through steel reducers such as a $10" \times 8"$ fitting.
   c. Pump manufacturing must have acceptable tolerances to achieve any reasonable production. These tolerances are basically -0 to $+8\%$ variation in pump head at rated flow and efficiency. Recognizing also that pumps are tested at specific suction pressures and temperatures and operated at other pressures and temperatures, it is obvious that tested pump performance is quite different from that achieved with the pump in operation on a HVAC water system.
   d. The better control of heating and cooling coils has brought a greater diversity. It is difficult for the designer to calculate accurately this diversity before the water system is placed in operation.

2. Realizing the above inexactness, the HVAC industry, in the past, resorted to balancing valves, pressure regulating valves, and complicated piping systems to destroy design overpressure and to make the systems function properly. The variable speed pump now eliminates design overflow and overpressure along with the many mechanical systems that were used in the past.

Along with these great changes has come the realization that energy conservation in HVAC water systems cannot be just the energy consumed in boilers or chillers. Great emphasis has been made on the improved efficiency of boilers and the reduced KW per ton of chillers; now it is understood that the energy consumption of pumps and central plant auxiliaries such as cooling towers must be included in an overall energy statement for chilled water plants. Coefficient of Performance and KW/Ton for the total central plant will be developed in the chapter on chillers.

With the development of digital electronics and the variable speed pump, we now have the tools to allow for the above inexactness and eliminate it in operation. We can remove much of the old mechanical complexity that was used to destroy excess pump pressure. Our motto should be the old admonition to the young engineer, KISS! KEEP IT SIMPLE STUPID!

In view of the great amount of detailed information that had to be gathered to produce this handbook, a number of people who are recognized as authorities in their field of endeavor have been called upon for their advice. They have reviewed portions of the book that pertain to their expertise and have been very helpful in the development of a HVAC pump handbook.

The following is a list of these knowledgeable people: Ahart, James R., PE, Consulting Engineer, Clayton, MO; Arnold, Charles G., PE, HDR Inc., Omaha, NE; Donley, Dallas E., ITT A-C Pump, Cincinnati, OH; Doolin, John H., Hydraulic Institute, Parsippany, NJ; Fediuck, Russell, General Electric Supply, Division of General Electric Company, Cincinnati, OH; Hanley, Allen J., M. Eng., Westmount, Quebec; Kincaid, Ben L., Keller-Rivest Inc., Indianapolis, IN; Lyons, Danial A., Systecon Inc., West Chester, OH; Nelson, Sidney W., Nelcor Inc., Cincinnati, OH; Patterson, Neil R., PE, Cincinnati, OH; Pilaar, Neil, Aerco International Inc., Northvale, NJ; Plummer, Robert W., Dean Brothers Pump Division, MET Pro Corporation, Indianapolis, IN; Staples, Lawrence S., L. S. Staples Company, Kansas City, MO; Sueker, Keith H., P.E., Consulting Engineer, Halmar Robicon Group, Pittsburgh, PA; Utesch, A. L., Cybernetic Systems Management Corporation, Argyle, TX; Ziel, Perry H., PE, Physical Plant Consultant,

Cincinnati, OH. Grateful acknowledgment is made to these engineers and authorities. This handbook would have been impossible without their assistance.

The author wishes to acknowledge also his debt to Engineering. This great profession, based upon natural law, has provided a field of work so rewarding in knowledge and personal relationships.

James B. Rishel

The Basic Tools

Digital Electronics and HVAC Pumps

1.1 Introduction

The emergence of digital electronics has had a tremendous impact on industrial societies throughout the world. In the heating, ventilating, and air-conditioning (HVAC) industry, the development of digital electronics has brought an end to the use of many mechanical devices; typical of this is the diminished use of mechanical controls for HVAC air and water systems. Today,s digital control systems, with built-in intelligence, more accurately evaluate water and system conditions and adjust pump operation to meet the desired water flow and pressure conditions.

Drafting boards and drafting machines have all but disappeared from the design rooms of heating, ventilating, and air-conditioning engineers and have been replaced by computer-aided drafting (CAD) systems. Tedious manual calculations are being done more quickly and accurately by computer programs developed for specific design applications. All this has left more time for creative engineering on the part of designers to the benefit of the client.

1.2 Computer-Aided Calculation of HVAC Loads and Pipe Friction

The entire design process for today's water systems, from initial design to final commissioning, has been simplified and improved as a result of the new, sophisticated computer programs. One of the most capable programs for sizing and analyzing flow in fluid systems is the piping systems analysis program developed by APEC, Inc. (Automated Procedures for Engineering Consultants), headquartered in Dayton,

Ohio. APEC, a nonprofit, worldwide association of consulting engineers and in-house design group, is dedicated to improving quality and productivity in the design of HVAC air and water systems through the development and application of advanced computer software.

The APEC PSA-1 program accurately calculates the friction losses and sizes of pipes as well as simulating flow under different operating conditions in either new or existing piping systems. Analyzing the fluid flow in systems with diversified loads, multiple pumps, and chillers or boilers is essential if engineers are to truly understand the real operating conditions of large HVAC water systems. This understanding can only be achieved through the use of a computer program capable of such thorough analysis.

1.2.1   Typical input for APEC piping system analysis program

The following is representative system data into an APEC's computer program for calculating pipe sizing, friction, and flow analysis, and typical output.

Master Data files

Pipe, fitting, and valve files.

Material	friction loss
Actual ID for nominal	copper type (M,L)
Steel schedule	other

Standard pipe pre-entered or custom. Cost estimation optional.

Insulation file

Type
K value
Thickness

Cost estimation optional.

Fluids. Provisions for all fluid types with:

Density	Temperature
Viscosity	Specific heat

System data

Pipe environment.   Required for heat/loss gain.

Space temperature	Outside air temperature
Soil conductivity	Burial depth

Program options

System sizing
Flow simulation
Cost estimate

Flow simulation options-typical entries

Maximum iterations	30
Intermediate results	Every 3 iterations
Temperature tolerance for convergence	0.50
Relaxation parameters	0.50°F
Fill pressure	46 ft of head (or H_2O)

Pump data

Variable *or* constant speed?
Points for pump curves:

Terminal data (coils, etc.)

Fluid flow
Pressure drop
Coil cfm (ft^3/min)
Inlet air temperature
Leaving air set point

Valve data

Valve coefficient
Trial setting
Valve control

1.2.2 Typical output for APEC piping system analysis program

Table 1.1 includes samples of output headings, with one line of output for only three output forms available. The output also has forms that mirror the input, so the designer has a complete record of the entire analysis.

This program is now being expanded to include many additional piping features and to accommodate contemporary computer practices such as Windows to speed the development and manipulation of project data.

TABLE 1.1 Sample Output Headings

Pressure-Drop Analysis*

Node		Pipe		Pipe PD	Terminal		CV PD	Fitting PD	Special PD	Total PD
From	To	Diameter	Length, ft		Flow	PD				
5	6	2.50	22.5	0.75	70	22	31.0	0.21		53.96

System Estimate†

Item	Size	Description	Quantity	Unit	Material		Labor		Total Cost
					Unit	Cost	Unit	Cost	
1	2.00	Schedule 40	115.0	LF	2.42	278	42	4830	5108

Final Simulation Results

Link		Pipe diameter	Flow (gpm)		Pressure head (ft)		Temperature, †F
Start	End		Input	Actual	At start	Node	
4	5	2.5	70	75.3	34.4	(79.37)	160

*Chiller or boiler pressure drop not included.
†Labor and cost unit are entered by user as master data for given localities. Cost estimates are not intended to give accurate costs for bidding purposes.

1.3 Pressure-Gradient Diagrams

The pressure-gradient diagram provides a visual description of the changes in total pressure in a water system. To date, these diagrams have been drawn manually; the actual drawing of the pressure-gradient diagram is now being evaluated for conversion to software; when this is completed, the diagram will appear automatically on the computer screen after the piping friction calculations are completed.

The pressure-gradient diagram has proved to be an invaluable tool in the development of a water system. It will appear throughout this book for various types of water systems. Its generation will be explained in Chap. 3. Clarification should be made between a pressure gradient and the hydraulic gradient of a water system. The hydraulic gradient includes the velocity head $V^2/2g$, of the water system, while the pressure gradient includes only the static and pressure heads. Velocity head is usually a number less than 5 ft and is not used to move water through pipe, as are static and pressure heads. Using the hydraulic gradient with the velocity head increases the calculations for developing these diagrams; therefore, the pressure gradient is used instead. Velocity head cannot be ignored, since it represents the kinetic energy of the water in the pipe. Velocity head will be emphasized in this book when it becomes a factor in pipe design, particularly in piping around chillers and in the calculation of pipe fitting and valve losses.

1.4   Speed and Accuracy of Electronic Design of Water Systems

The tremendous amount of time saved by electronic design enables the engineer to evaluate a water system under a number of different design constraints. The designer can load certain design requirements into a computer, and while the computer is doing all the detailed calculations for that program, the next program of design considerations can be set up for calculations by the computer. After all the programs have been run, the designer can select the one that provides the optimal system conditions that meet the specifications of the client. The designer now has time to play "what if" to achieve the best possible design for a water system. In the past, the engineer often was time driven and forced to utilize much of a past design to reach a deadline for a current project. Now the engineer can model pumping system performance under a number of different load conditions and secure a much more complete document on the energy consumption of proposed pumping systems.

The designer can compute the *diversity* of an HVAC system with much greater accuracy. Diversity is merely the actual maximal heating or cooling load on an HVAC system divided by the capacity of the installed equipment. For example, assume that the total cooling load on a chilled water system is 800 tons, but there are 1000 tons of cooling equipment installed on the system to provide cooling to all parts of that system. This disparity is caused by changing sun loads or differences in occupancy. The diversity in this case would be 800/1000, or 0.80 (80 percent).

This is a simpler and easier definition of diversity than a more technical definition that states that diversity is the maximal heating or cooling load divided by the sum of all the individual peak loads. For example, a 10-ton air handler might have a peak load of only 9.2 ton. The true diversity might be slightly less than that acquired by using the installed load.

1.4.1   Equation solution by computer

A number of equations are provided herein for the accurate solution of pressures, flows, and energy consumptions of HVAC water systems. These equations have been kept to the algebraic level of mathematics to aid the HVAC water system designer in the application of them to computer programs. Computer software is now available commercially to assist in the manipulation of these equations. Typical of them is the EES—Engineering Equation Solver program available from F-Chart Software, Middleton, Wis.

1.5   Databasing

After the designer has completed the overall evaluation of a water system, databasing can be used to search elements of past designs for use on a current project. Databasing is a compilation of information on completed designs in computer memory that can be recalled for use on future projects. To use it, the designer can enter key factors that would describe a current project and then allow the computer to search a database for similar completed designs that would have the same defining elements. For example, a project designed without databasing may have a total of 5000 design hours. After searching the database, a design might be found that could provide 3000 design hours from a previous project, leaving a requirement for 2000 new design hours. When this current project is completed, it would be entered into the database for similar future reference.

1.6   Electronic Communication

With the technical advances that are occurring in communications, rapid communication is available between various architectural and engineering offices. Databasing can be linked between main and branch offices of a multioffice firm so that job and data sharing can be established between the various offices as desired by the engineering management.

Interoffice communication also has been accelerated with the use of electronic mail such as E-Mail. Such mail can reduce the time for asking crucial questions and receiving responses. It reduces error with regard to documentation and maintains a file on the correspondence.

1.7   Electronic Design of the Piping and Accessories

Similar to load calculations and general system layout, digital electronics has invaded the actual configuration of the water system itself. This includes the methods of generating hot or cold water, storage of the same, and distribution of the water in the system. The distribution of water in an HVAC system is no longer dependent on mechanical devices such as pressure-regulating valves, balancing valves, crossover bridges, reverse-return piping, and other energy-consuming mechanical devices that force the water through certain parts of the system. Almost all mechanical devices are disappearing, other than temperature-control valves for heating and cooling coils. How this is done will be described in detail in Chaps. 10 and 15 during actual design of HVAC systems.

HVAC water systems are being reduced to major equipment such as boilers or chillers, heating and cooling coils, pumping systems, connecting piping, and electronic control. Simplicity of system design is ruling the day with very few flow- or pressure-regulating devices; this results in much higher overall system operating efficiencies.

1.8 Electronic Selection of HVAC Equipment

A major part of the designer's work is the selection of equipment for a water system. This includes, for example, chillers, cooling towers, boilers, pumping systems, heating and cooling coils, and control systems. In the past, designers depended on manufacturers' catalogs to furnish the technical information that provided the selection of the correct equipment for a water system. This had to do with the hope that the catalogs were current. Now comes the CD-ROM disc and on-line data services that provide current information and rapid selection of equipment that meets the designer's specifications. Many manufacturers are converting their technical catalogs to software such as CD-ROM discs, providing both performance and dimensional data. The day of the technical catalog is almost gone.

1.9 Electronic Control of HVAC Water Systems

Along with these changes in mechanical design, electronic control of HVAC water systems, in the form of direct digital control or programmable-logic controllers, has all but eliminated older mechanical control systems such as pneumatic control. The advent of universal protocols such as BACnet® has enabled most control and equipment manufacturers to interface together on a single installation. BACnet® description is available from ASHRAE headquarters in Atlanta, Georgia.

1.10 Electronics and HVAC Pumps

How do all these electronic procedures relate to HVAC pumps? Efficient pump selection and operation depend on the accurate calculation of a water system's flow and pump head requirements. Digital electronics has created greater design accuracy, which guarantees better pump selection. Incorrect system design will result in (1) pumps that are too small and incapable of operating the water system or (2) pumps that are too large with excess flow and head resulting in inefficient operation. The use of electronic design aids has improved the chances of selecting an efficient pumping system for each application. Accurate calculation of flow and head requirements of constant-volume HVAC systems has reduced the energy destroyed in balance valves that are used to eliminate excessive pressure.

1.11   Electronics and Variable-Speed Pumps

One of the greatest effects on HVAC water systems by electronics is the development of variable-frequency drives for fans and pumps. The day of the constant-speed pump with its fixed head-capacity curve is coming to an end, giving way to the variable-speed pump, which can adjust more easily to system conditions with much less energy and with smaller forces on the pump itself. Along with the constant-speed pump go the mechanical devices described earlier that overcame the excess pressures and flows of that constant-speed pump. The variable-frequency drive with electronic speed control and pump programming matches the flow and head developed by pumps to the flow and head required by the water system without mechanical devices such as balance valves.

1.12   Electronic Commissioning

Another great asset of electronics applied to water systems is its use during the commissioning process. There are always changes in drawings and equipment during the final stage of starting and operating a water system for the first time. Many of these changes in equipment and software can be recorded easily through the use of portable computers or other hand-held electronics. The agony of ensuring that "as-built" drawings are correct has been reduced greatly.

Electronic instrumentation and recording devices have accelerated the commissioning of water systems. Verification of compliance of the equipment of a water system is enhanced by these instruments.

1.13   Purpose of This Book

It is one of the basic purposes of this book to describe in detail all the preceding uses of electronics in the design and application of pumps to HVAC water systems. This must be done with recognition that the rapid development of new software and equipment is liable to relegate any description of digital electronics to obsolescence at the time of writing. The development of on-line data services is going to change even further the way we design these water systems.

The HVAC design engineers must understand where their offices are in the use of available electronic equipment and services; this ensures that they are providing current system design at a minimum cost to their company. The engineers who do not use electronic equipment, network the office, or subscribe to on-line data services as they come available will not be able to keep up with his or her contemporaries in design accuracy and speed.

One of the reasons for the writing of this book was to produce a handbook for HVAC pumps that would provide basic design and application data and embrace the many and rapid changes that have occurred in water system design and operation. This *Handbook* has been written to guide the student and inexperienced designer and, at the same time, provide the knowledgeable designer with some of the latest procedures for improving water system design and operation.

The advent of electronic control and the variable-speed pump has obsoleted many of the older designs of these water systems. We have the opportunity now to produce highly efficient systems and to track their performance electronically, ensuring that the projected design is achieved in actual operation.

1.14 Bibliography

Piping Systems Analysis–1, APEC, Inc., Dayton, Ohio, 1988.

2

Physical Data for HVAC System Design

2.1 Introduction

There can be confusion about the standards that exist for the design and operation of HVAC systems and equipment such as pumps. It is important for the designer to understand what these standards are, both for the HVAC equipment and for the water systems themselves. These standards can be established by technical societies, governmental agencies, trade associations, and as codes for various governing bodies. The designer must be aware of the standards and codes that govern each application.

Included in this chapter are standard operating conditions for HVAC equipment; also, this chapter has brought together much of the technical data on air, water, and electricity necessary for designing and operating these water systems. The only information on water not included in this chapter is pipe friction, as described in Chap. 3, and the specific heat of water at higher temperatures, as described in Chap. 21 for medium- and high-temperature water systems. Also included in this chapter are standard operating conditions for HVAC equipment.

It is hoped that most of the technical information needed by the HVAC system designer for pump application is included in this book. The cross-sectional area, in square feet, and the volume, in gallons, of commercial steel pipe and circular tanks have been included on a linear-foot basis. This is valuable information for the designer in calculating the liquid volume of HVAC water systems and energy storage tanks.

2.2 Standard Operating Conditions

Every piece of HVAC equipment available is based on some particular operating conditions such as maximum temperature or pressure; usually, these conditions are spelled out by the manufacturer. It is the responsibility of the design engineer to check these conditions and to ensure that they are compatible with the system conditions. It is very important that variations in electrical service as well as maximum ambient air temperature be verified for all operating equipment.

2.2.1 Standard air conditions

Standard air conditions must be defined for ambient and ventilation air. *Ambient air* is the surrounding air in which all HVAC equipment must operate. Standard ambient air is usually listed as 70°F, while maximum ambient air temperature is normally listed at 104°F. This temperature is the industry standard for electrical and electronic equipment. For some boiler room work, the ambient air may be listed as high as 140°F. It is incumbent on the designer to ensure that his or her equipment is compatible with such ambient air conditions.

Along with ambient air temperature, the designer must be concerned with the quality of *ventilation air*. This is the air that is used to cool the operating equipment as well as provide ventilation for the building. The designer must ensure that the equipment rooms are not affected by surrounding processes that contain harmful substances. This includes chemicals in the form of gases or particulate matter. Hydrogen sulfide is particularly dangerous to copper-bearing equipment such as electronics. Many sewage treatment operations generate this gas, so it is very important that any HVAC equipment installed in sewage treatment facilities be protected from ambient air that can include this chemical. Dusty industrial processes must be separate from equipment rooms to keep equipment clean. Dust that coats heating or cooling coil surfaces or electronics will have a substantial effect on the performance and useful life of that equipment. The designer must be aware of the presence of any such substances that will harm the HVAC equipment.

Ventilation air does not bother the operation of the pump itself, but it does affect the pump motor or variable-speed drive. This is the air that is used to cool this electrical equipment. Evaluating ventilation air is part of the design process for the selection of such equipment and is therefore very important in equipment selection. Outdoor air data including maximum wet bulb and dry bulb temperatures is listed in the American Society of Heating, Refrigerating, and Air-Conditioning Engineers' (ASHRAE's) *Systems and Equipment Handbook* for most of the principal cities. Indoor air quality must be verified as well, both from a chemical content basis as well as from a temperature basis. Heat generation in the equipment rooms must be removed by ventilation or mechanical cooling to ensure that the design standards of the equipment are not exceeded.

2.2.2 Operating pressures

Gauge pressure is that water or steam pressure that is measured by a gauge on a piece of HVAC equipment. Following is the basic equation for gauge, absolute, and atmospheric pressures.

$$psia = psig + P_e \qquad (2.1)$$

where psia = absolute pressure, lb/in^2 (psi)
    psig = gauge pressure, lb/in^2 (psi)
    P_e = atmospheric pressure, lb/in^2 (psi)

For example, if a water system is operating at 75 psig pressure at an altitude of 1000 ft, from Table 2.1, the atmospheric pressure is 14.2 lb/in^2, so the absolute pressure is 89.2 psia.

TABLE 2.1   Variation of Atmospheric Pressure with Altitude

Altitude, ft	Average Pressure P_e, PSIA	Average Pressure P_a, ftH$_2$O, Up to 85°F
0	14.7	34.0
500	14.4	33.3
1,000	14.2	32.8
1,500	13.9	32.1
2,000	13.7	31.6
2,500	13.4	31.0
3,000	13.2	30.5
4,000	12.7	29.3
5,000	12.2	28.2
6,000	11.8	27.3
7,000	11.3	26.1
8,000	10.9	25.2
9,000	10.5	24.3
10,000	10.1	23.3
15,000	8.3	19.2
20,000	6.7	15.5

SOURCE: *Cameron Hydraulic Data,* 15th ed., Ingersoll Rand, Woodcliff Lake, N.J., 1977, p. 7-4; used with permission.

The *atmospheric pressure* of outdoor air varies with the altitude of the installation of HVAC equipment and must be recognized in the rating of most HVAC equipment. Table 2.1 describes the variation of atmospheric pressure with altitude.[1] This table lists atmospheric pressure in feet of water as well as pounds per square inch. For water temperature in the range of 32 to 85°F, the feet of head can be used directly in the net positive suction head (NPSH) and cavitation equations found in Chap. 6 on pump performance. For precise calculations and higher-temperature waters, the atmospheric pressure in lb/in^2 absolute must be corrected for the specific volume of water at the operating temperature. See Eq. 6.10, which corrects the atmospheric pressure in feet of water to the actual operating temperature of the water.

2.3   Thermal Equivalents

There are some basic thermal and power equivalents that should be summarized for HVAC water system design. This book is based on 1 Btu (British thermal unit) being equal to 778.2 ft · lb (foot pounds). Other sources list 1 Btu as equal to from 778.0 to 778.26 ft · lb, which results in different thermal equivalents. For example, the ASHRAE *Systems and Equipment Handbook* lists 1 Btu as equal to 778.17 ft · lb, while Keenan and Keyes's *Thermodynamic Properties of Steam* defines 1 Btu as 778.26 ft · lb. The following thermal and power equivalents will be found in this book:

$$1 \text{ Btu (British thermal unit)} = 778.2 \text{ ft} \cdot \text{lb}$$

$$1 \text{ brake horsepower, bhp} = 33,000 \text{ ft} \cdot \text{lb/min}$$

$$1 \text{ brake horsepower hour, bhph} = 2544 \text{ Btu/h}$$

$$= 0.746 \text{ kWh (kilowatthour)}$$

$$1 \text{ kWh} = 1.341 \text{ bhp}$$

$$= 3412.0 \text{ Btu/h}$$

2.4   Water Data

Water is not as susceptible to varying atmospheric conditions as is air, but its temperature and quality must be measured. Standard water temperature can be stated as 32, 39.2 (point of maximum density), or 60°F. It is not very important which of these temperatures is used for HVAC pump calculations, since water has a density near 1.0 and a viscosity around 1.5 cSt (centistokes) at all these temperatures.

TABLE 2.2   Viscosity of Water

Temperature of water, °F	Absolute viscosity, cP	Kinematic viscosity, ft²/s
32	1.79	1.93×10^{-5}
40	1.55	1.67×10^{-5}
50	1.31	1.41×10^{-5}
60	1.12	1.21×10^{-5}
70	0.98	1.06×10^{-5}
80	0.86	0.93×10^{-5}
90	0.77	0.83×10^{-5}
100	0.68	0.74×10^{-5}
120	0.56	0.61×10^{-5}
140	0.47	0.51×10^{-5}
160	0.40	0.44×10^{-5}
180	0.35	0.39×10^{-5}
200	0.30	0.34×10^{-5}
212	0.28	0.32×10^{-5}
250	0.23	0.27×10^{-5}
300	0.19	0.22×10^{-5}
350	0.15	0.18×10^{-5}
400	0.13	0.16×10^{-5}
450	0.12	0.16×10^{-5}

SOURCE: *Engineering Data Book,* Hydraulic Institute, Parsippany, N.J., 1990, p. 19; and *Systems and Equipment Handbook,* ASHRAE, Atlanta, Ga., p. 14.3; used with permission.

Operations with water at temperatures above 85°F must take into consideration both the specific gravity and viscosity. Tables 2.2, 2.3, and 2.4 provide these data for water from 32 to 450°F.

2.4.1   Viscosity of water

There are two basic types of viscosity, dynamic, or absolute, and kinematic. *Dynamic viscosity* is expressed in force-time per square length terms and in the metric system usually as centipoise (CP). In most cases, the viscosity of water will be stated as *kinematic viscosity* in centistokes (cSt) in the metric system and in square feet per second in the English system. If the viscosity of a liquid is expressed as an absolute viscosity in centipoise, the conversion formula[2] to kinematic viscosity in square feet per second is

$$v = \frac{6.7197 \cdot 10^{-4} \cdot \mu}{\gamma} \tag{2.2}$$

where v = kinematic viscosity, ft²/s
μ = absolute viscosity, CP

TABLE 2.3 Vapor Pressures and Specific
Weights for Water for Temperatures from 32 to
212°F

Temperature, °F	Absolute pressure, ftH$_2$O	Specific weight γ, lb/ft^3
32	0.20	62.42
40	0.28	62.42
45	0.34	62.42
50	0.41	62.38
55	0.49	62.38
60	0.59	62.34
65	0.71	62.34
70	0.84	62.26
75	1.00	62.23
80	1.17	62.19
85	1.38	62.15
90	1.62	62.11
95	1.89	62.03
100	2.20	62.00
105	2.56	61.92
110	2.97	61.84
115	3.43	61.80
120	3.95	61.73
130	5.20	61.54
140	6.78	61.39
150	8.75	61.20
160	11.19	61.01
170	14.19	60.79
180	17.85	60.57
190	22.28	60.35
200	27.60	60.13
210	33.96	59.88
212	35.38	59.81

SOURCE: *Cameron Hydraulic Data,* 15th ed., Ingersoll
Rand, Woodcliff Lake, N.J., 1977; used with permission.

$\epsilon\gamma$ = specific weight, lb/ft^3

If the viscosity is expressed as the kinematic viscosity in the metric system in centistokes, the conversion formula[2] for kinematic viscosity in the English system is

$$v, \text{ft}^2/\text{s} = 1.0764 \cdot 10^{-5} \cdot v, \text{cSt} \qquad (2.3)$$

Kinematic viscosity in English units of square feet per second is the easiest expression of viscosity to use where other English units of length, flow, and head are used in HVAC pumping. This is the term required for computing the Reynolds number with English units.

TABLE 2.4   Vapor pressures and specific weights for
water, for Temperatures of 212 to 450°F

Temperature, °F	Absolute pressure, psia	Specific weight γ, lb/ft^3
212	14.70	59.81
220	17.19	59.63
230	20.78	59.38
240	24.97	59.10
250	29.83	58.82
260	35.43	58.51
270	41.86	58.24
280	49.20	57.94
290	57.56	57.64
300	67.01	57.31
320	89.66	56.66
340	118.01	55.96
360	153.03	55.22
380	195.77	54.47
400	247.31	53.65
420	308.83	52.80
440	381.59	51.92
450	422.6	51.55

SOURCE: Joseph H. Keenan and Frederick G. Keyes,
Thermodynamic Properties of Steam, Wiley, New York, 1936, p.
34; used with permission.

Contemporary computer programs for pipe friction automatically include these data for the water under consideration. Table 2.2 provides
the absolute viscosity in centipoise and the kinematic viscosity in
square feet per second.

2.4.2   Vapor pressure and specific weight
for water, 32 to 212°F

The *vapor pressure* of water for temperatures up to 450°F must be included, since this information is necessary in evaluating the possibilities of *cavitation* and in the calculation of net positive suction head
available for pumps, which is included in Chap. 6 on pump performance. Vapor pressure is the absolute pressure, psia, at which water
will change from liquid to steam at a specific temperature. For each
temperature of water, there is an absolute pressure at which water
will change from a liquid to a gas. Table 2.3 provides these vapor
pressures up to 210°F, as well as the specific weight of water at these
temperatures. The vapor pressures are shown in feet of water and not
pounds per square inch at these temperatures for NPSH calculations.
Specific weight γ is the density in pounds per cubic feet of water at a
particular temperature.

2.4.3   Vapor pressure and specific weight for water, 212 to 450°F

Vapor pressures for water from 212 to 450°F, along with its specific weight, are provided in Table 2.4; the values in the table, unlike Table 2.3, are expressed in absolute pressures for determining the minimum allowable pressures of hot water systems operating in this temperature range. These pressures are used to calculate and avoid cavitation at any point in these hot water systems.

2.4.4   Solubility of air in water

It is important to know the amount and source of air in an HVAC water system. Air is undesirable in pumps because of its great effect on the pump's performance and useful life.

Air enters an HVAC water system from the original filling of the system and from the make-up water that is required to keep the system full. Air should not enter the system from any other source. Air occurs naturally in water. Table 2.5 provides the basic data on the solubility of air in water.

As indicated in the table, the amount of air that can be dissolved in water decreases with temperature and increases with system pres-

TABLE 2.5   Solubility of Air in Water

Temperature, °F	System Gauge Pressure, psig						
	0	20	40	60	80	100	120
40	0.0258	0.0613	0.0967	0.1321	0.1676	0.2030	0.2384
50	0.0223	0.0529	0.0836	0.1143	0.1449	0.1756	0.2063
60	0.0197	0.0469	0.0742	0.1014	0.1296	0.1559	0.1831
70	0.0177	0.0423	0.0669	0.0916	0.1162	0.1408	0.1654
80	0.0161	0.0387	0.0614	0.0840	0.1067	0.1293	0.1520
90	0.0147	0.0358	0.0569	0.0750	0.0990	0.1201	0.1412
100	0.0136	0.0334	0.0532	0.0730	0.0928	0.1126	0.1324
110	0.0126	0.0314	0.0501	0.0689	0.0877	0.1065	0.1252
120	0.0117	0.0296	0.0475	0.0654	0.0833	0.1012	0.1191
130	0.0107	0.0280	0.0452	0.0624	0.0796	0.0968	0.1140
140	0.0098	0.0265	0.0432	0.0598	0.0765	0.0931	0.1098
150	0.0089	0.0251	0.0413	0.0574	0.0736	0.0898	0.1060
160	0.0079	0.0237	0.0395	0.0553	0.0711	0.0869	0.1027
170	0.0068	0.0223	0.0378	0.0534	0.0689	0.0844	0.1000
180	0.0055	0.0208	0.0361	0.0514	0.0667	0.0820	0.0973
190	0.0041	0.0192	0.0344	0.0496	0.0647	0.0799	0.0950
200	0.0024	0.0175	0.0326	0.0477	0.0628	0.0779	0.0930
210	0.0004	0.0155	0.0306	0.0457	0.0607	0.0758	0.0909

SOURCE: *Technical Bulletin 8-80*, Amtrol, Inc., West Warrwick, R.I., 1985, p. 14; used with permission.

sure. This table demonstrates Henry's law, which states that the amount of air dissolved in water is proportional to the pressure of the water system. This table should be used in place of similar charts for open tanks and deaerators where the only pressure is atmospheric pressure at 0 psig, and the amount of air dissolved in the water approaches zero at 212°F. It is evident from this table that make-up water that is supplied by the domestic water system can contain a great amount of air.

To demonstrate the release of air from water, assume that the return water has a temperature of 180°F and the system pressure is 40 psig. Make-up water entering the system at 50°F will have at least a 0.0836 ratio of air to water. It could have much more air than this, since it may have been reduced from a higher city water pressure. When the make-up water is heated to 180°F at 40 psig, the air content will drop to a ratio of 0.0361, which is less than half that of the cold make-up water.

An interesting and easy experiment to observe the release of air when water is heated is as follows:

1. Take a frying pan and fill it with potable water from the kitchen cold water faucet.

2. Place it on the stove, and heat the water to boiling.

3. Note that bubbles form as soon as the temperature begins to rise. This is air coming out of solution with the water, since the water cannot hold as much air with the higher temperature.

4. As the water approaches 212°F, the water begins to boil.

5. Allow the water to cool, and then reheat the water to boiling.

6. Note that this time bubbles do not appear until steam begins to form. This demonstrates that the water has been deaerated during the first boiling. It also provides a visual example of what happens to cold water when it is heated in an HVAC water system.

As shown in Table 2.5, when water passes through pumps and the pressure is increased, the water will increase its affinity for air. It is therefore imperative that the air in the make-up water be removed from the water as soon as it reaches system temperature by locating the water make-up near the air-elimination equipment such as an air separator. The optimal location for the air-elimination equipment depends on the configuration of the water system. Generally, it may be best to locate an air separator near the suctions of the distribution pumps with air vents at the high points where the system pressure is the lowest. For heating systems with high supply water temperatures, it may be advisable to add a dip tube and an air vent at the water discharge from the heater or boiler. These tables also demonstrate that it is wise to have manual or automatic air vents at the top of a building where the system pressure is the lowest. A detailed discussion of air removal from HVAC water systems is included in Chap. 9.

2.5   Glycol-Based Heat-Transfer Fluid (HTF) Solutions

Glycol-based heat-transfer solutions are prevalent in the HVAC industry. Glycol is used to (1) avoid freezing in equipment of an HVAC water system such as heating and cooling coils and (2) to transfer heat to and from energy storage tanks that use ice. There is a substantial variation in both viscosity and density of the solution as the percentage of glycol is varied with temperature. Information is provided on ethylene glycol–based heat-transfer fluids, since they have been the most used in the HVAC industry. Special applications of glycol-based heat-transfer fluids, such as contact with potable water or food, may require the use of propylene glycol–based solutions.

Figure 2.1 provides information on the viscosity and Fig. 2.2 on the specific gravity of ethylene glycol–based heat-transfer solutions. Care should be taken to avoid using glycol solutions near their freezing curves because slush occurs there that will change radically the pump's performance. Figure 2.3 provides the freezing curve for an ethylene glycol–based heat-transfer fluid. Verification of the minimum percentage of glycol to prevent slush formation at the minimum operating temperature should be sought from the supplier of the heat-transfer fluid.

There is an appreciable variation in the specific heat of glycol-based heat-transfer solutions. It is less than that of water for most percentages of glycol. Figure 2.4 provides the specific heat for ethylene glycol–based heat-transfer solutions. The pump flow, in gallons per minute, is calculated from Eq. 2.4:

$$\text{Pump flow, gal/min} = \frac{\text{Btu/h} \cdot 7.48 \text{ gal/ft}^3}{c_p \cdot \Delta T - {}^\circ\text{F} \cdot \gamma \cdot s \cdot 60 \text{ min/h}}$$

$$= \frac{0.125 \cdot \text{Btu/h}}{c_p \cdot \Delta T - {}^\circ\text{F} \cdot \gamma \cdot s} \tag{2.4}$$

where c_p = specific heat of ethylene glycol heat-transfer solution at constant pressure
s = specific gravity of the same solution

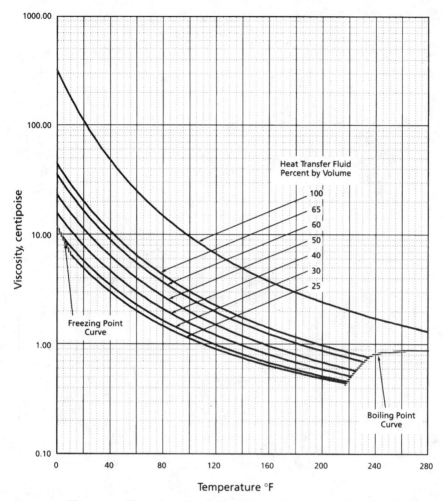

Figure 2.1 Viscosity of heat transfer fluids. (*From Engineering Data for Ethylene Glycol–Based Heat Transfer Fluids, Union Carbide Corporation, Danbury, Conn., 1993, p. 20.*)

ΔT = differential temperature
γ = specific weight of water

Note.   All these values must be at the operating temperature of the solution. For many glycol installations, this calculation should be run at several different operating temperatures. For example, assume that an ethylene glycol–based heat-transfer fluid has a heating load of 5 million Btu/h at 30°F with a differential temperature of 12°F. The

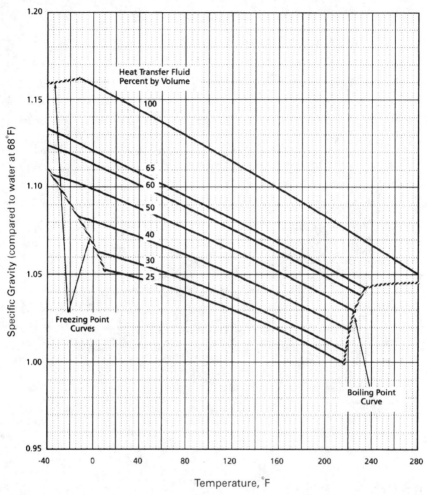

Figure 2.2 Specific gravity of heat transfer fluids. (*From Engineering Data for Ethylene Glycol–Based Heat Transfer Fluids, Union Carbide Corporation, Danbury, Conn., 1993, p. 18.*)

pumps will be operating with a 40% glycol solution. From Fig. 2.4, the specific heat of the glycol will be 0.84, and from Fig. 2.2, the specific gravity of the solution will be 1.074 based upon water at 68°F or 62.32 lb/ft^3.

$$\text{Pump flow, gal/min} = \frac{0.125 \cdot 5{,}000{,}000}{0.84 \cdot 12 \cdot 62.32 \cdot 1.074} = 926 \text{ gal/min}$$

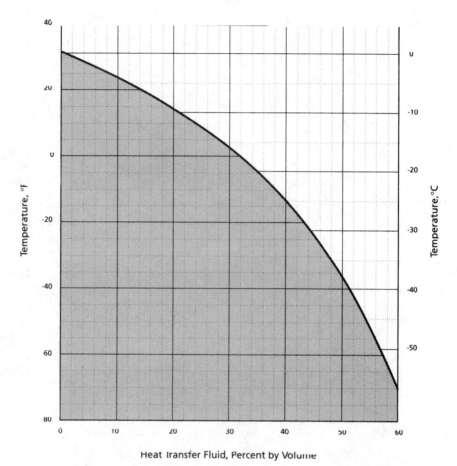

Figure 2.3 Freezing points of heat transfer fluids. (*From Engineering Data for Ethylene Glycol–Based Heat Transfer Fluids, Union Carbide Corporation, Danbury, Conn., 1993, p. 10.*)

2.6 Steam Data

Steam is used for many heating processes in HVAC. Most of the steam data come from one source, namely Keenan and Keyes' *Thermodynamic Properties of Steam*. This is a fundamental reference for any engineer working with steam. Table 2.6 provides the basic steam data, while vapor pressures are included in Tables 2.3 and 2.4 for the computation of net positive suction head available and the determination of cavitation pressures at various operating temperatures.

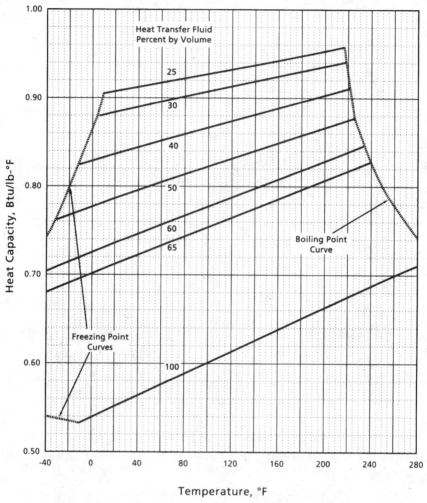

Figure 2.4 Specific heat of heat transfer fluids. (*From Engineering Data for Ethylene Glycol–Based Heat Transfer Fluids, Union Carbide Corporation, Danbury, Conn., 1993, p. 24.*)

In HVAC, there are two basic steam pressure ranges: (1) up to 15 psig (250°F) and (2) above 15 psig steam pressure. This is derived from the American Society of Mechanical Engineers' (ASME) boiler codes, (1) the *Heating Boiler Code* for steam pressures up to 15 psig and (2) the *Power Boiler Code* for steam pressures above 15 psig. See Chap. 19 for additional information on boilers.

TABLE 2.6 Basic Steam Data

Absolute pressure, lb/in^2	Steam temperature, °F	Enthalpy, Btu/lb		
		Saturated liquid	Evaporation	Saturated vapor
14.7	212.00	180.07	970.3	1150.4
16	216.32	184.42	967.6	1152.0
18	222.41	190.56	963.6	1154.2
20	227.96	196.16	960.1	1156.3
22	233.07	201.33	956.8	1158.1
24	237.82	206.14	953.7	1159.8
26	242.25	210.62	950.7	1161.3
28	246.41	214.83	947.9	1162.7
30	250.33	218.82	945.3	1164.1
35	259.28	227.91	939.2	1167.1
40	267.25	236.03	933.7	1169.7
45	274.44	243.36	928.6	1172.0
50	281.01	250.09	924.0	1174.1
60	292.71	262.09	915.5	1177.6
70	202.92	272.61	907.9	1180.6
80	312.03	282.02	901.1	1183.1
90	320.27	290.56	894.7	1185.3
100	327.81	298.40	888.8	1187.2
125	344.33	315.68	875.4	1191.1
150	358.42	330.51	863.6	1194.1
174	370.29	343.10	853.3	1196.4
200	381.79	355.36	843.0	1198.4

NOTE: Absolute pressure = gauge + atmospheric pressures.
SOURCE: Joseph H. Keenan and Frederick G. Keyes, *Thermodynamic Properties of Steam*, Wiley, New York, 1936, p. xxx.

2.7 Areas and Volumes of Pipe and Tanks

Table 2.7 provides the cross-sectional area, in equivalent square feet, and the volume, in gallons, of commercial steel pipe and circular tanks per linear foot of such pipe and tanks. The volume of the pipe or tank can be determined by multiplying the cross-sectional area by the length or height in feet. The volume in gallons per ft has been provided to simplify the calculations for HVAC water system volume and tank storage.

2.8 Electrical Data

Following is a brief review of electrical power supplies and use with HVAC pumps. Chapter 7 provides a detailed evaluation of electric motors. The standard frequency for electric power in the United States is 60 Hz (hertz), or cycles per second. Many foreign countries have stan-

TABLE 2.7   Areas and Volumes of Pipe and Tanks

Pipe Size	Inside Diameter, in	Area, ft²	Volume, gal/ft	Inside Diameter, in	Area, ft²	Volume, gal/ft
1¼	1.380	0.0104	0.078†	66	23.758	177.71†
1½	1.610	0.0141	0.106	72	28.274	211.49
2	2.067	0.0247	0.185	84	38.485	287.87
2½	2.469	0.0333	0.249	90	44.179	330.46
3	3.068	0.0513	0.384	96	50.266	375.99
4	4.026	0.0882	0.660	102	56.745	424.45
5	5.047	0.1389	1.039	108	63.617	475.86
6	6.065	0.2006	1.501	114	70.882	530.20
8	7.981	0.3474	2.599	120	78.540	587.48
10	10.02	0.5476	4.096	144	113.097	845.97
12	11.938	0.7773	5.814	168	153.938	1,151.46
14	13.124	0.9394	7.027	192	201.062	1,503.94
16	15.000	1.2272	9.180	216	254.469	1,903.43
18	16.875	1.5533	11.619	240	314.159	2,349.91
20	18.812	1.9302	14.438	288	452.389	3,383.87
24	22.624	2.7917	20.882	360	706.858	5,287.30
30	29.00*	4.5869	34.310	432	1,017.87	7,613.67
36	35.25†	6.8257	51.056	504	1,385.44	10,363.09
42	41.25†	9.2806	69.419	576	1,809.56	13,535.51
48	47.25†	12.1771	91.085	720	2,827.43	21,149.18

*All pipe sizes up to 24 in are schedule 40, while 30 in is schedule 20.
†Pipe sizes 36, 42, and 48 in are standard inside diameters.
NOTE: To convert the above volumes in gallons to pounds of water, multiply gallons by

$$\text{Pounds of water} = \frac{\gamma}{7.48}$$

where γ is the specific weight of the water at the operating temperature. For example, water at 45°F has a specific weight of 62.42 lb/ft³, so a 10-in schedule 40 steel pipe has 4.096 gal/ft³ or 34.18 lb/ft³.

dardized on 50 Hz; there also may be rural areas of the United States still operating on 50-Hz power. Tables 2.8 and 2.9 provide nominal power distribution voltages and standard nameplate voltages for motors operating at both 60 and 50 Hz. Electric power utilities are allowed a variation of ±5 percent from the distribution system voltages listed in these tables.

The most popular power for HVAC applications is 480-V, three-phase. Single-phase power is seldom used above 7½ hp. The 208-V service is derived from a Y-connected transformer in the building being served; three-phase motors as high as 60 hp are available for this voltage. The higher voltages of 2400 and 4160 V are used generally on motors of 750 hp and larger.

Electrical machinery such as motors and variable-speed drives have specified voltage tolerances that exceed those of the electrical utility. The electrical design engineer must develop the building power distribution to ensure that its voltage drop does not exceed the

TABLE 2.8   Standard 60-Hz Voltages

Nominal distribution system voltage	Motor nameplate voltage	
	Below 125 hp	125 hp and up
Polyphase		
208	200	—
240	230	—
480	460	460
600	575	575
2400	2300	2300
4160	4000	4000
Single-phase		
120	115	—
208	200	—
240	230	—

SOURCE: *AC Motor Selection and Application Guide,* Bulletin GET = 6812B, General Electric Company, Fort Wayne, Ind., 1993, p. 2; used with permission.

TABLE 2.9   Standard 50-Hz Voltages

Nominal distribution system voltage	Motor nameplate voltage	
	Below 125 hp	125 hp and up
Polyphase (see note) 200	200	
	220	—
	380	380
	415	415
	440	440
	550	550
	3000	3000
Single-phase (see note)	110	—
	200	—
	220	—

NOTE: Distribution system voltages vary from country to country; therefore, motor nameplate voltage should be selected for the country in which the motor will be operated.

SOURCE: *AC Motor Selection and Application Guide,* Bulletin GET = 6812B, General Electric Company, Fort Wayne, Ind., 19xx, p. 2; used with permission.

voltage tolerances of the electrical equipment. Typically, the voltage tolerance of most electric motors is ±10%, and those for most variable-speed drives appear to be +10 percent and −5 percent. The actual tolerances for this equipment should be verified by the HVAC water system designer. For example, the utility voltage at a building transformer may be 480 V ±5 percent, or 456 to 504 V. A 460-V variable-speed drive has an allowable voltage variation of 437 to 506 V. Therefore, the building power distribution system must be designed so that the power supply to the variable-speed drive does not drop below 437 V under any load condition.

Power factor correction equipment can be required by public utilities or state law above a certain size of motor. This should be checked by the designer at the beginning of the development of a specific project. Generally, public utilities do not require power factor correction at most places in their electrical distribution until the load approaches 500 kVa.

The popularity of the variable-frequency drive has created a problem for public utilities. This is the *harmonic distortion* caused by the alteration of the sine wave by the variable-frequency drive. The public utility furnishing power on a project may have a specification on the maximum allowable harmonic distortion. Also, the owner of the facility may have tolerances on harmonic distortion.

More information on power factor correction and harmonic distortion is included in Chap. 7.

2.9   Efficiency Evaluations of HVAC Water Systems

Several expressions of efficiency will be provided in the following chapters that relate to the effectiveness of pump selection and application. These will include

1. System efficiency, which determines the quality of use of pump head in a water system. This will be expressed as a percentage, coefficient of performance, kW/ton, or kW/1000 mbh.

2. Wire-to-water efficiency of a pumping system, which demonstrates the use of energy in a pumping system.

3. Kilowatts per ton efficiency for an entire chilled-water plant, which includes the energy consumption of the pumps and cooling towers, not just the efficiency of the chillers themselves.

4. Boiler efficiency as a percentage as related to entering water temperature.

These efficiencies are possible now that digital computers are available to make the calculations rapidly and accurately. The equations for HVAC systems and equipment included herein enable the HVAC operator to observe these water systems and ensure that they are functioning at optimal efficiency.

2.10   Additional Reading

It is important that the HVAC designer be well versed in the basic fluids and services available at the point of installation of each project.

Local codes and services must be checked for compatibility with the final design. The manuals of the technical societies are excellent sources for additional reading, particularly those of the American Society of Heating, Refrigerating, and Air-Conditioning Engineers (ASHRAE) and the Institute of Electrical and Electronics Engineers (IEEE).

2.11 Bibliography

Cameron Hydraulic Data, 15th ed., Ingersoll Rand, Woodcliff, N.J., 1966.

Engineering Data Book, 2d. ed., Hydraulic Institute, Parsippany, N.J., 1990.

Eshbach, Ovid W., *Eshbach's Handbook of Engineering Fundamentals,* John Wiley & Sons, New York, 1990.

Grimm, Nils R., and Robert C. Rosaler, *Handbook of HVAC Design,* McGraw-Hill, New York, 1990.

Handbook of Essential Engineering Information and Data, McGraw-Hill, New York, 1991.

Handbook of Fundamentals, American Society of Heating, Refrigerating, and Air-Conditioning Engineers, Atlanta, Ga, 1993.

Keenan, Joseph H., and Frederick G. Keyes, *Thermodynamic Properties of Steam,* John Wiley & Sons, New York, 1936.

Mark's Standard Handbook for Mechanical Engineers, Ninth ed., McGraw-Hill, New York, 1978.

3

Piping System Friction

A comprehensive chapter on pipe friction has been included in this *Handbook* for HVAC pumps because the sizing of pumps is determined principally on pump capacity and head. A poor computation of system friction will have a disastrous effect on pump selection and operation. There is not a more critical subject facing HVAC water system designers than the development of better procedures for calculating pump head for these systems.

As pointed out in the introduction to this book, pipe friction analysis is, at best, an inexact science. Much needs to be done to achieve better information on pipe and fitting friction. Research work on reducing pipe friction through the use of additives to water is being carried out; surfactants are one class of chemicals that are being studied to reduce piping friction. The increase in cost of energy will provide the driving force to achieve better piping friction data and better piping design.

Good piping design always balances *first cost* against *operating cost,* taking into consideration all factors that exist on each installation. These are the two basic parameters that influence pipe sizing in the HVAC industry, since excessive corrosion or fouling should not exist in these water systems.

Obviously, piping costs increase and power costs decrease with increases in pipe diameter. The American Society of Heating, Refrigerating, and Air-Conditioning Engineers (ASHRAE) has information that indicates that *velocities in the range of 10 to 17 ft/s in HVAC systems do not create erosion or noise in the larger sizes of pipe.* The overall controlling factor is *friction,* which increases exponentially with velocity. Friction in piping is the principal source of increased operating costs for these water systems.

3.1 Maximum Velocity in Pipe

There are a number of conflicting tables on the maximum allowable water velocity in HVAC piping. The failure of many of these tables of maximum velocity is their lack of consideration of the hydraulic radius of commercial pipe. The *hydraulic radius* of a pipe is the cross-sectional area of a pipe divided by the circumference of its inner surface. It is calculated as follows:

$$\text{Area} = \frac{\pi d^2}{4}$$

$$\text{Circumference} = \pi d$$

$$\text{Hydraulic radius} = \frac{\text{area}}{\text{circumference}} = \frac{d}{4} \qquad (3.1)$$

where d = inside diameter, in

Obviously, the hydraulic radius increases with pipe diameter, and therefore, the allowable velocity should increase with the pipe diameter. Hydraulic radii for commercial pipe are shown in Table 3.1. It is quite clear that 36-in inside diameter (ID) pipe with a hydraulic radius of 9.0 in must be rated velocity-wise differently than 3-in schedule 40 pipe with a hydraulic radius of 0.8 in.

Hydraulic radius may introduce a new guideline for the reevaluation of friction for flow of water in piping and pipe fittings. The current information on pipe friction and recommended velocities in pipe are too dependent on testing done on small pipe; often the data are then extrapolated for larger pipe. It is very difficult to test large pipe fittings such as those with diameters greater than 20 in.

There are several recommendations for allowable velocity in HVAC pipe; some are based on a maximum friction loss per 100 ft. Actually, as indicated above, final pipe velocity is within the province of the designer who is responsible for first cost as well as operating costs. Here is an excellent point at which the designer can use computer capability in sizing piping. Several computer runs at different pipe sizes can be done to achieve the economically desirable pipe size. This should be done for the major piping such as loops and headers. The size of smaller branches and coil connections will fall more into the realm of the designer's experience. Table 3.2 provides an elementary example of this program comparing 12-, 14-, and 16-in diameter commercial pipe.

The operating cost decreases while maintenance and amortization of the first cost increase with the pipe size. The economic pipe size is at the minimum point of the sum of the two values or curves.

TABLE 3.1   Maximum Water Capacities of Steel Pipe (in gal/min)

Size, in	Schedule	Maximum flow, gal/min	Velocity, ft/s	Loss, ft/100 ft	Hydraulic radius, in
2	40	45	4.3	3.85	0.5
2½	40	75	5.0	4.10	
3	40	130	5.6	3.92	0.8
4	40	260	6.6	4.03	1.0
6	40	800	8.9	4.03	1.5
8	40	1,600	10.3	3.82	2.0
10	40	3,000	12.2	4.06	2.5
12	40	4,700	13.4	3.98	3.0
14	40	6,000	14.2	3.95	3.3
16	40	8,000	14.5	3.49	3.8
18	40	10,000	14.3	2.97	4.2
20	40	12,000	13.8	2.44	4.5
24	40	18,000	14.4	2.10	5.7
30	20	30,000	14.6	1.61	7.3
36	36-in ID	45,000	14.1	1.18	9.0
42	42-in ID	60,000	13.9	0.95	10.5

TABLE 3.2   Total Owning Cost of Piping

Pipe size, in	Amortized first cost per year	Annual operating cost	Total annual owning cost
12	$12,000	$16,000	$28,000
14	14,000	12,000	26,000
16	17,000	10,000	27,000

Obviously, the total owning costs of the piping system should be generated for each application. The derivation of these data is beyond the scope of this book, but there are programs available for computing these costs in detail.

3.1.1   Maximum capacities and velocities of actual piping

Table 3.2 is a general recommendation to designers for the maximum capacity and pipe velocity for standard sizes of steel pipe. One basis for this information comes from ASHRAE Design Study RP-450, which researched the literature available on pipe velocities and resulting friction. This research paper is an excellent document for reviewing the history of and literature on pipe friction and maximum allowable water velocities.

It is obvious that Table 3.2 is but a preliminary road map for the knowledgeable piping designer. With the present information available, the pipe designer must rely on actual, personal experience.

3.1.2 Pipe velocity is the designer's responsibility

It is also very clear from Table 3.2 that sizing all pipe, particularly large pipe in the range from 20 to 42 in in diameter, requires a detailed analysis of the entire piping system to achieve the economical size for a particular installation. It cannot be based on a rule that limits pipe velocity. Reiterating, it is the designer's responsibility to determine pipe size and maximum velocity. There are so many judgment calls in the final selection of pipe diameter that it is not a simple process. For a hypothetical example, if you have 12,000 gal/min flowing in a chiller header in a central energy plant, you could use 20-in-diameter steel pipe if the header is only 100 ft long. This would reduce the cost of the piping and tees where the chillers are connected. On the other hand, if a chilled water supply main runs for 1000 ft to a group of buildings, you might use 24-in-diameter pipe to reduce the overall friction loss. The *cost of piping accessories and the length of pipe involved* affect the decision on the final pipe size. These are the evaluations that a good pipe designer must make.

The physical pressure that the pipe must operate under and the possibilities of corrosion as well as availability determine the schedule or wall thickness of steel pipe. The designer should make the velocity calculation and, therefore, the friction calculations based on the actual inside diameter of the pipe to be used in the water system. The designer should check the actual project conditions to ensure that the pipe inside diameter to be used for each pipe size is available at the job site at the time of construction.

3.2 Pipe and Fitting Specifications

Elements of an HVAC water system are connected together by means of piping. In most cases, this piping is steel, although various types of plastic piping are now appearing in this industry.

Most steel piping used in the HVAC industry for low-temperature applications conforms to American Society for Testing and Materials (ASTM) Specifications A53 or A120. Higher-temperature applications such as high-pressure steam and high-temperature water may require seamless piping per ASTM A106; local and ASME codes should be checked for detailed pipe, flange, bolting, and fitting specifications for particular applications such as high-temperature water and high-pressure steam. Steel fittings follow American National Standards Institute (ANSI) Specification B16.5, whereas threaded cast iron fittings comply with ANSI Specification B16.4 and flanged cast iron fittings with ANSI B16.1.

TABLE 3.3   Rating of Cast Iron Pipe Fittings (Threaded Fittings, ANSI B16.4)

| Temperature, °F | Working pressures, nonshock, psig | |
	Class 125	Class 250
−20 to 150	175	400
200	165	370
250	150	340
300	140	310
350	125	280
400	NA	250

NOTE: NA = not acceptable.

TABLE 3.4   Rating of Cast Iron Pipe Fittings (Flanged Fittings, ANSI B16.1)

| Temperature, °F | Working pressures, nonshock, psig | | | | |
| | Class 125 | | | Class 250 | |
	1–12 in	14–24 in	30–48 in	1–12 in	14–24 in*
−20 to 150	200	150	150	500	300
200	190	135	115	460	280
225	180	130	100	440	270
250	175	125	85	415	260
275	170	120	65	395	250
300	165	110	50	375	240
325	155	105	NA	355	230
350†	150	100	NA	335	220
375	145	NA	NA	315	210
400‡	140	NA	NA	290	200
425	130	NA	NA	270	NA
450	125	NA	NA	250	NA

NOTE: NA = not acceptable.
*For liquid service, these ratings are for class 250 flanges only, not for class 250 fittings.
†353°F to reflect 125-psig steam pressure.
‡406°F to reflect 250-psig steam pressure.

The pressure and temperature ratings of steel pipe, fittings, and flanges are beyond the scope of this book, since there are a number of codes for specific applications. Tables 3.3 and 3.4 list the temperature and pressure ratings for cast iron fittings that are in common use in HVAC water systems. The cast iron data are included in this book because of cast iron's greater reduction in allowable pressure with higher temperatures.

3.3 Steel Pipe Friction Analysis

As water flows through pipe, friction is generated that resists the flow. Energy is required to overcome this friction, and this energy must be derived from (1) pumps, (2) reduction in system pressure, or (3) changes in static head. How this is done in actual practice requires an evaluation of the basic equation for fluid systems: the *Bernoulli theorem*. The total hydraulic head at any point in a piping system can be computed by this theorem:

$$H = Z + hg + hv \qquad (3.2)$$

where H = total system head, ft
 Z = static head, ft
 hg = system pressure, ft H_2O
 $hv = V^2/2g$, velocity head, ft

For example, assume the following:

1. The pipe is 10 ft above the ground, which, in this case, is assumed to be the datum for all energy measurements. (Often, this datum is the elevation above sea level.)
2. The pressure in the pipe is 40 psig.
3. 200 gal/min of water at 50°F is flowing in a 4-in-diameter pipe. At this flow, from Table 3.5, the velocity head $V^2/2g$ equals 0.4 ft. The total head H in the 4-in pipe is 10 + 40 × 2.31 + 0.4 = 102.8 ft. This is the hydraulic gradient at this particular point in the piping.

Equation 3.2 is for a frictionless system. For practical applications of the Bernoulli equation, the friction of the system from one point to another must be included in the equation. This is usually expressed as an additional term Hf in feet.

Bernoulli's theorem must be studied carefully to ensure that it is understood fully. *This theorem states simply that the total energy is a constant in a system and that all energy must be accounted for in any analysis.* A typical application of this theorem is in the use of a hot or chilled water distribution system referred to as *distributed pumping,* which will be demonstrated in several places in this book.

Distributed pumping is based on the Bernoulli theorem, which states that energy for pipe friction can come from three sources: (1) elevation, (2) system pressure, and (3) velocity head. Distributed pumping derives its system distribution friction head from the second source, namely, system pressure. Distributed pumping appears to be

TABLE 3.5   Flow of Water in Steel Pipe

Flow, gal/min	1¼-in schedule 40, ID = 1.380 in			1½-in schedule 20, ID = 1.610 in			2-in schedule 40, ID = 2.067 in		
	V, ft/s	$\frac{V^2}{2g}$, ft	Hf, ft/100 ft	V, ft/s	$\frac{V^2}{2g}$, ft	Hf, ft/100 ft	V, ft/s	$\frac{V^2}{2g}$, ft	Hf, ft/100 ft
3	0.64	0.006	0.21						
4	0.86	0.011	0.34						
5	1.07	0.018	0.51	0.79	0.010	0.242			
6	1.29	0.026	0.71	0.95	0.014	0.333	0.57	0.005	0.10
7	1.50	0.035	0.93	1.10	0.019	0.439	0.67	0.007	0.13
8	1.72	0.046	1.18	1.26	0.025	0.558	0.77	0.009	0.17
9	1.93	0.058	1.46	1.42	0.031	0.689	0.86	0.012	0.21
10	2.15	0.072	1.77	1.58	0.039	0.829	0.96	0.014	0.25
12	2.57	0.103	2.48	1.89	0.056	1.16	1.15	0.021	0.34
14	3.00	0.140	3.28	2.21	0.076	1.53	1.34	0.028	0.45
16	3.43	0.183	4.20	2.52	0.099	1.96	1.53	0.036	0.58
18	3.86	0.232	5.22	2.84	0.125	2.42	1.72	0.046	0.72
20	4.29	0.286	6.34	3.15	0.154	2.94	1.91	0.057	0.87
22	4.72	0.346	7.58	3.47	0.187	3.52	2.10	0.069	1.03
24	5.15	0.412	8.92	3.78	0.222	4.14	2.29	0.082	1.20
26	5.58	0.483	10.37	4.10	0.261	4.81	2.49	0.096	1.39
28	6.01	0.561	11.9	4.41	0.303	5.51	2.68	0.111	1.60
30	6.44	0.644	13.6	4.73	0.347	6.26	2.87	0.128	1.82
35	7.51	0.877	18.5	5.52	0.472	8.52	3.35	0.174	2.42
40	8.58	1.14	23.5	6.30	0.618	10.79	3.82	0.227	3.10
45	9.65	1.44	29.7	7.09	0.782	13.7	4.30	0.288	3.85
50	10.7	1.79	36.0	7.88	0.965	16.4	4.78	0.355	4.67
55	11.8	2.16	43.2	8.67	1.17	19.7	5.26	0.430	5.59
60	12.9	2.57	51.0	9.46	1.39	23.2	5.74	0.511	6.59
65	13.9	3.02	59.6	10.24	1.63	27.1	6.21	0.600	7.69
70	15.0	3.50	68.8	11.03	1.89	31.3	6.69	0.696	8.86
75	16.1	4.02	78.7	11.8	2.17	35.8	7.17	0.799	10.1
80	17.2	4.58	89.2	12.6	2.47	40.5	7.65	0.909	11.4
85	18.2	5.17	100.2	13.4	2.79	45.6	8.13	1.03	12.8
90	19.3	5.79	112.0	14.2	3.13	51.0	8.60	1.15	14.2
95				15.0	3.48	56.5	9.08	1.28	15.8
100				15.8	3.86	62.2	9.56	1.42	17.4
120							11.5	2.05	24.7
140							13.4	2.78	33.2
160							15.3	3.64	43.0
180							17.2	4.60	54.1
200							19.1	5.68	66.3

Note: V = velocity, feet per second; $V^2/2g$ = velocity head, feet; Hf = friction loss, feet per 100 feet of pipe. No aging factor, manufacturer's tolerance, or any factor of safety has been included in the friction losses Hf.

TABLE 3.5 Flow of Water in Steel Pipe (*Continued*)

Flow, gal/min	2½-in schedule 40, ID = 2.067 in			3-in schedule 40, ID = 3.068 in			4-in schedule 40, ID = 4.026 in		
	V, ft/s	$\frac{V^2}{2g}$, ft	Hf, ft/100 ft	V, ft/s	$\frac{V^2}{2g}$, ft	Hf, ft/100 ft	V, ft/s	$\frac{V^2}{2g}$, ft	Hf, ft/100 ft
14	0.94	0.014	0.09						
16	1.07	0.018	0.24						
18	1.21	0.023	0.30						
20	1.34	0.028	0.36	0.87	0.012	0.13	0.50	0.004	0.03
22	1.47	0.034	0.43	0.95	0.026	0.15	0.55	0.005	0.04
24	1.61	0.040	0.50	1.04	0.036	0.18	0.60	0.006	0.05
26	1.74	0.047	0.58	1.13	0.047	0.21	0.66	0.007	0.06
28	1.88	0.055	0.66	1.22	0.059	0.25	0.71	0.008	0.07
30	2.01	0.063	0.75	1.30	0.073	0.26	0.76	0.009	0.07
35	2.35	0.086	1.00	1.52	0.089	0.35	0.88	0.012	0.10
40	2.68	0.112	1.28	1.74	0.105	0.44	1.01	0.016	0.12
45	3.02	0.141	1.60	1.95	0.124	0.55	1.14	0.020	0.15
50	3.35	0.174	1.94	2.17	0.143	0.66	1.26	0.025	0.18
55	3.69	0.211	2.32	2.39	0.165	0.79	1.39	0.030	0.21
60	4.02	0.251	2.72	2.60	0.187	0.92	1.51	0.036	0.25
65	4.36	0.295	3.16	2.82	0.211	1.07	1.64	0.042	0.29
70	4.69	0.342	3.63	3.04	0.237	1.22	1.76	0.048	0.33
75	5.03	0.393	4.13	3.25	0.264	1.39	1.89	0.056	0.37
80	5.36	0.447	4.66	3.47	0.293	1.57	2.02	0.063	0.42
85	5.70	0.504	5.22	3.69	0.354	1.76	2.15	0.071	0.47
90	6.03	0.565	5.82	3.91	0.421	1.96	2.27	0.080	0.52
95	6.37	0.630	6.45	4.12	0.495	2.17	2.40	0.089	0.57
100	6.70	0.698	7.11	4.34	0.574	2.39	2.52	0.099	0.62
110	7.37	0.844	8.51	4.77	0.659	2.86	2.77	0.119	0.74
120	8.04	1.00	10.0	5.21	0.749	3.37	3.02	0.142	0.88
130	8.71	1.18	11.7	5.64	0.846	3.92	3.28	0.167	1.02
140	9.38	1.37	13.5	6.08	0.948	4.51	3.53	0.193	1.17
150	10.05	1.57	15.4	6.51	1.06	5.14	3.78	0.222	1.32
160	10.7	1.79	17.4	6.94	1.17	5.81	4.03	0.253	1.49
170	11.4	2.02	19.6	7.38	1.42	6.53	4.28	0.285	1.67
180	12.1	2.26	21.9	7.81	1.69	7.28	4.54	0.320	1.86
190	12.7	2.52	24.2	8.25	1.98	8.07	4.79	0.356	2.06
200	13.4	2.79	26.7	8.68	2.29	8.90	5.04	0.395	2.27
220	14.7	3.38	32.2	9.55	2.63	10.7	5.54	0.478	2.72
240	16.1	4.02	38.1	10.4	3.00	12.6	6.05	0.569	3.21
260	17.4	4.72	44.5	11.3	3.38	14.7	6.55	0.667	3.74
280	18.8	5.47	51.3	12.2	3.79	16.9	7.06	0.774	4.30
300	20.1	6.28	58.5	13.0	4.23	19.2	7.56	0.888	4.89
320				13.9	4.68	22.0	8.06	1.01	5.51
340				14.8		24.8	8.57	1.14	6.19
360				15.6		27.7	9.07	1.28	6.92
380				16.5		30.7	9.58	1.43	7.68
400				17.4		33.9	10.1	1.58	8.47
420							10.6	1.74	9.30
440							11.1	1.91	10.2
460							11.6	2.09	11.1
480							12.1	2.27	12.0
500							12.6	2.47	13.0
550							13.9	2.99	15.7
600							15.1	3.55	18.6
650							16.4	4.17	21.7
700							17.6	4.84	25.0

Note: V = velocity, feet per second; $V^2/2g$ = velocity head, feet; Hf = friction loss, feet per 100 feet of pipe. No aging factor, manufacturer's tolerance, or any factor of safety has been included in the friction losses Hf.

TABLE 3.5 Flow of Water in Steel Pipe (*Continued*)

Flow, gal/min	5-in schedule 40, ID = 5.047 in			6-in schedule 40, ID = 6.065 in			8-in schedule 40, ID = 7.981 in		
	V, ft/s	$\dfrac{V^2}{2g}$, ft	Hf, ft/100 ft	V, ft/s	$\dfrac{V^2}{2g}$, ft	Hf, ft/100 ft	V, ft/s	$\dfrac{V^2}{2g}$, ft	Hf, ft/100 ft
30	0.48	0.004	0.024	0.33	0.002	0.010	0.19	0.001	0.003
40	0.64	0.006	0.040	0.44	0.003	0.016	0.26	0.001	0.004
50	0.80	0.010	0.059	0.56	0.005	0.024	0.32	0.002	0.007
60	0.96	0.014	0.081	0.67	0.007	0.034	0.39	0.002	0.009
70	1.12	0.020	0.108	0.78	0.009	0.045	0.45	0.003	0.012
80	1.28	0.026	0.137	0.89	0.012	0.056	0.51	0.004	0.015
90	1.44	0.032	0.169	1.00	0.016	0.070	0.58	0.005	0.019
100	1.60	0.040	0.204	1.11	0.019	0.084	0.64	0.006	0.022
120	1.92	0.058	0.286	1.33	0.028	0.118	0.77	0.009	0.031
140	2.25	0.078	0.380	1.55	0.038	0.155	0.90	0.013	0.041
160	2.57	0.102	0.487	1.78	0.049	0.198	1.03	0.016	0.052
180	2.89	0.129	0.606	2.00	0.062	0.246	1.15	0.021	0.064
200	3.21	0.160	0.736	2.22	0.077	0.299	1.28	0.026	0.078
220	3.53	0.193	0.879	2.44	0.093	0.357	1.41	0.031	0.093
240	3.85	0.230	1.035	2.66	0.110	0.419	1.54	0.037	0.109
260	4.17	0.270	1.20	2.89	0.130	0.487	1.67	0.043	0.126
280	4.49	0.313	1.38	3.11	0.150	0.560	1.80	0.050	0.144
300	4.81	0.360	1.58	3.33	0.172	0.637	1.92	0.058	0.163
320	5.13	0.409	1.78	3.55	0.196	0.719	2.05	0.066	0.184
340	5.45	0.462	2.00	3.78	0.222	0.806	2.18	0.074	0.206
360	5.77	0.518	2.22	4.00	0.240	0.898	2.31	0.083	0.229
380	6.09	0.577	2.46	4.22	0.277	0.993	2.44	0.092	0.253
400	6.41	0.639	2.72	4.44	0.307	1.09	2.57	0.102	0.279
420	6.74	0.705	2.98	4.66	0.338	1.20	2.70	0.113	0.308
440	7.06	0.774	3.26	4.89	0.371	1.31	2.83	0.124	0.338
460	7.38	0.846	3.55	5.11	0.405	1.42	2.96	0.135	0.369
480	7.70	0.921	3.85	5.33	0.442	1.54	3.08	0.147	0.402
500	8.02	1.0	4.16	5.55	0.479	1.66	3.21	0.160	0.424
550	8.82	1.21	4.98	6.11	0.580	1.99	3.53	0.193	0.507
600	9.62	1.44	5.88	6.66	0.690	2.34	3.85	0.230	0.597
650	10.4	1.69	6.87	7.22	0.810	2.73	4.17	0.271	0.694
700	11.2	1.96	7.93	7.77	0.939	3.13	4.49	0.313	0.797
750	12.0	2.25	9.05	8.33	1.08	3.57	4.81	0.360	0.907
800	12.8	2.56	10.22	8.88	1.23	4.03	5.13	0.409	1.02
850	13.6	2.89	11.5	9.44	1.38	4.53	5.45	0.462	1.15
900	14.4	3.24	12.9	9.99	1.55	5.05	5.77	0.518	1.27
950	15.2	3.61	14.3	10.5	1.73	5.60	6.09	0.577	1.41
1000	16.0	4.00	15.8	11.1	1.92	6.17	6.41	0.639	1.56
1100	17.6	4.84	19.0	12.2	2.32	7.41	7.05	0.773	1.87
1200	19.2	5.76	22.5	13.3	2.76	8.76	7.70	0.920	2.20
1300	20.8	6.75	26.3	14.4	3.24	10.2	8.34	1.08	2.56
1400	22.5	7.83	30.4	15.5	3.76	11.8	8.98	1.25	2.95
1500	24.1	8.99	34.8	16.7	4.31	13.5	9.62	1.44	3.37
1600	25.7	10.2	39.5	17.8	4.91	15.4	10.3	1.64	3.82
1700	27.3	11.6	44.5	18.9	5.54	17.3	10.9	1.85	4.29
1800	28.8	12.9	49.7	20.0	6.21	19.4	11.5	2.07	4.79
1900	30.5	14.4	55.2	21.1	6.92	21.6	12.2	2.31	5.31
2000	32.1	16.0	61.0	22.2	7.67	23.8	12.8	2.56	5.86
2200				24.4	9.27	28.8	14.1	3.09	7.02
2400				26.6	11.1	34.2	15.4	3.68	8.31
2600							16.7	4.32	9.70
2800							18.0	5.01	11.20
3000							19.2	5.75	12.8

Note: V = velocity, feet per second; $V^2/2g$ = velocity head, feet; Hf = friction loss, feet per 100 feet of pipe. No aging factor, manufacturer's tolerance, or any factor of safety has been included in the friction losses Hf.

41

TABLE 3.5 Flow of Water in Steel Pipe (*Continued*)

Flow, gal/min	10-in schedule 40, ID = 10.020 in			12-in schedule 40, ID = 11.938 in			14-in schedule 40, ID = 13.124 in		
	V, ft/s	$\dfrac{V^2}{2g}$, ft	Hf, ft/100 ft	V, ft/s	$\dfrac{V^2}{2g}$, ft	Hf, ft/100 ft	V, ft/s	$\dfrac{V^2}{2g}$, ft	Hf, ft/100 ft
120	0.49	0.004	0.010						
140	0.57	0.005	0.014						
160	0.65	0.007	0.017						
180	0.73	0.008	0.022						
200	0.81	0.010	0.026	0.57	0.005	0.011	0.47	0.003	0.007
220	0.90	0.013	0.031	0.63	0.006	0.013	0.52	0.004	0.009
240	0.98	0.015	0.036	0.69	0.007	0.016	0.57	0.005	0.010
260	1.06	0.017	0.042	0.75	0.009	0.018	0.62	0.006	0.012
280	1.14	0.020	0.048	0.80	0.010	0.021	0.66	0.007	0.014
300	1.22	0.023	0.054	0.86	0.012	0.023	0.71	0.008	0.015
350	1.42	0.032	0.072	1.00	0.016	0.031	0.83	0.011	0.019
400	1.63	0.041	0.092	1.15	0.020	0.039	0.95	0.014	0.025
450	1.83	0.052	0.114	1.29	0.026	0.049	1.07	0.018	0.032
500	2.03	0.064	0.138	1.43	0.032	0.059	1.19	0.022	0.037
550	2.24	0.078	0.164	1.58	0.039	0.070	1.31	0.026	0.045
600	2.44	0.093	0.192	1.72	0.046	0.082	1.42	0.031	0.052
650	2.64	0.109	0.224	1.86	0.054	0.095	1.54	0.037	0.061
700	2.85	0.126	0.256	2.01	0.063	0.109	1.66	0.043	0.068
750	3.05	0.145	0.291	2.15	0.072	0.124	1.78	0.049	0.078
800	3.25	0.165	0.328	2.29	0.082	0.140	1.90	0.056	0.087
850	3.46	0.186	0.368	2.44	0.092	0.156	2.02	0.063	0.098
900	3.66	0.208	0.410	2.58	0.103	0.173	2.13	0.071	0.108
950	3.87	0.232	0.455	2.72	0.115	0.191	2.25	0.079	0.120
1000	4.07	0.257	0.500	2.87	0.128	0.210	2.37	0.087	0.131
1100	4.48	0.311	0.600	3.15	0.154	0.251	2.61	0.106	0.157
1200	4.88	0.370	0.703	3.44	0.184	0.296	2.85	0.126	0.185
1300	5.29	0.435	0.818	3.73	0.216	0.344	3.08	0.148	0.215
1400	5.70	0.504	0.940	4.01	0.250	0.395	3.32	0.171	0.247
1500	6.10	0.579	1.07	4.30	0.287	0.450	3.56	0.197	0.281
1600	6.51	0.659	1.21	4.59	0.327	0.509	3.79	0.224	0.317
1700	6.92	0.743	1.36	4.87	0.369	0.572	4.03	0.252	0.355
1800	7.32	0.834	1.52	5.16	0.414	0.636	4.27	0.283	0.395
1900	7.73	0.929	1.68	5.45	0.461	0.704	4.50	0.315	0.438
2000	8.14	1.03	1.86	5.73	0.511	0.776	4.74	0.349	0.483
2200	8.95	1.25	2.23	6.31	0.618	0.930	5.21	0.422	0.584
2400	9.76	1.48	2.64	6.88	0.735	1.093	5.69	0.503	0.696
2600	10.6	1.74	3.08	7.45	0.863	1.28	6.16	0.590	0.816
2800	11.4	2.02	3.56	8.03	1.00	1.47	6.64	0.684	0.947
3000	12.2	2.32	4.06	8.60	1.15	1.68	7.11	0.786	1.04
3200	13.0	2.63	4.59	9.17	1.31	1.90	7.58	0.894	1.18
3400	13.8	2.97	5.16	9.75	1.48	2.13	8.06	1.01	1.34
3600	14.6	3.33	5.76	10.3	1.65	2.37	8.53	1.13	1.50
3800	15.5	3.71	6.40	10.9	1.84	2.63	9.01	1.26	1.67
4000	16.3	4.12	7.07	11.5	2.04	2.92	9.48	1.40	1.81
4500	18.3	5.21	8.88	12.9	2.59	3.65	10.7	1.77	2.27
5000	20.3	6.43	10.9	14.3	3.19	4.47	11.9	2.18	2.78
5500				15.8	3.86	5.38	13.1	2.64	3.36
6000				17.2	4.60	6.39	14.2	3.14	3.95
6500				18.6	5.39	7.47	15.4	3.69	4.64
7000				20.1	6.26	8.63	16.6	4.28	5.32
8000							19.0	5.59	6.90
9000							21.3	7.08	8.7

Note: V = velocity, feet per second; $V^2/2g$ = velocity head, feet; Hf = friction loss, feet per 100 feet of pipe. No aging factor, manufacturer's tolerance, or any factor of safety has been included in the friction losses Hf.

TABLE 3.5 Flow of Water in Steel Pipe (*Continued*)

Flow, gal/min	16-in schedule 40, ID = 15.000 in			18-in schedule 40, ID = 16.876 in			20-in schedule 40, ID = 18.812 in		
	V, ft/s	$\frac{V^2}{2g}$, ft	Hf, ft/100 ft	V, ft/s	$\frac{V^2}{2g}$, ft	Hf, ft/100 ft	V, ft/s	$\frac{V^2}{2g}$, ft	Hf, ft/100 ft
300	0.55	0.005	0.008	0.43	0.003	0.004	0.35	0.002	0.003
400	0.73	0.008	0.013	0.57	0.005	0.007	0.46	0.003	0.004
500	0.91	0.013	0.019	0.72	0.008	0.011	0.58	0.005	0.006
600	1.09	0.018	0.027	0.86	0.012	0.015	0.69	0.007	0.009
700	1.27	0.025	0.036	1.00	0.016	0.020	0.81	0.010	0.012
800	1.45	0.033	0.045	1.15	0.021	0.026	0.92	0.013	0.015
900	1.63	0.042	0.056	1.29	0.026	0.032	1.04	0.017	0.019
1000	1.82	0.051	0.068	1.43	0.032	0.039	1.15	0.021	0.023
1200	2.18	0.074	0.095	1.72	0.046	0.054	1.38	0.030	0.032
1400	2.54	0.100	0.127	2.01	0.063	0.072	1.62	0.041	0.042
1600	2.90	0.131	0.163	2.30	0.082	0.092	1.85	0.053	0.054
1800	3.27	0.166	0.203	2.58	0.104	0.114	2.08	0.067	0.067
2000	3.63	0.205	0.248	2.87	0.128	0.139	2.31	0.083	0.081
2500	4.54	0.320	0.377	3.59	0.200	0.211	2.89	0.129	0.123
3000	5.45	0.461	0.535	4.30	0.288	0.297	3.46	0.186	0.174
3500	6.35	0.627	0.718	5.02	0.392	0.397	4.04	0.254	0.232
4000	7.26	0.820	0.921	5.74	0.512	0.511	4.62	0.331	0.298
4500	8.17	1.04	1.15	6.45	0.647	0.639	5.19	0.419	0.372
5000	9.08	1.28	1.41	7.17	0.799	0.781	5.77	0.517	0.455
6000	10.9	1.84	2.01	8.61	1.15	1.11	6.92	0.745	0.645
7000	12.7	2.51	2.69	10.0	1.57	1.49	8.08	1.014	0.862
8000	14.5	3.28	3.49	11.5	2.05	1.93	9.23	1.32	1.11
9000	16.3	4.15	4.38	12.9	2.59	2.42	10.39	1.68	1.39
10000	18.2	5.12	5.38	14.3	3.20	2.97	11.5	2.07	1.70
12000	21.8	7.38	7.69	17.2	4.60	4.21	13.8	2.98	2.44
14000				20.1	6.27	5.69	16.2	4.06	3.29
16000				22.9	8.19	7.41	18.5	5.30	4.26
18000							20.8	6.71	5.35
20000							23.1	8.28	6.56

Note: V = velocity, feet per second; $V^2/2g$ = velocity head, feet; Hf = friction loss, feet per 100 feet of pipe. No aging factor, manufacturer's tolerance, or any factor of safety has been included in the friction losses Hf.

difficult to understand until this simple fact is realized. Figure 3.1 has been generated to demonstrate the use of this theorem.

It should be noted in the preceding example that velocity head $V^2/2g$ is so small that it is seldom used in water distribution calculations. Therefore, it is not included in the pressure gradients described in this *Handbook*. The total hydraulic gradient for a water system does include the velocity head. Velocity head should not be ignored totally, since it does come into importance when determining the flow in pipe around chillers and boilers. Also, it is the correct basis for computing friction loss in pipe fittings.

TABLE 3.5 Flow of Water in Steel Pipe (*Continued*)

Flow, gal/min	24-in schedule 40, ID = 22.624 in			30-in schedule 20, ID = 29.000 in			36-in inside diameter		
	V, ft/s	$\dfrac{V^2}{2g}$, ft	Hf, ft/100 ft	V, ft/s	$\dfrac{V^2}{2g}$, ft	Hf, ft/100 ft	V, ft/s	$\dfrac{V^2}{2g}$, ft	Hf, ft/100 ft
300	0.24	0.001	0.001						
400	0.32	0.002	0.002	0.19	0.001	0.001			
500	0.40	0.002	0.003	0.24	0.001	0.001			
600	0.48	0.004	0.004	0.29	0.001	0.001			
700	0.56	0.005	0.005	0.34	0.002	0.001			
800	0.64	0.006	0.006	0.39	0.002	0.002			
900	0.72	0.008	0.008	0.44	0.003	0.002			
1000	0.80	0.010	0.009	0.49	0.004	0.003	0.32	0.002	0.001
1200	0.96	0.014	0.013	0.58	0.005	0.004	0.38	0.002	0.001
1400	1.12	0.019	0.017	0.68	0.007	0.005	0.44	0.003	0.002
1600	1.28	0.025	0.022	0.78	0.009	0.007	0.50	0.004	0.002
1800	1.44	0.032	0.027	0.87	0.012	0.008	0.57	0.005	0.003
2000	1.60	0.040	0.033	0.97	0.015	0.010	0.63	0.006	0.003
2500	1.99	0.062	0.050	1.21	0.023	0.015	0.79	0.010	0.005
3000	2.39	0.089	0.070	1.46	0.033	0.021	0.95	0.014	0.007
3500	2.79	0.121	0.093	1.70	0.045	0.028	1.10	0.019	0.010
4000	3.19	0.158	0.120	1.94	0.059	0.035	1.26	0.025	0.012
4500	3.59	0.200	0.149	2.19	0.074	0.044	1.41	0.031	0.015
5000	3.99	0.247	0.181	2.43	0.092	0.054	1.58	0.039	0.019
6000	4.79	0.356	0.257	2.91	0.132	0.075	1.89	0.056	0.026
7000	5.59	0.485	0.343	3.40	0.180	0.100	2.21	0.076	0.035
8000	6.38	0.633	0.441	3.89	0.235	0.129	2.52	0.099	0.044
9000	7.18	0.801	0.551	4.37	0.297	0.161	2.84	0.125	0.055
10000	7.98	0.989	0.671	4.86	0.367	0.196	3.15	0.154	0.067
12000	9.58	1.42	0.959	5.83	0.528	0.277	3.78	0.222	0.094
14000	11.2	1.94	1.29	6.80	0.719	0.371	4.41	0.303	0.126
16000	12.8	2.53	1.67	7.77	0.939	0.478	5.04	0.395	0.162
18000	14.4	3.21	2.10	8.74	1.19	0.598	5.67	0.500	0.203
20000	16.0	3.96	2.58	9.71	1.47	0.732	6.30	0.618	0.248
22000	17.6	4.79	3.10	10.7	1.78	0.886	6.93	0.746	0.300
24000	19.2	5.70	3.67	11.7	2.12	1.05	7.56	0.888	0.357
26000	20.7	6.69	4.29	12.6	2.48	1.24	8.20	1.04	0.419
28000				13.6	2.88	1.43	8.83	1.21	0.486
30000				14.6	3.30	1.61	9.46	1.39	0.540
35000				17.0	4.49	2.17	11.03	1.89	0.724
40000				19.4	5.87	2.83	12.6	2.47	0.941
45000				21.9	7.42	3.56	14.1	3.13	1.18
50000				24.3	9.17	4.38	15.8	3.86	1.45
55000							17.4	4.70	1.75
60000							18.9	5.56	2.07
70000							22.1	7.56	2.81

Note: V = velocity, feet per second; $V^2/2g$ = velocity head, feet; Hf = friction loss, feet per 100 feet of pipe. No aging factor, manufacturer's tolerance, or any factor of safety has been included in the friction losses Hf.

TABLE 3.5   Flow of Water in Steel Pipe (*Continued*)

Flow, gal/min	V, ft/s	$\frac{V^2}{2g}$, ft	h, ft per 100 ft of pipe
		42-in inside diameter	
1000	0.232	0.00083	0.00047
1500	0.347	0.00187	0.00098
2000	0.463	0.00333	0.00164
2500	0.579	0.00521	0.00246
3000	0.695	0.00750	0.00343
3500	0.811	0.0102	0.00454
4000	0.926	0.0133	0.00580
4500	1.042	0.0169	0.00720
5000	1.16	0.0208	0.00874
6000	1.39	0.0300	0.0122
7000	1.62	0.0408	0.0162
8000	1.85	0.0533	0.0208
9000	2.08	0.0675	0.0258
10000	2.32	0.0833	0.0314
12000	2.78	0.120	0.0441
14000	3.24	0.163	0.0591
16000	3.71	0.213	0.0758
18000	4.17	0.270	0.0944
20000	4.63	0.333	0.115
25000	5.79	0.521	0.176
30000	6.95	0.750	0.250
35000	8.11	1.02	0.334
40000	9.26	1.33	0.433
45000	10.42	1.69	0.545
50000	11.6	2.08	0.668
60000	13.9	3.00	0.946
70000	16.2	4.08	1.27
80000	18.5	5.33	1.66
90000	20.8	6.75	2.08
100000	23.2	8.33	2.57

Note: V = velocity, feet per second; $V^2/2g$ = velocity head, feet; Hf = friction loss, feet per 100 feet of pipe. No aging factor, manufacturer's tolerance, or any factor of safety has been included in the friction losses Hf.

Source: Data in this table have been derived from the *Engineering Data Book,* 2d ed., Hydraulic Institute, Parsippany, N.J., 1990.

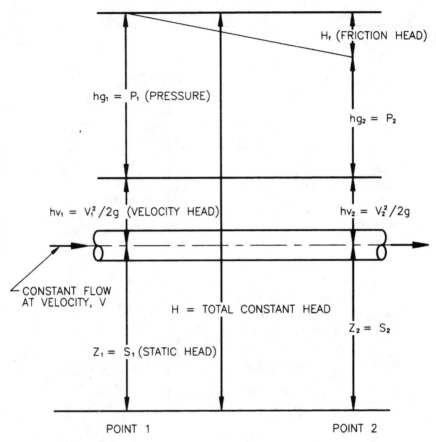

Figure 3.1 Bernoulli diagram.

3.3.1 Pipe friction formulas

The amount of friction that is created by flow of water in piping has been determined by a number of sources. Today, there are two principal formulas for determining pipe friction. These are the Darcy Weisbach (Fanning) and the Williams and Hazen formulas:

Darcy Weisbach formula (Basis for Table 3.5)

$$Hf = f \cdot \frac{L}{D} \cdot \frac{V^2}{2g} \tag{3.3}$$

where Hf = friction loss, ft of liquid
L = pipe length, in ft
D = average inside diameter, ft

f = friction factor

The friction factor f is usually derived from the Colebrook equation:[1]

$$\frac{1}{\sqrt{f}} = -2\log_{10}\left(\frac{\epsilon}{3.7D} + \frac{2.51}{R\sqrt{f}}\right) \tag{3.4}$$

where R = Reynolds number
  ϵ = absolute roughness parameter (typically 0.00015 for steel pipe)

For practical purposes, the friction factor f is calculated from the Moody diagram described below.

Williams and Hazen formula

$$Hf = 0.002083 \cdot L \cdot \left(\frac{100}{C}\right)^{1.85} \cdot \frac{\text{gal/min}^{1.85}}{d^{4.8655}} \tag{3.5}$$

where C = a design factor determined for various types of pipe
  d = inside diameter of pipe, in

There are a number of sources for securing the data for the aforementioned equations in either tabular or software form. *Before any data on pipe friction are used, either in tabular or computer software form, be sure that the pipe under consideration has the same inside diameter as that in the tables or computer software!* The following pages demonstrate some of the sources for pipe friction data in tabular form. A principal source is the Hydraulic Institute's *Engineering Data Book*. This book is based on the Darcy Weisbach formula, and Table 3.5 has been developed from Hydraulic Institute data for steel pipe. It is strongly recommended that this data book be acquired by anyone who is involved in piping design.

3.3.2 Reynolds number and the Moody diagram

The Hydraulic Institute's *Engineering Data Book* contains some very practical information on the generation and use of Reynolds number. Reynolds number is a dimensionless number that simplifies the calculation of pipe friction under varying velocities and viscosities.

$$\text{Reynolds number } R = \frac{V \cdot D}{v} \tag{3.6}$$

where V = velocity, ft/s
  D = pipe diameter, ft

v = kinematic viscosity, ft^2/s

For example, assume that 50°F water is flowing through 4-in schedule 40 steel pipe at 200 gal/min. From Table 3.5, the velocity is 5.04 ft/s, and the diameter of the pipe is 4.026 in, or 0.336 ft. From Table 2.2, the kinematic viscosity for 50°F water is 1.41×10^{-5} ft^2/s. The Reynolds number is

$$\frac{5.04 \times 0.336}{1.41 \times 10^{-5}} = 1.20 \times 10^5$$

From the Moody diagram shown in Fig. 3.2, the friction factor f is 0.0198.

The Moody diagram, named after its originator, is described in Fig. 3.2. This diagram generates the friction factor f of the Colebrook equation (Eq. 3.5) for a broad variety of Reynolds numbers and, therefore, velocities and viscosities. This particular diagram accepts a constant relative roughness factor ϵ of 0.00015 and provides curves for all popular sizes of steel pipe. This is adequate for most HVAC applications of steel pipe.

Another diagram that is useful for work with a number of liquids where the kinematic viscosity is known is shown in Fig. 3.3. The Reynolds number can be selected from this diagram once the pipe diameter, water velocity, and viscosity of the liquid are developed. Also, if the temperature of the liquid is determined, the kinematic viscosity of that liquid can be determined from this figure.

The practical use of Reynolds number and the Moody diagram can be demonstrated by the following example. Assume that

1. A condenser water system is used for cooling computer air-conditioners employing a propylene glycol solution of 50% where the liquid temperature can vary from 30 to 130°F.

2. The flow rate is 750 gal/min at 30°F in a 6-in-diameter steel pipe.

Question. What is the variation in friction loss in feet per 100 feet of length as the liquid varies from 30 to 130°F?

Factual data

1. From Fig. 2.1, the viscosity of the glycol solution is 8.5 cP at 30°F and 1.4 cP at 130°F.

2. From Fig. 2.3, the specific gravity of the glycol solution is 1.09 × water at 60°F, or 67.95 lb/ft^3. Likewise, the specific gravity at 130°F is 1.06 × water at 60°F, or 66.08 lb/ft^3.

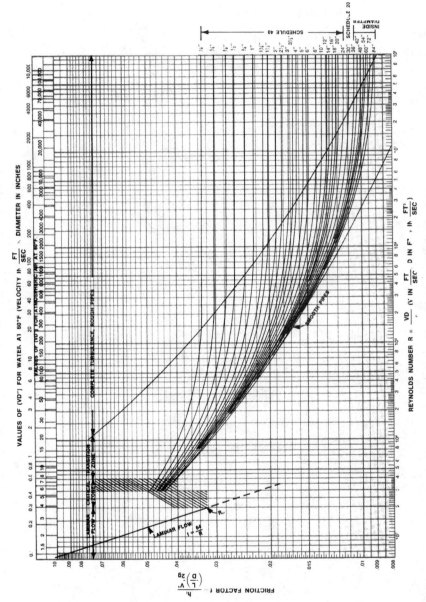

Figure 3.2 Moody diagram. (*From Engineering Data Book, 2d ed., Hydraulic Institute, Parsippany, N.J., 1990, p. 37.*)

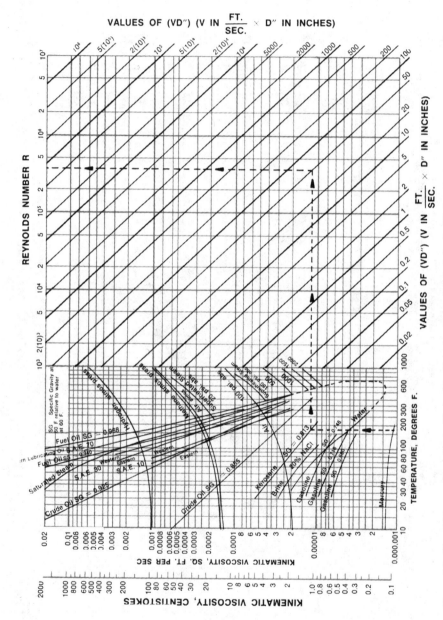

Figure 3.3 Reynolds number chart. *(From Engineering Data Book, 2d ed., Hydraulic Institute, Parsippany, N.J., 1990, p. 42.)*

3. From Table 3.5, the inside diameter of 6-in steel schedule 40 pipe is 6.065 in, or 0.505 ft. Also from this table, with 750 gal/min flowing through this pipe, the velocity is 8.33 ft/s, and the velocity head is 1.08 ft.

Computing the kinematic viscosity first in square feet per second from Eq. 2.1. At 30°F,

$$v = \frac{6.7197 \times 8.5 \times 10^{-4}}{67.95} = 0.841 \times 10^{-4}$$

At 130°F,

$$v = \frac{6.7197 \times 1.4 \times 10^{-4}}{66.08} = 0.142 \times 10^{-4}$$

The data have now been collected to compute the Reynolds number at the two temperatures. At 30°F,

$$R = \frac{V \times D}{v} = \frac{8.33 \times 0.505}{0.841 \times 10^{-4}} = 5.00 \times 10^{4}$$

At 130°F,

$$R = \frac{8.33 \times 0.505}{0.142 \times 10^{-5}} = 2.96 \times 10^{5}$$

The friction factors can now be selected from the Moody diagram (Fig. 3.2). At 30°F and a Reynolds number of 5.00×10^{4}, the friction factor f is 0.002. At 130°F and a Reynolds number of 2.96×10^{5}, the friction factor f is 0.017.

The friction in feet per 100 ft can now be calculated from the Darcy Weisbach equation (Eq. 3.3). At 30°F,

$$Hf = \frac{0.022 \times 100 \times 1.08}{0.505} = 4.71 \text{ ft/100 ft}$$

At 130°F,

$$Hf = \frac{0.017 \times 100 \times 1.08}{0.505} = 3.64 \text{ ft/100 ft}$$

This example demonstrates the use of Reynolds number and the Moody diagram. It also emphasizes the variation in pipe friction with viscosity. In this case, the friction at 130°F was 77 percent of that at 30°F. Also, this demonstrates that the friction for water at 60°F from

Table 3.5 of 3.57 ft/100 ft should not be used in calculating friction losses for this glycol solution.

3.3.3   Use of the Darcy Weisbach equation

For those who wish to study pipe friction further or use the Darcy Weisbach equation for generating their own computer program for pipe friction, the preceding example provides a guide for doing so. Also, the two figures from the Hydraulic Institute's *Engineering Data Book* (Figs. 3.2 and 3.3) should clarify the use of Reynolds number for the calculation of pipe friction.

Another source for Darcy Weisbach data is *Cameron Hydraulic Data,* published by the Ingersoll Rand Company. Like the data book of the Hydraulic Institute, it is an excellent source for pipe and water data and is a necessary reference manual for any serious designer of piping.

Both these sources do not include any allowance in their tables for pipe aging, variation in pipe manufacture, or field assembly. The Hydraulic Institute recommends that 15 percent allowance be made for these factors; I consider this factor adequate for loop-type systems such as hot or chilled water systems. It is not adequate for cooling tower water, which is exposed to air in the water; it is recommended that a factor of 20 percent be added for this service if steel pipe is used. This possible increase in friction for steel pipe may demonstrate the application for plastic pipe on cooling tower water; its use is limited by size and structural capability.

3.3.4   Use of the Williams-Hazen formula

The Williams-Hazen formula is very popular in the civil engineering field and can be used for HVAC piping design if it is understood properly. This formula gives accurate values for liquids that have a viscosity of around 1.1 cSt, such as water at 60°F. It is therefore acceptable for chilled water and even condenser water. It will show an error as much as 20 percent high if used for hot water over 200°F.

This formula is based on design factors that relate to the roughness of the pipe involved. These design factors are called *C factors* that appear in the preceding equation and range from 80 to 160, 80 being for the roughest pipe and 160 for the smoothest pipe.

Table 3.6 is taken from an older edition of *Cameron Hydraulic Data* and includes the *C* factors for various types of pipe. If a *C* factor of 140 is used with the Williams-Hazen formula, it will yield friction data for steel pipe that is somewhat comparable with the Darcy Weisbach formula with the 15 percent aging factor as recommended above. This is adequate for closed piping such as chilled water but should not be used with cooling tower piping. Cooling tower piping

TABLE 3.6   Values of *C for the Williams-Hazen Formula*

Type of Pipe	Values of C		
	Range	Average value	Design value
Cement = asbestos	160–140	150	150
Fiber	—	150	150
Bitumastic- or enamel-lined iron or steel, centrifugally applied	160–130	148	140
Cement-lined iron or steel, centrifugally applied	—	150	140
Copper, brass, or glass piping or tubing	150–120	140	150
Welded or seamless steel	150–120	140	140
Wrought iron	150–80	130	100
Cast iron	150–80	130	100

SOURCE: *Cameron Hydraulic Data,* 15th ed., Ingersoll Rand Company, Woodcliff, N.J., 1977; used with permission.

should be calculated with a C factor of 130 for steel pipe when using the Williams-Hazen formula.

For values of C,

1. Range: High is for smooth and well-installed pipe; low is for poor or corroded pipe.

2. The average value is for good clean and new pipe.

3. The design value is the common value used for design purposes.

Note. The American Iron and Steel Institute recommends a C factor of 140 for steel pipe for hot and chilled water systems.

There is some confusion in the industry about the use of the Williams-Hazen formula. It is very adequate for cold water such as chilled water. A number of organizations publish Williams-Hazen data for various pipe sizes; organizations such as the American Iron and Steel Institute provide manuals on piping that include Williams-Hazen friction data on the water flow in the commercial sizes of steel pipe. It is very adequate for cold water such as chilled water. Since it cannot easily accommodate viscosity or specific gravity corrections, it should not be used for hot water systems.

3.3.5   Review of pipe friction and velocities

The preceding recommendations are the result of my experience; every designer should develop parameters based on his or her experience. It should be pointed out that pipe friction is not an exact science. A study of the available literature will demonstrate the variations that can exist with pipe friction when manufacturing tolerances and specific aging processes are included. This has caused some designers to use much larger safety factors in their design of piping.

The figures and tables of this chapter provide data that can be referred to during a piping design procedure. Software is also available that implements the formulas and eliminates much of the drudgery in calculating pipe friction. Software packages are available from various sources at reasonable costs. It is urged that the design basis for such software be fully understood before it is used on a project. Designers with knowledge of software and computers can develop some of their own computer programs for pipe friction by using the aforementioned equations.

3.4   Steel and Cast Iron Pipe Fittings

A significant part of the friction loss for HVAC piping is caused by the various fittings that are used to connect the piping. There have been some very unacceptable practices used for the calculation of fitting losses. For example, some people recommend that the pipe friction be calculated and a percentage of that loss be added for the pipe fitting and valve losses. This is a very poor and inexact method of computing friction losses for valves and fittings. The proper method of computing fitting loss is to determine that loss for each and every fitting and valve. Again, the Hydraulic Institute's *Engineering Data Book* and Ingersol Rand's *Cameron Hydraulic Data* are excellent sources for fitting and valve losses. Interesting data on the possible variations in the friction losses for fittings and valves are included in Table 32(c) of the Hydraulic Institute's *Engineering Data Book*; variations from ± 10 percent to as high as ± 50 percent are described.

Most fitting losses are referenced to the velocity head $V^2/2g$ of the water flowing in the pipe. Table 3.5 for steel pipe friction lists the velocity head at various flows in the pipe. A K factor has been developed for many of the popular pipe fittings so that the loss through fittings Hf is

$$Hf = K \cdot \frac{V^2}{2g} \tag{3.7}$$

The Hydraulic Institute provides various K factors for valves and fittings in their *Engineering Data Book*; these are listed in Tables 3.7 through 3.12 as approximate values. Unfortunately, at this writing, many popular fittings such as reducing elbows or tees have no data. Also, welded steel pipe reducers have significant losses, and there are no reliable data on them.

TABLE 3.7   *K Factors for Inlets from Open Tanks*

Type of inlet	K factor
Bell-mouth inlet	0.05
Square-edged inlet	0.5
Inward-projecting pipe	1.0*

*K decreases with increases in wall thickness and rounding of edges.

TABLE 3.8   *K Factors for Elbows, 90°*

Size, in	Regular screwed	Long-radius screwed	Regular flanged	Long-radius flanged
½	2.1	—	—	—
¾	1.7	0.85	—	—
1	1.5	0.76	0.43	—
1¼	1.3	0.65	0.40	0.37
1½	1.2	0.53	0.39	0.35
2	1.0	0.43	0.37	0.31
2½	0.85	0.36	0.35	0.27
3	0.76	0.29	0.33	0.25
4	0.65	0.23	0.32	0.22
6	—	—	0.29	0.18
8	—	—	0.27	0.15
10	—	—	0.25	0.14
12	—	—	0.24	0.13
14	—	—	0.23	0.12
16	—	—	0.23	0.11
18	—	—	0.22	0.10
20	—	—	0.22	0.09
24	—	—	0.21	—

Miscellaneous *K factors*

Couplings and unions. Couplings and unions depend on the quality of manufacture. Assume an average K factor of 0.05 for them.

Reducing bushings and couplings. The K factor for reducing bushings can vary from 0.05 to 2.0. Replace them wherever possible with threaded tapered fittings.

Sudden enlargements. Sudden enlargements such as reducing flanges can have a K factor as high as 1.0. Replace them with tapered fittings wherever possible.

3.4.1   Tapered fittings

There are a number of tests and equations for computing the K factors for tapered fittings. Like other fitting calculations, they are, at

TABLE 3.9   *K Factors for Elbows, 45°*

Size, in	Regular screwed	Long-radius flanged
$\frac{1}{2}$	0.36	—
$\frac{3}{4}$	0.35	—
1	0.33	0.21
$1\frac{1}{4}$	0.32	0.20
$1\frac{1}{2}$	0.31	0.18
2	0.30	0.17
$2\frac{1}{2}$	0.29	0.15
3	0.28	0.14
4	0.27	0.14
6	—	0.13
8	—	0.12
10	—	0.11
12	—	0.10
14	—	0.09
16	—	0.08

TABLE 3.10   *K Factors for Tees*

Size, in	Screwed-type, branch flow	Flanged, line flow	Flanged, branch flow
$\frac{1}{2}$	2.4	—	—
$\frac{3}{4}$	2.1	—	—
1	1.9	0.27	1.0
$1\frac{1}{4}$	1.7	0.25	0.94
$1\frac{1}{2}$	1.6	0.23	0.90
2	1.4	0.20	0.83
$2\frac{1}{2}$	1.3	0.18	0.80
3	1.2	0.17	0.75
4	1.1	0.15	0.70
6	—	0.12	0.61
8	—	0.10	0.58
10	—	0.09	0.54
12	—	0.09	0.52
14	—	0.08	0.51
16	—	0.08	0.47
18	—	0.07	0.44
20	—	0.07	0.42

NOTE: For screwed-type, line flow tees, a *K* factor of 0.90 is used for all sizes.

best, approximate. Two of the most accurate calculations seem to be those included in *Cameron Hydraulic Data.*

Tapered reducing fittings

$$K = 0.8 \sin \frac{\theta}{2} \left(1 - \frac{d_1^{\,2}}{d_2^{\,2}} \right) \tag{3.8}$$

TABLE 3.11 *K Factors for Globe and Gate Valves (100 percent open)*

Size, in	Globe valves		Gate valves	
	Screwed	Flanged	Screwed	Flanged
$\frac{1}{2}$	14	—	0.32	—
$\frac{3}{4}$	10	—	0.28	—
1	8.7	14	0.24	—
$1\frac{1}{4}$	8.0	12	0.22	—
$1\frac{1}{2}$	7.5	10	0.20	—
2	6.9	8.5	0.17	—
$2\frac{1}{2}$	6.5	7.9	0.15	—
3	6.0	7.0	0.14	0.22
4	5.8	6.4	0.12	0.16
6	—	5.9	—	0.11
8	—	5.7	—	0.08
10	—	5.7	—	0.06
12	—	5.6	—	0.05
14	—	5.6	—	0.05
16	—	5.6	—	0.05
18	—	5.6	—	0.04
20	—	5.6	—	0.03

TABLE 3.12 *K Factors for Miscellaneous Valves (100 percent open)*

Size, in	Screwed check	Angle valve		Ball valve*	Butterfly valve
		Screwed	Flanged		
$\frac{1}{2}$	5.6	—	—	—	—
$\frac{3}{4}$	3.7	5.5	—	0.11	—
1	3.0	4.7	5.0	0.10	—
$1\frac{1}{4}$	2.7	3.8	3.7	0.09	—
$1\frac{1}{2}$	2.5	3.0	3.0	0.09	—
2	2.3	2.2	2.5	0.08	—
$2\frac{1}{2}$	2.2	1.7	2.3	0.07	3.5
3	2.2	1.4	2.2	0.07	3.0
4	2.2	1.0	2.2	0.06	1.5
6	—	—	2.1	0.05	1.0
8	—	—	2.1	0.05	0.8
10	—	—	2.1	0.05	0.7
12	—	—	2.1	0.04	0.7
14	—	—	2.1	0.04	0.6
16	—	—	2.1	0.04	0.6
18	—	—	2.1	—	—
20	—	—	2.1	—	—

NOTE: Flanged check valves are assumed to have a K factor of 2.0 for all sizes.

where θ = total angle of reduction
$\quad d_1$ = smaller, leaving pipe diameter
$\quad d_2$ = larger, entering pipe diameter

Tapered enlarging fittings

$$K = 2.6 \sin \frac{\theta}{2} \left(1 - \frac{d_1^{\,2}}{d_2^{\,2}}\right)^2 \tag{3.9}$$

where θ = total angle of enlargement
$\quad d_1$ = smaller, entering pipe diameter
$\quad d_2$ = larger, leaving pipe diameter

Note: *Do not use these formula for steel reducing fittings!* Steel reducing fittings are not tapered; they have a reverse, ogee curve that causes a much greater friction loss. Until better data are available, use the following equation for the friction loss in steel reducing or increasing fittings:

$$Hf = \frac{V_1^{\,2} - V_2^{\,2}}{2g} \tag{3.10}$$

where V_1 = velocity in the smaller pipe
$\quad V_2$ = velocity in the larger pipe

Table 3.13 is for commercially available cast iron tapered fittings, both reducing and increasing, using Eqs. 3.8 and 3.9.

The use of equivalent feet of pipe as a means of calculating friction in fittings and valves does not seem to provide as accurate results as the K factors of the Hydraulic Institute. Data on fitting and valve losses, as indicated above, are at best approximate.

Following is a list of the fittings and valves that are encountered in HVAC systems:

1. Tees, straight or reducing, with flow through the run and through the branch

2. Elbows, straight or reducing, of various radii

3. Valves, ball, gate, globe, or butterfly

4. Reducers and increasers, cast iron taper type

5. Reducers and increasers, steel type

6. Strainers

7. Flow meters

8. Entrance and exit losses from tanks

TABLE 3.13   *K Factors for Cast Iron Tapered Fittings*

Pipe size, in		K Factor	
Large end	Small end	As an increaser	As a reducer
$2\frac{1}{2}$	$1\frac{1}{2}$	0.11	0.05
$2\frac{1}{2}$	2	0.02	0.01
3	$1\frac{1}{2}$	0.18	0.07
3	2	0.07	0.04
3	$2\frac{1}{2}$	0.01	0.01
4	$1\frac{1}{2}$	0.34	0.12
4	2	0.21	0.08
4	$2\frac{1}{2}$	0.10	0.05
4	3	0.04	0.02
5	$2\frac{1}{2}$	0.23	0.09
5	3	0.13	0.06
5	4	0.02	0.02
6	$2\frac{1}{2}$	0.34	0.13
6	3	0.24	0.10
6	4	0.09	0.05
6	5	0.01	0.01
8	4	0.26	0.11
8	5	0.13	0.07
8	6	0.05	0.03
10	6	0.18	0.08
10	8	0.03	0.02
12	6	0.31	0.13
12	8	0.11	0.06
12	10	0.02	0.02
14	6	0.42	0.16
14	8	0.22	0.10
14	10	0.08	0.05
14	12	0.01	0.01
16	10	0.16	0.08
16	12	0.05	0.04
16	14	0.01	0.01

NOTE: Computed from Eqs. 3.8 and 3.9.

It should be noted that the following devices are not included in the preceding list of fittings. All these devices waste energy and should be avoided if possible and used only as a last resort:

1. Sudden enlargements such as screwed bushings on small pipe or reducing-type flanges on larger pipe

2. Balance valves, manual or automatic

3. Pressure-reducing valves

4. Pressure-regulating valves

5. Bull-head tee connection

6. Multiple-duty valves

Item 5 refers to connecting a pipe tee with a supply on each run of the tee and the discharge on the branch connection. This is a wasteful connection and should be avoided. None of the usual references for friction losses through tees includes any data for this type of connection, which is known to have a sizable friction loss.

Item 6, multiple-duty valves, refers to a combination check, balancing, and shutoff valve. These valves can be wasteful of energy, since they can be set at an intermediate position and left there at a high friction loss. These valves should not be used without downstream shutoff valves, since the check-valve portion cannot be repaired without draining the system. In most cases, the cost of this multiple-duty valve is so much more than that for an ordinary check valve and shutoff valve that they are not economical. Also, their friction loss is usually greater than that for an individual check valve and butterfly valve (see Fig. 5.8).

Certain crossover bridges can be piping practices that waste energy by using return pressure-reducing valves. The additional piping losses also can increase the energy consumption of the total piping system.

3.4.1 Effect of fabrication on steel fitting loss

The great variation in steel pipe fitting friction loss is due to some extent to the method of fabricating steel pipe assemblies. The fitting losses listed previously are for factory-formed fittings such as tees and elbows. Figure 3.4a describes the three ways that a steel tee can be fabricated: (1) a factory-fabricated tee with dimensions conforming to ANSI Specification B16.5, (2) a "fishmouth" fabrication in which the end of the branch pipe is cut to a curve equal to the outside diameter of the main pipe, and (3) an unacceptable fabrication that should never be used. Unfortunately, if the inspection is of questionable quality, this type of fabrication may result. A fourth method that is available is an acceptable practice and is called a *saddle tee.*

It is obvious that the friction loss for the factory-fabricated tee will be less than that for the field-fabricated tees, particularly for the type 3 tee, where the branch pipe projects into the main pipe. The loss for the latter may be two to three times that for the factory-fabricated tee. The smooth radii of the factory-fabricated tee reduce appreciably the friction of the sharp corners of the type 2 tee and the reentrant loss of the type 3 tee.

Another case of fabricated fittings is the 45- and 90-degree elbows. Figure 3.4b describes (1) a standard production elbow and (2) a typical field-fabricated mitered elbow. If a sizable radius is maintained on the mitered elbow, its friction loss will be similar to that of the factory-fabricated elbow. Often designers reduce their fitting losses through elbows by specifying long-sweep elbows with a greater radius.

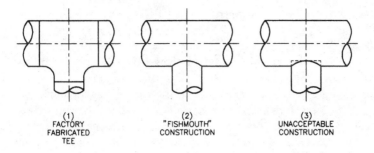

(1)
FACTORY
FABRICATED
TEE

(2)
"FISHMOUTH"
CONSTRUCTION

(3)
UNACCEPTABLE
CONSTRUCTION

a. Methods of fabricating a steel tee.

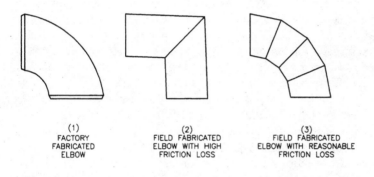

(1)
FACTORY
FABRICATED
ELBOW

(2)
FIELD FABRICATED
ELBOW WITH HIGH
FRICTION LOSS

(3)
FIELD FABRICATED
ELBOW WITH REASONABLE
FRICTION LOSS

b. Methods of fabricating a steel elbow.

Figure 3.4   Fabrication of steel pipe fittings.

Field fabrication and selection of fittings have a great impact on the overall friction loss in the piping of an HVAC water system. Continued vigilance should be provided to maintain the quality of field-fabricated fittings and to eliminate the use of screwed bushings and reducing flanges.

3.5 Thermoplastic Pipe

Thermoplastic pipe is being used in the HVAC industry for cold water services such as cooling tower and chilled water systems. It can offer distinct advantages cost-wise on some installations. It has a decided advantage on cooling tower water that may be laden with oxygen and where steel pipe is susceptible to rusting. The pressure and temperature ratings of plastic pipe and fittings are available from the principal manufacturers of this pipe. Most thermoplastic pipe as used on

HVAC water systems is manufactured in PVC (polyvinyl chloride), schedule 40 and 80 pipe, and CPVC (chlorinated polyvinyl chloride), schedule 80 pipe.

Thermoplastic pipe is manufactured in sizes up to 16 in, but many common pipe fittings may not be readily available over 12-in pipe size. This is its principal drawback for larger cooling tower installations.

Thermoplastic pipe offers a lower resistance to water flow than steel pipe. It is the plastic pipe industry's standard to use the Williams-Hazen formula for calculating pipe friction with a C factor of 150. This is acceptable, since most applications of plastic pipe are for water near 60°F. On warmer water, the manufacturer of the plastic pipe under consideration should provide friction loss data comparable with those secured from the use of Reynolds numbers and the Darcy Weisbach equation. Fitting losses for plastic pipe are calculated in equivalent lengths of pipe comparable with those for steel pipe fittings. It is my opinion that careful research should be conducted on plastic pipe fittings to determine their losses at higher velocities and temperatures. The sharper radii in many of the fittings may generate fitting losses that are greater than comparable fittings for steel pipe.

There are a number of types of plastic pipe now available to the HVAC industry, from PVC, schedule 40, to Fiberglas. ASHRAE's 1992 issue of the *System and Equipment Handbook* provides information on the various types of plastic pipe. Table 10, "Properties of Plastic Pipe," page 42.13, provides a summary of this important type of piping. In assessing plastic material for a particular application, the following factors should be evaluated:

1. *Temperature.*   The temperature and pressure ratings of some types of plastic pipe and fittings decrease appreciably above 100°F.

2. *Expansion.*   Most plastic pipe has a coefficient of expansion greater than steel pipe; expansion must be taken into consideration on HVAC systems where variations occur in both ambient air temperature and the liquid being transported.

3. *Hydraulic shock.*   Plastic pipe does not have the withstandability of steel pipe. The actual configuration of the plastic pipe installation, including methods of support, should be approved by the manufacturer of the plastic pipe.

4. *Maximum allowable velocity.*   Table 3.2 outlines maximum velocities for steel pipe. These velocities may be excessive for certain types of plastic pipe. Actual maximum velocities under consideration for a specific installation of plastic pipe should be approved by the manufacturer of the plastic pipe.

Most manufacturers of plastic pipe have very good application manuals that provide most of the answers to the preceding installation variables.

3.6 Glycol Solutions

Few special liquids are used in the HVAC industry, but both ethylene glycol and propylene glycol are found in these water systems. The friction for ethylene glycol based, heat transfer solutions can be computed using Fig. 2.1, which shows the variation in viscosity for various temperatures and percent of solution. Once the viscosity is known, the Reynolds number can be determined, and then the f factor can be computed for the Darcy Weisbach equation. The pump horsepower is increased due to the higher specific gravity; this will be discussed in Chap. 6. The specific gravity of ethylene glycol based heat transfer solutions is provided in Fig. 2.2. For heat transfer work, Fig. 2.4 provides specific heat data.

Care should be taken in selecting the percentage of glycol used in a water system. This percentage should be such that the pumps never operate near the freezing curve shown in Figs. 2.1 and 2.2. Slush can form as the freezing temperature is approached, causing poor pump performance.

3.7 Pressure-Gradient Diagrams

The pressure-gradient diagram was first mentioned in Chap. 1. It will be found throughout this book as a tool for describing the various pumping and circuiting procedures for HVAC water systems. It should be pointed out that pressure-gradient diagrams include only static and physical pressures. Velocity head is a dynamic head and is therefore not included in pressure-gradient diagrams.

The pressure-gradient diagram is an excellent method of checking energy transformations in a building. It is somewhat a graphic representation of Bernoulli's theorem. The vertical dimension of these diagrams is to scale and is shown normally in feet of head (see Eq. 4.1 for computing feet of head from known pressures and temperatures). The horizontal dimension is not to scale and is used to separate the specific changes in static and friction losses in various parts of the system.

Pump head is usually shown vertically upward, while friction losses can be shown vertically downward or diagonally. Computer programs are now being evaluated to facilitate the development of diagrams similar to those shown here.

Figure 3.5b illustrates the generation of a pressure-gradient dia-

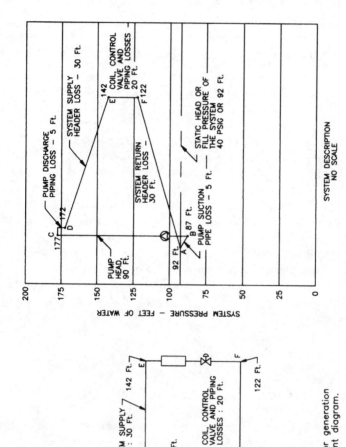

b. Generation of pressure gradient diagram.

a. Elementary system for generation os a pressure gradient diagram.

Figure 3.5 Generation of a pressure-gradient diagram.

gram for the simple secondary system of Fig. 3.5a. This diagram is for a building of horizontal development. The static pressure, or fill pressure, of the system is shown as a horizontal line; in this case, it is 40 psig or 92 ft. The pump head in this figure is 90 ft H_2O, while the pressure drop across the coil, its control valve, and piping, is 20 ft. The nodes, A through F, indicate the pressure, in feet, at each of them. These pressures are for the gauge pressure in the building. For campus-type operations, where there are buildings at different elevations, the nodes must indicate elevations above sea level to demonstrate changes in static pressure.

Typical buildings with their pressure-gradient diagrams are shown in Figs. 3.6 and 3.7 for both horizontal and vertical developments. A very important use of the pressure gradient is demonstrated in Fig. 3.7. Note that the circulating pump is located on the discharge of the generator; this reduces the pressure on the generator. If the pump were located on the inlet to the generator, it would have to withstand 160 psig of pressure. If the generator were a chiller, it might require high-pressure water boxes.

The pressure-gradient diagram should be developed with no flow on the building with the pumps running at shutoff head or no-flow condition. This condition provides the maximum possible head on the system. The second condition for the pressure-gradient diagram should be at the full flow or design situation for the system. The maximum pressure for the system can be computed easily by adding the pump shutoff head to the operating pressure of the building. In Fig. 3.7b, if the water system is 231 ft high from the highest point to the lowest point and the desired minimum pressure is 10 lb/in^2 at the highest point, the operating pressure is 231 divided by 2.31 + 10, or 110 psig. If the pump shutoff head (no-flow condition) is 115 ft or 50 psig, the maximum pressure that can be exerted on any part of the building is 110 + 50, or 160 psig. It is important that this zero-flow condition be computed to secure the maximum possible pressure that can be exerted on a water system. Just because a pump is variable-speed does not mean that this calculation should not be made. If the variable-speed pump controls fail, the pump could be operated at maximum speed.

Examples of the pressure-gradient diagram will be included throughout this book. It is a very useful tool for calculating operating pressures, as well as in determining system overpressure that may occur due to incorrect piping or pumping design. The overpressure indicated by the pressure-gradient diagram demonstrates energy-saving possibilities that can be achieved by revision of the pumps or piping.

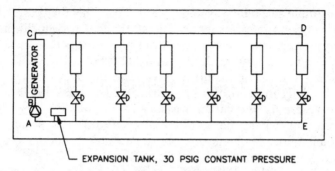

EXPANSION TANK, 30 PSIG CONSTANT PRESSURE

a. Typical building — horizontal development.

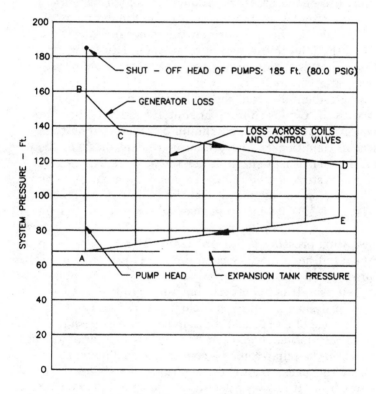

b. Pressure gradient — horizontal development.

Figure 3.6 Pressure-gradient diagram for a horizontal building.

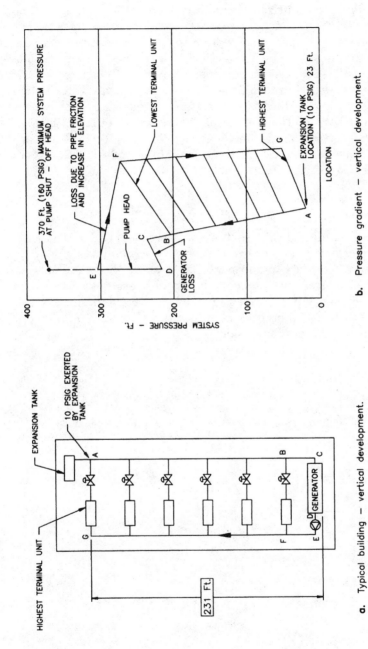

a. Typical building — vertical development.

b. Pressure gradient — vertical development.

Figure 3.7 Pressure-gradient diagram for a vertical building.

3.8 Piping Network Analyses

Multibuilding or multizone installations create friction analysis problems due to the various flow rates that can occur in parallel and series piping loops. Special software has been developed to compute the friction for such complex chilled or hot water systems. This software is now available for use on personal computers. Calculation of friction for piping networks was originated in the municipal water field; typical of this was the Hardy Cross method developed by Hardy Cross at the University of Illinois in 1936. Development of such software for contemporary computers has been completed by several universities and software companies. An example of this software is that produced by Donald Wood and James Funk at the University of Kentucky.

3.9 Summary

Care in determining pipe friction for a prospective HVAC water system cannot be overemphasized! There are so many systems in operation where too much pump head has been applied, and this excess head is destroyed by balance valves and pressure-reducing valves. System efficiency is described in Chap. 8; it emphasizes the need for care in computing pipe friction and in the avoidance of piping accessories that waste pumping energy.

3.10 Bibliography

Cameron Hydraulic Data, 15th ed., Ingersoll Rand, Woodcliff, N.J., 1977.
Engineering Data Book, 2d ed., Hydraulic Institute, Parsippany, N.J., 1990.
Mohinder, L. Nayyar, *Piping Handbook,* 6th ed., McGraw-Hill, New York, 1992.
The Handbook of Steel Pipe, Bulletin AISI SP 229-884-15M-SP, American Iron and Steel Institute, Washington, D.C., 1984.

HVAC Pumps and Their Performance

4

Basics of Pump Design

4.1 Introduction

A pump is a wheel rotating on stick. Some wheel; some stick! So much has been written on pumps, but often this material had to be in technical terms that made it difficult for the HVAC designer to understand how a pump functions. It is the intent of this *Handbook* to present pumps so that it is easy for the HVAC water system designer or operator to understand their operation. This can be done because most HVAC systems use water, not hydrocarbons or other liquids with broad ranges of viscosity and specific gravity. Several chapters are devoted to pumps to clarify (1) the basics of pump design, (2) the pump's physical arrangement, (3) pump performance, (4) motors and drives for pumps, and finally, (5) applying pumps to HVAC systems.

The centrifugal pump creates pressure on a water stream that can be used to move that water through a system of piping and equipment. The amount of pressure that the pump can create depends on the flow through it. Its fundamental performance curve is called a *head-capacity curve* and is described in Fig. 4.1a. The flow, in gallons per minute, is plotted horizontally, while the head, in feet, is plotted vertically. Pump head is always shown in feet of liquid rather than pounds of pressure. With head in feet, the pump curve is applicable to any liquid of any specific gravity. Only when calculating the brake horsepower of the pump does specific gravity become involved.

The feet of head per pound of pressure for any liquid is

$$\text{Feet of head } H = \frac{144 \text{ ft}^2}{\gamma \text{ ft/lb}} \tag{4.1}$$

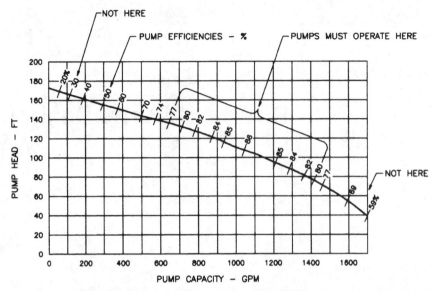

a. Typical pump head — capacity curve for one pump size.

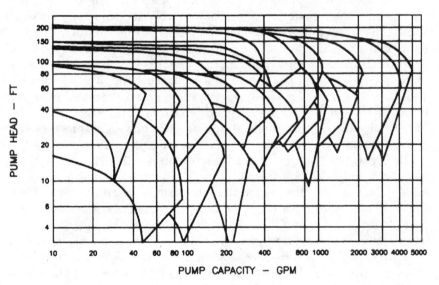

b. Family of pump head — capacity curves.

Figure 4.1 Typical pump head-capacity curves.

where γ is the specific weight of the liquid in pounds per cubic foot. For example, from Table 2.3, water at 180°F has a specific weight of 60.57 lb/ft³. Thus H, for 180°F water, is 144/60.57, or 2.38 ft/lb of pressure.

The words *pump duty* are often used when describing a pump's performance. Pump duty is usually defined as a certain flow in gallons per minute at a specific head in feet. This is the point at which the pump has been selected for a particular application. Usually, this point is selected as closely as possible to the best efficiency of a pump impeller, which, for Fig. 4.1a, is 86 percent.

Other common terms that are encountered in the pump industry are

1. *Carry out.* This means that the pump is operating at the far right of its curve, where the efficiency is poor.

2. *Shutoff head.* This is the head produced by a pump when it is running at the "no flow," or zero capacity, point.

3. *Churn.* A pump is said to be in churn when it is operating at shutoff head or no flow.

4.2   Centrifugal Pump Impeller Design

Centrifugal pumps, as their name implies, depend on centrifugal force to produce flow through the pumps. When an impeller rotates, the water in the impeller rotates as well, so there are two forces acting on that water, namely centrifugal and rotational. The impeller of the pump rotates in a body to utilize these forces which are imposed on the water in the impeller. These forces are aided by the shaping of the impeller to produce the correct passages in the impeller that will use this centrifugal force as well as the rotating velocity of the water in the impeller. Internal vanes are designed into the impeller that direct the flow through and out of the impeller. The relatively high efficiency of HVAC pumps, up to 90 percent, is due to the fact that the pump designer can design these vanes for maximum efficiency without concern for dirty or stringy material and without consideration for corrosive or erosive liquids. Figure 4.2a illustrates a cross section of a centrifugal pump impeller with vanes that direct the water through the impeller.

This figure provides the basic vector diagram for a centrifugal pump. It describes the forces imparted on a molecule of water leaving the impeller. $\mathbf{W}_2$ is the centrifugal force and $\mathbf{V}_2$ is the rotational force developed by the rotating energy of the impeller. These two forces produce the pump head and flow. The net force acting on the water is the vector diagram for these two forces. The vector diagram that the pump designer works with during the actual design of an efficient impeller is

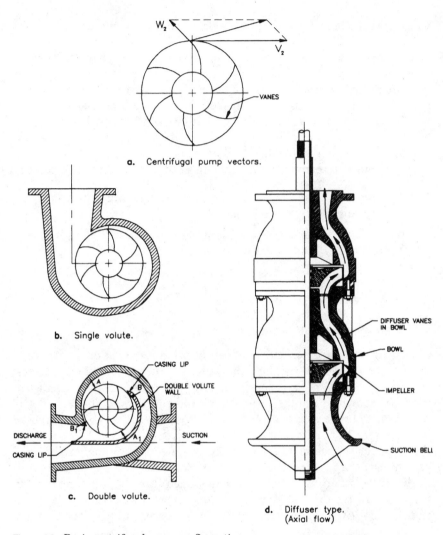

a. Centrifugal pump vectors.

b. Single volute.

c. Double volute.

d. Diffuser type.
    (Axial flow)

Figure 4.2 Basic centrifugal pump configurations.

much more complicated, since it must take into consideration entrance losses and other factors such as friction losses inside the impeller. The net angle of this vector diagram is the angle at which the water leaves the impeller and is particularly important to the pump designer because it plays a great part in determining the efficiency of a pump.

Inspection of an HVAC centrifugal pump impeller will reveal slots on the periphery of the impeller. These slots are narrow for low-capacity pumps and wide for higher-capacity pumps. The angle of the vane as it reaches the periphery of the impeller reveals the angle that the pump designer has selected for the vector $\mathbf{W}_2$ (Fig. 4.2a).

Pumps are highly specialized devices designed for very particular applications in order to achieve maximum efficiency and ease of use. HVAC pumps come under the broadest classification of centrifugal pumps—those for clear water service. Since corrosion and erosion are not factors in this pump service, the designer can concentrate on first cost, efficiency, and ease of operation and maintenance, the three driving forces in any product design. Much work is done by the pump designer in developing the internal passages of the impeller to achieve maximum, economical smoothness that will provide a minimum of friction to the water passing through the pump. Here lies a simple evaluation that can be made by any buyer of pumps. How smooth and how well are the internal passages of a pump impeller developed? How uniform are the internal vanes and leaving ports of the impeller? How uniform and smooth are the vanes at the suction port of the impeller? How smooth are the internal surfaces of the pump body, namely, the volute, or the bowl. All these factors have an impact on the overall efficiency of the pump.

4.3   General Performance of a Centrifugal Pump

Figure 4.1a provides general performance of a centrifugal pump, as well as typical hydraulic efficiencies for that pump. It should be noted that this efficiency drops to zero at zero flow and then increases to a point called the *best efficiency point* for the pump. As flow increases through the pump beyond this point, friction increases internally, and the efficiency starts to fall again. The best efficiency point of a pump will be emphasized in many parts of this book. Every water system designer should know where the best efficiency point occurs for each of the pumps involved in an HVAC water system.

The pump designer develops a basic configuration of a pump impeller and then selects various pump impeller diameters and speeds to create the vector diagrams that will produce the flow-head relationships needed for the broad application of that pump to HVAC water services. The capacity or flow range for a volute-type pump is determined by the size of its discharge connection. Therefore, the pump designer works with the vector diagrams and discharge connection areas to produce a number of pumps that will cover this broad range of flows and heads. The result is a number of head-capacity curves, as shown in Fig. 4.1b. This is often called a *family* of pump curves and provides a quick guide to the water system designer as to which size of pump may fit the needed flow and head for a particular application.

4.4 Sizing Centrifugal Pumps

As will be shown later, two types of centrifugal pumps are used in HVAC service, namely, volute and axial-flow pumps. Volute pump size is designated by the sizes of the suction and discharge connections, as well as by the pump diameter. A typical sizing for such a pump would be 6 × 4 × 8 in. This indicates that the pump has a 6-in diameter suction, a 4-in discharge, and an 8-in nominal size impeller. The impeller may have been trimmed to a specific diameter such as $7\frac{3}{4}$ in to achieve the best efficiency at the specified flow and head. Many pump companies would call this a 4-in pump, since they use the discharge connection size to classify their pumps.

Axial-flow pump sizing is determined by the suction and discharge connections of the body or bowl. The suction and discharge connections are almost always the same size. A 6-in suction and discharge would indicate a 6-in pump or bowl.

4.5 Specific Speed of a Pump

The experienced pump designer uses a formula called *specific speed N_s* to develop a pump impeller for a specific range of flows and heads, knowing that a certain specific speed will produce the desirable ratio between pump head and pump capacity. This formula (Eq. 4.2) is of great importance to the designer but of little interest to the user of pumps.

$$N_s = \frac{S \cdot \sqrt{Q}}{h^{3/4}} \tag{4.2}$$

where S = pump speed, in rev/min
      Q = pump flow, in gal/min
      h = pump head, in ft

For example, assume that a pump has a capacity of 1500 gal/min at 100 ft of head and is rotating at 1760 rev/min. Then,

$$\text{Specific speed} = \frac{1750 \cdot \sqrt{1500}}{100^{3/4}} = 2143$$

With the aid of this formula and a great amount of experience and testing, the designer can develop the above-mentioned family of pump curves.

Most HVAC pumps are of moderate specific speed in the range of 1000 to 5000; some large cooling tower pumps may have specific speeds as high as 8000. High-specific-speed pumps, as the preceding formula would indicate, are high capacity with low head, while low-specific-speed pumps are high head with low capacity. This is interesting background information for the HVAC water system designer, but it is not used in the design of these systems.

4.6 Critical Speed of a Pump

Every rotating device has a natural frequency at which vibrations become pronounced; the vibrations can increase until noise becomes objectionable, and there is danger in a pump that the rotating equipment will become damaged. The speed at which this occurs is called the *critical speed* of that rotating element. Some pumps will actually have more than one critical speed. Fortunately, for HVAC pumps, the impeller diameters are relatively small and the rotational energy low. All pump designers must take critical speed into consideration in the structural design of their pumps, but it is of little concern to the HVAC system designer in the selection of pumps.

The diameter or stiffness of the impeller shaft determines the critical speed of a pump. Therefore, the pump designer ensures that the critical speed of the pump is out of the normal range of pump operation. If a pump has a normal operating range of up to 1800 rev/min, the pump designer will ensure that the critical speed of that pump is at some speed such as 2500 rev/min. Therefore, in most cases, critical speed of HVAC pumps is not part of the pump selection process.

4.7 Minimum Speed for a Variable-Speed Pump

Minimum speed for a variable-speed HVAC pump will be discussed here, where information is also included on specific and critical speeds. *There is no minimum speed for these HVAC pumps.* Misinformation about specific, critical, and minimum speeds for HVAC pumps has caused misapplication of them along with the installation of unnecessary speed controls. As will be reviewed in Chap. 6, variable-speed centrifugal pumps can be operated down to any speed required by the water system. Since they are variable-torque machines, very little energy is required to turn them at low speeds. Information on this subject is included in Chap. 7 on pump drivers and drives as to the minimum speed required by motors for their cooling.

The HVAC water system designer really has no need to be concerned with the specific, critical, or minimum speeds of these pumps. Only on extremely low speed operations where variable-speed pumps are connected in series does the designer need to verify pump motor performance with respect to heating.

4.8   Minimum Flow for HVAC Pumps

The question often arises as to the minimum flow that is required through a pump to prevent it from overheating. The following equation provides this flow:

$$\text{Minimum flow } Q_m = \frac{\text{pump bhp} \cdot 2544}{\Delta T \cdot 60(\gamma/7.48)}$$

$$= \frac{317.2 \cdot \text{pump bhp}}{\Delta T \cdot \gamma} \tag{4.3}$$

where Pump bhp = brake horsepower of the pump near the shutoff or no-flow condition.

ΔT = allowable rise in temperature in °F of the water.

γ = specific weight of the water at the inlet water temperature

The difficulty with this equation is getting the actual brake horsepower consumed by the pump at very low flows. Some pump companies do publish pump brake horsepower curves down to zero flow. Other equations that use pump efficiency have the same problem, since the efficiency of a centrifugal pump approaches zero at very low flows. If this information is not available, the pump company should provide the minimum flow that will hold the temperature rise to the desired maximum. The obvious advantage of the variable-speed pump over the constant-speed pump appears here because there is much less energy imparted to the water at minimum flows and speeds.

For most HVAC applications, there should not be temperature rises in a pump higher than 10°F. *Do not bypass water to a pump suction! This continues to elevate the suction temperature* (see Fig. 10.14). Bypass water should be returned to a boiler feed system or deaerator on boiler feed systems or to the cooling tower, boiler, or chiller on condenser, hot, or chilled water systems. Since it is desired to maintain supply temperature in hot and chilled water mains, flow-control valves should be located at the far ends of these mains to maintain their water temperature. Usually, the heat loss or heat gain for these mains requires a flow that is greater than that required to maintain temperature in the pump itself.

4.9   Two Types of Centrifugal Pumps for HVAC Service

As indicated earlier, there are two basic types of centrifugal pumps for HVAC water systems. These are volute and axial-flow pumps. Volute pumps are found in most of the low-flow applications, while axial-flow pumps are used in the very high volume or low-flow, high-head applications. Chapter 5 will describe the various applications in detail.

4.9.1 Volute-type centrifugal pumps

The pump impeller produces the pump head, but it must be housed in a body that collects the water and delivers it to the system piping. In volute-type pumps, the body collects the water from the impeller and moves it around to the pump discharge connection (Fig. 4.2b, c). This body is called the *volute*. Unlike the axial-flow pump, normally, there are no collecting vanes in the body or volute to aid the flow. The cost of adding these volute vanes or diffusers would be prohibitive in volute-type pumps.

Most small HVAC centrifugal pumps are volute pumps, since this type of pump lends itself to the many different configurations that will be described in Chap. 5. Larger volute pumps may be equipped with what is called a *double volute* (Fig. 4.2c). This pump has a second volute cast into its casing. Radial thrust on a pump shaft can be reduced by means of the double-volute construction. The second passage in the volute should not be confused with the diffusers of axial-flow pumps. There is very little need for double-volute pumps in HVAC work, particularly with variable-speed pumps, in which the radial thrust is very low under most operating conditions. Radial thrust will be described in detail in Chap. 6.

4.9.2 Axial-flow centrifugal pumps

Axial flow pumps are arranged to pass water through a body that is called a *bowl*. This bowl has suction and discharge ports and contains diffuser vanes that aid the flow of water through the bowl. Because of the vanes, these pumps are often called *diffuser pumps*. Water flows out uniformly 360 degrees around the impeller; Figure 4.2d illustrates this flow. The pump body or bowl vanes are similar to the impeller vanes, and they aid the diffusion of water out of the impeller into the bowl and then into the discharge pipe or another bowl.

4.10 Open or Closed Impellers

There are several types of impellers available for clear service (clean water): closed, semiopen, and open. Most of the impellers of pumps in the HVAC industry are of the closed type. Closed-type impellers have a shroud and utilize case wear rings to impede bypassing. Other impellers are open type without shrouds or case wear rings. Usually, they are found in larger, axial-flow pumps. For most HVAC applications, the closed impeller with case wear rings provides the highest efficiency; this efficiency is sustained longer due to the ability of the pump designer to develop the smallest possible clearance between the impeller and the case wear ring. This reduces the flow of water from the discharge area with its higher pressure back to the suction connection with its lower pressure. This unwanted water flow is called *bypassing* or *suction recirculation* in the pump industry. Certified testing, which is discussed later, proves the quality of the pump designer's work and the pump manufacturing to produce the lowest amount of water bypassing at the case rings and therefore the highest pump efficiency.

4.11   General Pump Design Information

This chapter offers some very preliminary information on pump design. The pump types used in the HVAC industry are very standard designs that have been available for many years. Much has been done to secure maximum efficiency from these pumps. Like any other industry, there are practices that reduce the quality of pump design and manufacturing. Poor casting and finishing of impellers and volutes as well as eliminating casing wear rings on some designs reduces the possibilities of sustained, high pump efficiencies.

There is considerable additional information available in the pump industry for the HVAC design engineer who desires to study pump design further. Most of the pump manufacturers provide extensive information on their pump designs. Hydraulic Institute maintains pump design and testing standards that are available to water system designers.

4.12   Bibliography

Pump Handbook, McGraw-Hill, New York, 1996.
Centrifugal Pump Standards, Hydraulics Institute, Parsippany, N.J., 1996.

Physical Description of HVAC Pumps

5.1 Introduction

A number of different types of centrifugal pumps are used to handle all the various pump applications in the HVAC industry. HVAC pumps can vary from small in-line circulators with a capacity of 20 gal/min at 10 ft of head and fractional horsepower motors to large cooling tower pumps with capacities of 50,000 gal/min at 100 ft of head and equipped with 1500-hp motors. These pumps can be constant-speed or variable-speed. This chapter provides a physical description of the various pumps that are popular with HVAC designers for chilled, hot, and condenser water service.

5.2 Physical Description of HVAC Pumps

Centrifugal pumps account for most of the pump applications in the HVAC industry. Some positive-displacement pumps are found on HVAC systems, but they are used for specific applications with small capacities, not for distributing water in hot, chilled, and condenser water or cooling tower applications. A specific duty for positive-displacement pumps in the HVAC field is the metering of chemicals into water systems.

Table 5.1 provides a chart that describes the various types of centrifugal pumps found on HVAC systems. There is not universal agreement on some of the generic names applied to these pumps. For example, the double-suction pumps listed are often called *horizontal split case pumps.* The words *diffuser, axial flow,* and *turbine* are used interchangeably for one class of centrifugal pumps; axial flow appears

TABLE 5.1   Classification of HVAC Pumps

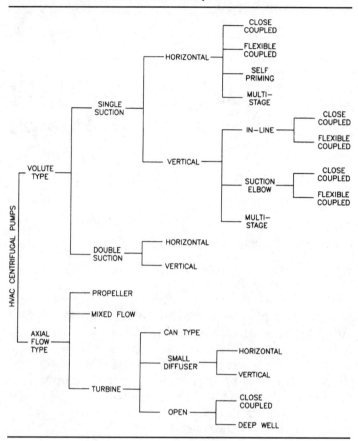

to be the least confusing and will be the name used herein. Hopefully, this chart will prove to be a guide for the designer to the most popular types of pumps used on these water systems.

5.3   Two Basic Types of Centrifugal Pumps

As demonstrated in Chap. 4, there are two principal types of centrifugal pumps for HVAC water systems; these are *volute* and *axial-flow pumps.* The volute pump is offered in many configurations and handles much of the pump duty on HVAC systems.

Although not as popular as the volute pump, the axial-flow pump handles some very important services. The words *axial flow* best describe these pumps, because water is diffused around the entire periphery of the impeller and flows axially through the pump. There is no volute to collect the water from the impeller. Instead, the impeller is housed in a bowl assembly that has a suction connection and a discharge section. Water enters the pump bowl and passes through the impeller and the bowl vanes to the discharge connection. With this arrangement, it is very easy to make up these pumps in what is called *stages*. A two-stage pump merely indicates that two bowls and impellers are installed in series together, as shown in Fig. 4.2d. How pumps operate in series or parallel is described in detail in Chap. 6.

5.3.1 Volute-type pumps

Volute-type pumps are provided in many different configurations and sizes to accommodate all the HVAC applications. They can be broken down into two fundamental types, single suction and double suction. These pumps are shown in Fig. 5.1a and b. There are no clearly defined conditions where each of these pump types are applied. Generally, the use of double-suction pumps begins with capacities around 500 gal/min and up to many thousands of gallons per minute. Single-suction pumps are applied in the smaller capacities, although there are excellent single-suction pumps that function with capacities as high as 4000 gal/min.

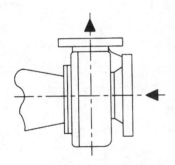

a. Single suction.

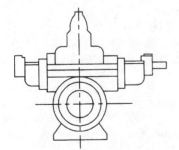

b. Double suction.

Figure 5.1 Types of volute pumps.

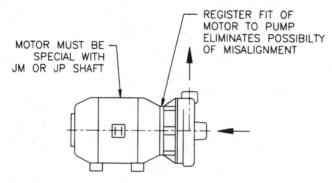

MOTOR MUST BE
SPECIAL WITH
JM OR JP SHAFT

REGISTER FIT OF
MOTOR TO PUMP
ELIMINATES POSSIBILTY
OF MISALIGNMENT

a. Close—coupled single suction pump.

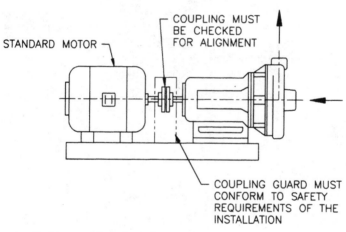

STANDARD MOTOR

COUPLING MUST
BE CHECKED
FOR ALIGNMENT

COUPLING GUARD MUST
CONFORM TO SAFETY
REQUIREMENTS OF THE
INSTALLATION

b. Flexible coupled single suction pump.

Figure 5.2   Horizontally mounted end-suction pumps.

The actual factors that determine which type of pump is to be selected
for a particular application are discussed in Chap. 6.

Single-suction pumps are furnished in a number of configurations
for the many small applications in HVAC; these are as follows:

1. *Close coupled* (Fig. 5.2a). This configuration is so named because
the motor is connected directly through an end bell flange to the
volute casing of the pump. The motor is special because it must have
a type JM extended shaft on pumps with mechanical seals and a type
JP extended shaft on pumps with packing.

The advantage of this configuration is the fact that the motor and pump cannot be misaligned with each other. Many designers prefer this pump over the flexibly coupled single-suction pump because of this. Another advantage of this pump is its physical space requirements, which are less than those required by the flexibly coupled single-suction pump. As indicated above, the motor is special because of the end bell mounting and the shaft extension and is not stocked as readily as the standard motor.

2. *Flexibly coupled* (Fig. 5.2*b*). This pump is mounted on a base and connected through a coupling to a standard motor. It can be equipped with a spacer-type coupling that allows removal of the pump bracket and impeller without moving the motor or its electrical connections.

This arrangement requires more equipment room space and must be aligned carefully after the pump base is set in place. The flexible coupling always must have a coupling guard in place before the pump is operated. The principal advantage of this pump is the fact that it uses a standard motor. In large motor sizes, it is less expensive than the close-coupled pump.

3. *In-line* (Fig. 5.3*b*). This configuration is so named because the pump can be inserted directly in a pipeline. The suction and discharge connections are in the same line, so there is no need for offsets or elbows in the connecting piping.

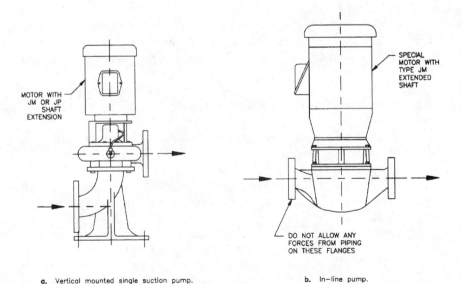

MOTOR WITH
JM OR JP
SHAFT
EXTENSION

SPECIAL
MOTOR WITH
TYPE JM
EXTENDED
SHAFT

DO NOT ALLOW ANY
FORCES FROM PIPING
ON THESE FLANGES

a. Vertical mounted single suction pump.   b. In-line pump.

Figure 5.3 Vertically mounted, end suction pumps.

This pump was built in small motor sizes initially, but it is now available in motor sizes as large as 100 to 150 hp. Great care must be taken with these pumps to ensure that the piping does not impose stress on the pump connections. If this happens, excessive maintenance on bearings and mechanical seals will result.

Because of the suction throat that must be built into the pump, its efficiency may not be a high as that for other single-suction pumps of similar size. The manufacturer of these pumps should verify pump performance. This performance also can be limited by the fact that the suction and discharge connections are almost always the same size. The pump designer of other types of single-suction pumps can have larger sizes for the suction connection, which may improve the overall efficiency of the pumps.

This pump can save considerable floor space in an HVAC equipment room. It can be mounted directly in the pipeline, or it can be mounted on a base like any other pump. The latter is preferred in the larger sizes to ensure that there are no pipe stresses on its suction and discharge connections.

4. *Vertical mount* (Fig. 5.3b). The vertically mounted single-suction pump is a close-coupled pump mounted on a pedestal-type long-sweep elbow that provides the suction connection as well as the support for the pump and motor. This pump, like the in-line pump, can save equipment room floor space. Because of the improved suction conditions in size and configuration, its efficiency should be greater than that for the in-line pump.

5. *Multistage.* This single-suction pump is often misnamed as a multistage horizontal split case pump, a name that does not recognize that it is a single-suction pump. This pump has two to five stages with internal passages that carry the water from one stage to the next. The pump is flexibly coupled to a standard motor when mounted horizontally. It can be mounted vertically and still be flexibly coupled to a vertical, solid-shaft motor.

This pump is not seen often on HVAC water systems. Its principal use is in pumping condenser water in tall buildings with indoor sumps that require high pump heads.

6. *Self-priming.* The self-priming pump is always a single-suction type. It can be close coupled or flexibly coupled to the motor like other single-suction pumps.

The principal feature that these pumps have is the ability to prime themselves. Water can be lifted from below-grade tanks into the pump without special priming equipment. Typical applications are chiller sumps and open energy storage tanks such as are used in cold water or ice storage systems.

Because of their self-priming requirements, the efficiency of these pumps is lower than the efficiency for other single-suction pumps of equal size. The impellers are usually of the open type without casing rings, which reduces the overall, lasting efficiency of this type of pump.

This pump is designed to operate on reasonable NPSH available conditions and can handle gases; therefore, it can be self-priming. The maximum dry-priming lift conditions for these pumps are usually shown on the head-capacity curves for them. The maximum allowable suction lift is the allowable lift or height of the pump above the minimum tank water surface. Under this condition, the pump is assumed to be dry. This means that within these lift conditions, the pump will prime itself as long as the NPSHR conditions are not exceeded. Manufacturers of self-priming pumps should provide information on the priming, repriming, and suction lift capabilities of their pumps before they are applied to an actual installation.

The suction pipe for most self-priming pumps should be the same size as the pump suction. This adds friction loss to the pumping installation and must be charged to the friction loss of the pump fittings. Due to this and the overall efficiency of self-priming pumps, the wire-to-water efficiency for such pumping systems is much lower than that for other centrifugal pumping systems of equal capacity and pump head.

5.3.2 Double-suction pumps

The workhorse of the HVAC field is the double-suction pump. It is by far the preference of most designers when the flow in a water system exceeds 1000 gal/min. The principal reasons for this popularity are its relatively high efficiency and its ability to be opened, inspected, and serviced without disturbing either the pump rotor, motor, or the connecting piping.

The latter feature is available if the pump casing is split parallel to the pump rotor or shaft. With this construction, called *axially split,* the top half of the casing can be removed, affording internal inspection of the impeller, case rings, and internal surfaces of the casing. Usually, there is no need for realignment of the pump and motor after it has been inspected, since the rotating element has not been disturbed.

The double-suction pump can be furnished in a casing that is split vertically or perpendicularly to the pump shaft. This construction requires pulling the rotating element out the end of the pump to inspect the impeller interior; the pump should be realigned with its motor after being inspected and reassembled. This pump is almost always mounted horizontally. Contrary to claims, this pump does not save

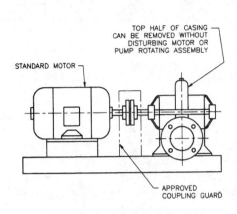

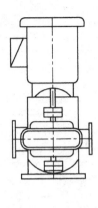

a. Horizontally mounted double suction pump.

b. Vertically mounted double suction pump.

Figure 5.4 Axially split, double suction pumps.

equipment room space in most cases because of the distance required at the pump end for removal of the rotating element.

The double-suction pump (Fig. 5.4), that has a casing split parallel to the pump shaft can be furnished mounted horizontally or vertically. This is why the popular name *horizontal split case* should not be used. The double-suction pump is always flexibly coupled to the electric motor, never close coupled. The vertically mounted double-suction pump uses a vertical, solid-shaft motor.

5.3.3 Axial-flow pumps

Axial-flow pumps are furnished in a number of different configurations. They are classified by impeller design and by their general construction and arrangement. An axial-flow pump consists of four basic subassemblies: (1) the bowl assembly that houses the impellers, (2) the column assembly that consists of pump shafting and piping to transfer the water from the bowl assembly to the discharge head, (3) the discharge head that receives water from the column assembly and delivers it to the system piping, and (4) the motor or other type of driver that powers the pump.

The three principal impeller classifications are *turbine, mixed flow,* and *propeller.*

1. *Turbine.* Turbine pumps are designed with lower specific speeds and are the type usually found in HVAC applications. Most of these pumps are equipped with enclosed impellers, although some manufacturers provide semiopen impellers. These impeller configurations are shown in Fig. 5.5a and b.

2 Impeller
6 Shaft, pump
8 Ring, impeller
10 Shaft, head
12 Shaft, line
13 Packing
17 Gland
29 Ring, lantern
39 Bushing, bearing
40 Deflector
55 Bell, suction
63 Bushing, stuffing-box
64 Collar, protecting
66 Nut, shaft-adjusting
70 Coupling, shaft
77 Lubricator
79 Bracket, lubricator
83 Stuffing-box
84 Collet, impeller lock
85 Tube, shaft-enclosing
101 Pipe, column
103 Bearing, lineshaft, enclosing
183 Nut, tubing
185 Plate, tension, tube
187 Head, surface discharge
189 Flange, top column
191 Coupling, column pipe
193 Retainer, bearing, open line shaft
197 Case, discharge
199 Bowl, intermediate
203 Case, suction
209 Strainer (optional)
211 Pipe, suction

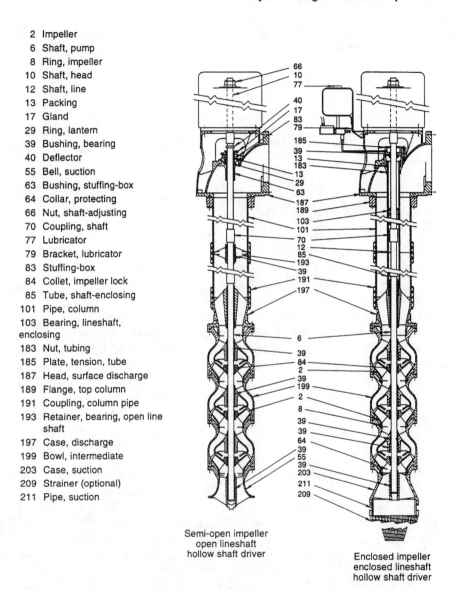

Semi-open impeller
open lineshaft
hollow shaft driver

Enclosed impeller
enclosed lineshaft
hollow shaft driver

Figure 5.5 Turbine type, axial-flow pumps. (*From Hydraulics Institute Standards.*)

Turbine-type pumps with enclosed impellers and bowl rings usually offer the highest sustained pump efficiency due to the ability to control more closely the amount of bypassing that occurs between the impeller and the bowl. Semiopen impellers must be checked periodically for the right clearance between the impeller and the bowl. The correct clearance between the impeller and the bowl is called the *lateral setting*. This setting can be accomplished by adjusting the top drive coupling of a vertical, hollow-shaft motor or by a special spacer coupling below a vertical, solid-shaft motor.

2. *Mixed flow.* Mixed-flow pumps are used on high-capacity, moderate-head applications; they are found in the HVAC industry on large cooling tower pump installations. These pumps have medium to high specific speeds. The impellers are almost always semiopen type.

3. *Propeller.* These impellers, as the name implies, look like multibladed propellers. They have high specific speeds and are designed for very high capacity, low-head applications. They are not found very often in the HVAC field.

Axial-flow pumps also can be classified by the method of lubricating the pump bearings. The pumps can be of open- or enclosed-line shafting, as shown in Fig. 5.5*a* and *b*.

Most HVAC axial-flow pumps are of the open-line shafting, where the water cools and lubricates the bearings. There should be no corrosive or erosive elements in HVAC water that require enclosed-line shafting. Enclosed-line shafting separates the water from the bearings by an internal tube. This tube is filled with oil, grease, or clear water that lubricates the bearings. Enclosed-line shafting requires an oil lubricator, grease cup, or source of clear water for lubrication.

Axial-flow pumps can be furnished in open construction, as shown in Fig. 5.2*a* and *b,* for installation in a tank or sump; they also can be supplied in a metal pipe with support, which is called a *can.* This construction is illustrated in Fig. 5.6.

The preceding description of axial-flow pumps is for the larger pumps used in cooling tower service with capacities in excess of 500 gal/min. Small multistage turbine pumps are being applied in the HVAC field on low-flow, high-head applications such as feeding high-pressure boilers. The four preceding subassemblies for this type of pump are furnished in a compact assembly and vertically mounted, which saves equipment room space. The smaller pumps are flexibly coupled through an integral bracket to a vertical, solid-shaft electric motor. A slightly larger pump of this configuration can be furnished mounted vertically with the same motor, or it can be supplied mounted horizontally and connected through a flexible coupling to a standard, horizontally mounted motor. The motor and pump are mounted on a structural base, as is the case with many volute-type pumps. These small turbine pumps are equipped with high-temperature mechanical seals for 250°F water.

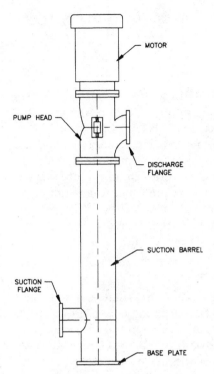

Figure 5.6 Axial-flow pump can type.

5.4   Positive-Displacement Pumps

Positive-displacement pumps as used in HVAC applications are small in size because, typically, they are used only for chemical feed service or other applications where a positive amount of material is required. Most of these pumps are of the *rotary* or *reciprocating* type. Since these pumps seldom enter into the overall performance of HVAC water systems, no further discussion of them will be included. Hydraulic Institute's *Centrifugal Pump Standards* is an excellent reference manual for these pumps.

5.5   Regenerative Turbine Pumps

A specialty pump that cannot be classified as a centrifugal pump or a positive-displacement pump is the regenerative turbine pump. This pump is used in the HVAC field primarily for low-flow, high-head applications such as boiler feed service. It has very tight clearances and, therefore, must be equipped with a very good suction strainer. Its efficiency is lower than other centrifugal pumps of similar size. It has been replaced on many applications by the small, multistage, turbine-type centrifugal pump.

5.6   Pump Construction

Most centrifugal pumps for the HVAC industry do not require the rugged construction found in industrial and power plant installations. Compared with other pump applications, HVAC pumping is easy pump duty. All the major liquids used in this field should be noncorrosive and nonabrasive. Therefore, these pumps must offer long life with very little in the way of repairs. Chapter 25 on maintenance of pumps will review, in detail, the causes for abnormal wear in pumps.

The description that follows pertains to most of the HVAC pumps. Understanding the basic parts of a pump, regardless of the type, will be helpful in selecting and observing the operation of a pump.

5.6.1   Typical parts of a centrifugal pump

Figure 5.7 describes a typical cross section of a single-suction, flexibly connected pump. The purpose of this drawing is to familiarize designers with the generic parts of a pump. This pump drawing illustrates the case rings mentioned below, and it points out most of the parts found in HVAC pumps; some of the more important parts are as follows:

Part name	Part number
Casing, body, or volute	1
Impeller	2
Shaft	6
Inboard bearing	16
Outboard Bearing	18
Frame or bracket	19
Case rings or covers	25 and 27

Other pump parts relating to axial-flow pumps can be found in Fig. 5.5, the drawings for the turbine pump.

5.6.2   Materials of construction

Since most HVAC pumps do not operate under corrosive or erosive operating conditions and are not of extremely high pressure, they are normally cast iron, bronze fitted. This means that the pump body is cast iron and the internal trim is bronze. The impeller and case wear rings are bronze, while the pump shaft is either carbon steel or stain-

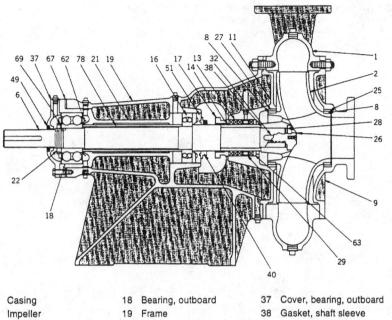

1	Casing	18	Bearing, outboard	37	Cover, bearing, outboard
2	Impeller	19	Frame	38	Gasket, shaft sleeve
6	Shaft, pump	21	Liner, frame	40	Deflector
8	Ring, impeller	22	Locknut, bearing	49	Seal, bearing cover, outboard
9	Cover, suction	25	Ring, suction cover	51	Retainer, grease
11	Cover, stuffing box	26	Screw, impeller	62	Thrower (oil or grease)
13	Packing	27	Ring, stuffing-box cover	63	Bushing, stuffing-box
14	Sleeve, shaft	28	Gasket, impeller screw	67	Shim, frame liner
16	Bearing, inboard	29	Ring, lantern	69	Lockwasher
17	Gland	32	Key, impeller	78	Spacer, bearing

Figure 5.7 Cross section of an end-suction, volute-type pump. (*From Hydraulics Institute Standards.*)

less steel. *Case wear rings* (see part numbers 25 and 27 in Fig. 5.7) are mounted in the pump casing at the point where the impeller turns in the casing. This provides a closer tolerance between the impeller and the casing and reduces the bypassing that can occur between the pump discharge and the pump suction. Bypassing is the water that escapes from the high-pressure, or discharge, side of the impeller to the suction side. Obviously, the smaller the amount of bypassing, the greater is the efficiency of the pump.

Other materials such as cast steel, stainless steel, and ductile iron are found in HVAC applications where specific conditions require their use. Some manufacturers of small pumps are now using stainless steel fabrications for impellers because they have found them to

TABLE 5.2   Typical Pressure-Temperature Limitations for HVAC
Centrifugal Pumps (Pressure in psig)

Casing and nozzle designation	0–150°F	175°F	200°F	250°F
A	510	487	467	426
B	450	430	412	375
L	425	406	390	360
K	400	384	367	336
J	350	336	325	300
C	300	290	280	270
H	275	265	255	240
D	250	240	232	215
E	175	170	164	150
F	150	143	137	125

NOTE: The temperature limitation for all these casings and nozzles is 250°F.

be less expensive than bronze castings. Stainless steel shafts are used
by some pump manufacturers when mechanical seals are furnished
without any sleeves under them.

Higher-temperature operations above 250°F, such as medium- and
high-temperature water, will require ductile iron, cast steel, or fabricated steel casings to withstand the system temperatures and pressures (see Chap. 21). The working pressures and temperatures of
standard cast iron casings are always rated by the pump manufacturer. They will provide a pressure-temperature chart similar to Table
5.2 for a specific class of pumps and will state whether the pump nozzles are furnished with 125- or 250-lb flanges. In the case of Table 5.2,
the various letters from A to F are for different sizes of nozzles with
different ratings for each. Some of the higher ratings are for ductile
iron casings. It is very important, even on a small, low-pressure project, that the design engineer check the pressure-temperature ratings
of the pumps. It is wise to have a rating chart similar to Table 5.2 in
the job file for each application.

5.6.3   Mechanical seals or packing?

Most HVAC pumps today are equipped with mechanical seals. Many
different types of mechanical seals are offered by a number of manufacturers. The books listed in the Bibliography at the end of this chapter have much information on the types of seals and their application.
Although the trend is toward mechanical seals, there are still places
where packing may be a better answer for ease of maintenance.

Any application that has a possibility of dirt or other foreign matter
may be an application for packing. Typical of these are condenser
pumps taking water from cooling tower sumps. Also, large facilities

with full-time maintenance personnel who are experienced in the technique of installing packing in all types of rotating elements may be candidates for packed pumps. Properly installed packing can eliminate waste water caused by continuous dripping.

Why use packing? A mechanical seal, when it fails, may throw water around the equipment room, while packing, as it starts to fail, just leaks water from the gland. Its failure is much more gradual than that of mechanical seals. The other reason may be because the client's personnel prefer packing; therefore, this is one question that the designer should ask when considering the pumps on a particular installation.

5.7 Mechanical Devices for Pumps

Most HVAC pumps do not require many special mechanical devices for them to sustain long life. Their useful life is much more dependent on their selection and operation. All of this will be covered in detail in the remaining chapters of this book. Air is a bothersome entity to pumps, so it never hurts to find methods to get air out of a pump. The best way, of course, is to never let it in the pump. A useful and inexpensive device is a *manual petcock* installed on top of the pump volute that will allow manual removal of air that collects there. If air continually enters the pump and cannot be stopped, this pet cock can be replaced with an *automatic air vent*. The automatic air vent should have a discharge line to near a floor drain, since they are notorious for leaking water. If the water carries grit or other small particulate matter, a *cyclone separator* can be installed on each flushing line to the seals or packing. Methods of venting air from water systems are included in Chap. 9.

There is much argument over the value of *strainers* on HVAC systems other than condenser water flowing to chillers. Many similar industrial applications are not equipped with strainers. The normal location of the strainer is on the suction of the pump. In the case of cooling tower water feeding a chiller, the strainer should be installed on the discharge of the pump, as shown in Chap. 11. The strainer is there to protect the tubes of the condenser, not the pump. The cooling tower sump strainer is adequate to keep rocks and other debris out of the pump. Installing the strainer on the pump discharge prevents the strainer from being obliterated with algae and ruining the pump. If the strainer becomes clogged when located on the pump suction, a vacuum will be generated on the pump suction, and it will be destroyed by steam and heat. For other HVAC applications, some engineers prefer to use strainers for start-up service only. After the system has been flushed and operated for a while, the strainer screens or baskets are removed.

Coupling guards conforming to Occupational Safety and Health Administration (OSHA) or particular state codes should be located over the rotating coupling and pump shaft where it is exposed. OSHA's requirements should be checked from time to time to ensure that their requirements are met.

Balance valves should never be installed on the discharge of any pump, let alone variable-speed pumps. Some designers feel that they are a safeguard to prevent overpressuring a system; this is a terrible waste of energy. On constant-speed pumps, the amount of overpressure should be determined and the impeller trimmed to eliminate as much of the overpressure as possible. See Chap. 29 on the procedure for trimming an impeller. On variable-speed pumps, the pump controls should be designed to prevent this overpressuring by limiting the maximum speed of the pump.

Multiple-duty valves that include the features of a check valve, a balance valve, and a shutoff valve should never be used on the discharge of any pump and particularly on the discharge of a variable-speed pump (Fig. 5.8). Some multiple-duty valves can be closed partially without any indication of this condition. This creates a terrible waste of pumping energy. Also, some multipurpose valves must have another valve downstream of them to repair the check-valve feature

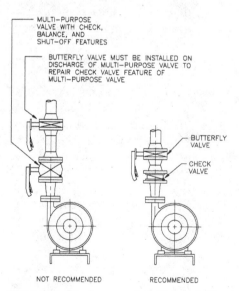

Figure 5.8 Discharge valves for variable-speed pumps.

of the multipurpose valve. A separate check valve and shutoff valve should be provided instead. They are lower in cost and usually have a lower friction loss than comparable multiple-duty valves.

On continuously running pumps taking suction from a tank below the pumps, the greater efficiency of the standard centrifugal pump should give it priority over a self-priming pump. It will be necessary to provide a *priming means* to get the pump started. This can be a priming tank or a separate source of water. This tank holds enough water to keep the pump flooded until it has primed the suction line. With a properly sized priming system, the standard centrifugal pump evacuates the air in the suction line and continues to pump water from the lower tank. There are manufacturers of priming systems for centrifugal pumps, and there is considerable information in the pump references on the actual design of a priming system. The NPSHR characteristics of the pump must be checked carefully when taking suction from lower tanks.

5.8 Bibliography

Additional centrifugal pump descriptions and performance can be found in the catalogs of (1) the AC Pump Division of the ITT Fluid Technology Corporation, Cincinnati, OH; (2) Aurora Pump, a Unit of General Signal, Houston, TX; (3) Goulds Pumps, Seneca Falls, NY; (4) Ingersoll Rand, Woodcliff Lakes, NJ; (5) Paco Pumps, a Flow Technologies Company, Aurora, IL; and (6) Peerless Pumps, Indianapolis, IN.

ANSI/HI 1.1-1.5-1994, *Centrifugal Pumps,* Hydraulic Institute, Parsippany, N.J., 1994.
ANSI/HI 2.1-2.5-1994, *Vertical Pumps,* Hydraulic Institute, Parsippany, N.J., 1994.
Pump Handbook, McGraw-Hill, New York, 1996.

6

HVAC Pump Performance

6.1 Introduction

This evaluation of HVAC pump performance is limited to centrifugal pumps. The use of positive-displacement and regenerated turbine pumps in HVAC systems is for highly specialized low-volume, high-head applications. They are not used on the water services that consume most of the pumping energy in HVAC systems. Their application is so specialized that the performance required of these pumps, when encountered, should be sought from the manufacturer of the pump under consideration.

This review of HVAC pump performance will attempt to provide a general approach that is applicable to all pumps from small circulators to very large condenser pumps.

6.2 Pump Head-Capacity Curves

A study of centrifugal pump operations must begin with an evaluation of the basic performance of such a pump. The pressure or head that is developed by any pump depends on the flow through that pump. This basic head-flow relationship is called the pump's *head-capacity curve*. This was shown in elementary form in Chap. 4 and is now demonstrated with other accompanying curves in Fig. 6.1. Figure 6.1*a* is the head-capacity curve for a medium-sized constant-speed pump, and it is in the form normally found in pump manufacturers' catalogs. Included in this figure are the efficiency curves, the brake horsepower curves, and a net positive suction head (NPSH) curve. The NPSH required curve will be discussed later in this chapter.

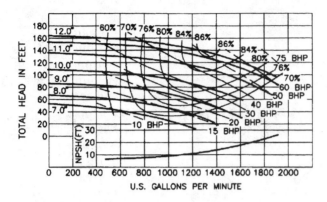

a. Head – capacity curves for a constant speed pump.

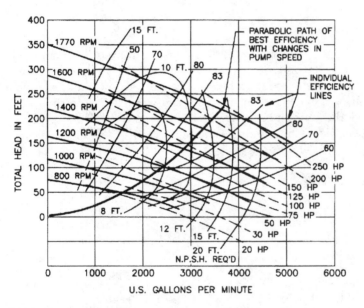

b. Head – capacity curves for a variable speed pump.

Figure 6.1 Typical head-capacity curves for most HVAC pumps.

6.2.1 Constant-speed head-capacity curves

Most HVAC pump head-capacity curves provided by the pump manufacturers are constant-speed curves for a particular speed such as 875, 1150, or 1750 rev/min. These curves are derived from dynamometer tests in the pump manufacturer's plant.

Head-capacity curves for pumps can have several shapes depending on the specific speed of the pump. Figure 6.1*a* is for a pump of moderate specific speed, 1560 rev/min, and is called a *continuously rising characteristic curve* because the head increases continuously to the shutoff, or no-flow, condition. High-specific-speed pumps such as mixed-flow or propeller pumps can have a looped head-capacity curve, as shown in Fig. 6.2*a*. Normally, manufacturers of such pumps do not provide a complete head-capacity curve in their catalogs. Only the right-hand portion of the curve where the pump is to be operated is provided, as shown in Fig. 6.2*b*. These pumps require some care in their application due to the fact that they can operate at several capacities at the same head; this is shown in Figure 6.2*a*, where the pump at 34 ft of head can operate at 800, 1700, and 2700 gal/min. The pump can shift from one capacity to another, causing shock and vibration in the pump itself and in the water system. Only experienced pump application engineers should handle the selection, installation, and operation of these large pumps.

The head-capacity curve describes the net head available for useful pumping after all internal losses of the pump are deducted from the theoretical head that could be produced by an impeller of a particular diameter and operating at a certain speed. The efficiency curves superimposed on the head-capacity curve demonstrate how efficient the pump is when operated at a specific point on its head-capacity curve. Some pump manufacturers provide efficiency curves similar to those in Fig. 6.1. Others provide a separate efficiency curve from zero efficiency at zero flow to maximum flow. This is a rising curve to the point of maximum efficiency at the optimal flow. This maximum efficiency point on such a curve is called the *best efficiency point.* Great emphasis will be placed on the importance of this best efficiency point in Chap. 10. Some pump manufacturers, particularly of small pumps, do not provide efficiency curves; instead, they provide a performance curve for a certain impeller diameter and with a specific size of electric motor. *Such pump curves should not be used for selection of HVAC pumps. If the manufacturer cannot provide an efficiency curve for a pump, another manufacturer should be consulted.*

6.2.2 Typical constant-speed pump head-capacity curves

Figure 6.3*a* through *d* describes typical pump head-capacity curves for a broad range of HVAC applications. No particular type of centrifugal pump is represented. These curves are for pumps with good formation and have best efficiencies that are at the level desired for these water systems. Almost all HVAC applications should have

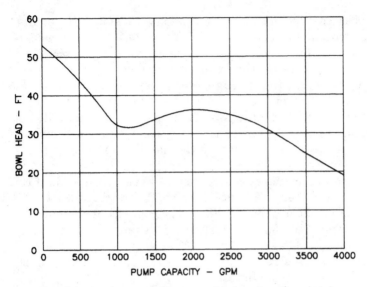

a. Complete head — capacity curve for impeller diameter A.

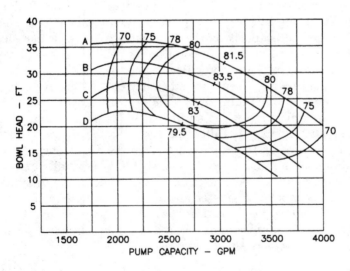

b. Operating range.

Figure 6.2 Head-capacity curves for a high specific speed, axial-flow pump.

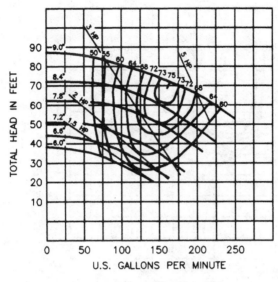

a.

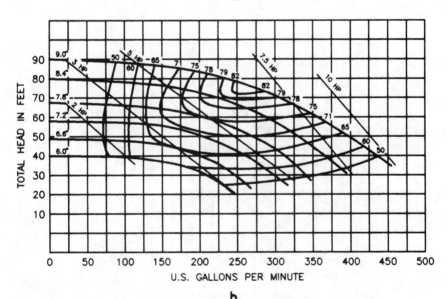

b.

Figure 6.3 Examples of good pump performance.

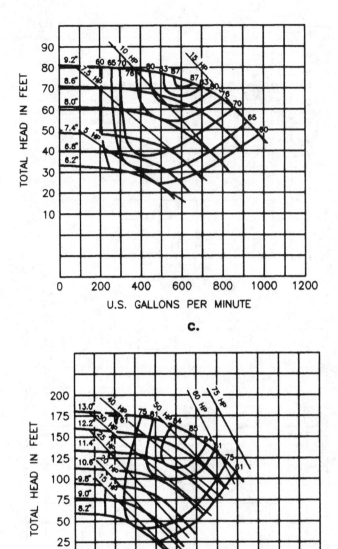

Figure 6.3 (*Continued*)

pumps with the best efficiency above 80 percent, excepting small pumps or pumps in low-flow, high-head applications.

Velocity head $V^2/2g$ of the pump connections, must be considered when the head-capacity curve is developed. On pumps with suction pipes, the difference between the velocity head for the water flowing out of the discharge connection and for the water flowing into the pump suction is included in these head-capacity curves. On axial-flow pumps taking water directly from a sump, there is no inlet velocity head deducted from the discharge velocity head. There may be an efficiency reduction for single-impeller performance that does account for some entrance losses.

Normal pump head-capacity curves are provided by manufacturers at particular induction motor speeds such as 875, 1150, and 1750 rev/min. A range of impeller diameters will be shown on each curve. A pump is considered to be *trimmed* to a specific diameter for each particular application. The range of impeller trimming is shown on a typical head-capacity curve, from minimum to maximum diameter. For example, the pump curve described in Fig. 6.3d has an unusual impeller diameter range, from 8.2 to 13.0 in diameter. Although this much trimming is allowed, cutting the impeller to such an extent has a great effect on pump efficiency. This particular pump has a peak efficiency of 87 percent; cutting the impeller to its minimum diameter would reduce the efficiency to 61 percent. Great care must be taken in trimming pump impellers; an alternative is to install a variable-speed drive and operate the pump at a lesser speed. Pumps must never be trimmed below the minimum diameter shown by the pump manufacturer on the published head-capacity curves.

Note. All manufacturers' pump head-capacity curves are dynamometer curves at a fixed speed such as 1750 rev/min. Actual field performance will be different from these curves when the pumps are operated by induction motors whose speed varies from 1750 rev/min fully loaded to 1800 rev/min unloaded. Manufacturers' pump curves should not be used to program pumps; they can be used for estimating pump performance and initial calculations required for wire-to-water efficiency and energy consumption. See Fig. 23.1 for a typical variation in the pump head-capacity curve due to induction motor speed variation.

6.2.3   Variable-speed pump head-capacity curves

Since speed variation is an added factor, the variable-speed pump's head-capacity curves must be provided for at least one diameter of impeller. If another impeller diameter is desired, a second set of curves should be provided for that diameter. The speed variation is usually shown from around 45 to 100 percent speed, or typically from 800 to 1750 rev/min. Below 45 percent speed, pump performance becomes a variable, and the pump manufacturers do not like to certify pump performance at such low speeds for most HVAC pumps. Figure 6.1b describes a typical set of pump curves for a large centrifugal pump operating at a maximum speed of 1750 rev/min. This pump has a specific speed of 1860 rev/min. It must be remembered that this group of curves is for one impeller diameter, while Fig. 6.1a is for another pump at constant speed and various impeller diameters.

Due to the many variable-speed pumps that are now used in the HVAC field, it is important that the HVAC water system designer become familiar with what happens when the speed of a pump is varied from minimum to maximum speed. Figure 6.1b is one of the most important figures in this book. Of particular importance in this figure is the curve describing the parabolic path of the best efficiency point as the pump changes speed. This curve will help the water system designer understand how the efficiency of a pump varies as the speed changes. The variations in efficiency at different points on the pump head-capacity curves are also provided in this figure as lines of constant efficiency. Lastly, this figure is a graphic representation of the pump affinity laws that are so important.

6.2.4   Steep versus flat head-capacity curves

There is much discussion in the industry as to when a pump head-capacity curve is steep or flat. If a pump head-capacity curve rises more than 25 percent from the design point to the shutoff head or no-flow point, it is considered to be steep; it is considered flat if this rise is less than 25 percent. For example, if the design condition of a pump is 500 gal/min at 100 ft of head and the shutoff (no-flow) condition is 120 ft, it would be considered a flat-curved pump. If the shutoff condition were 135 ft, the pump would be considered steep-curved. The pump curves in Fig. 6.4 are considered to be flat. The shape of the pump curve was of great concern when constant volume-water systems were popular. It was desired to have relatively steep pump curves for these operations so that a minor change in system head would not make a great change in system flow. Steep curves were preferred, and flat-curved pumps were avoided. Small, constant-speed pumping systems with three-way control valves on the heating or cooling coils must be concerned with the shape of the pump curve. A flat-curved pump may create instability in the operation of the three-way valves.

With the advent of variable-volume water systems utilizing variable-speed pumps and system differential pressure controls, there is very little need to be concerned about the shape of the pump curve.

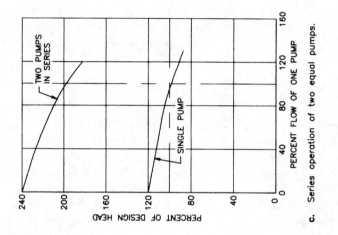

c. Series operation of two equal pumps.

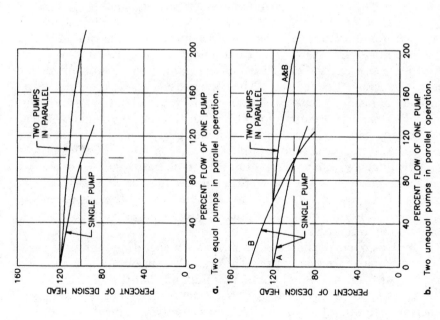

a. Two equal pumps in parallel operation.

b. Two unequal pumps in parallel operation.

Figure 6.4 Series/parallel performance of HVAC pumps. *(From the Water Management Manual, Systecon, Inc., West Chester, Ohio, 1992, Fig. 3-18.)*

Peak efficiency is what is sought, not the shape of the curve. In fact, flat-curved pumps are desired. Pumps with flat-curved head-capacity characteristics should be sought for variable-speed pumping applications, provided there is no loss in pump efficiency. There is less speed reduction with a flat-curved pump than with a steep-curved pump. With less speed reduction, the wire-to-shaft efficiency of the variable-speed drive and motor is greater because the speed reduction is less.

6.3 Series/Parallel Operation of Centrifugal Pumps

The broad variation in water requirements in HVAC systems usually requires more than one pump to handle all the system flows from minimum to maximum capacities. Likewise, condenser water service in some tall buildings requires pump heads that cannot be handled economically by a single-stage pump. The first condition requires parallel operation of more than one pump, and the second condition requires more than one impeller operating in series. By far most HVAC applications consist of multiple pumps in parallel with few of them in series operation. Figure 6.4 describes the various resulting head-capacity curves when operating pumps in parallel and in series.

Parallel operation enables the water system designer to select a number of pumps that will produce efficient operation from minimum to maximum system flow. Operating pumps of equal size and head in parallel results in the capacities of the pumps being multiplied together, as shown in Fig. 6.4a. This figure describes two pumps with such equal head-capacity curves operating in parallel; the head is the same for one or two pumps operating, but the capacity is doubled for two-pump operation. Parallel operation of pumps requires great care to ensure that the number of pumps running is the most efficient for a particular flow and head condition. When operating pumps in parallel, the total capacity and head of the pumps must be that required by the uniform system head curve or any point within the system head area. (See Chap. 9 for a description of uniform system head curves and system head areas.)

The great mistake that is often made with operating pumps in parallel is to measure the capacity of one existing pump at a certain head, install another equal pump, and find out that the two pumps do not produce twice the capacity. Pump operation cannot be predicted without computing the system head curve or system head area of the water system. This is pointed out here because of the danger of trying to run pumps in parallel without evaluating the head of the water system. This will be covered, in detail, in Chap. 10.

Considerable care should be exercised in trying to operate pumps in parallel that have different head-capacity curves; this is described in Fig. 6.4b. There is great danger that the pump with the lower head-capacity curve, pump A in this figure, can be operated at the shutoff head condition and cause heating in the pump. This will happen if these pumps are operated together at any less than 62 percent capacity of one pump. Also, it may be very difficult to program and operate the pumps together at an adequate wire-to-water efficiency.

Pumps are operated in series when the flow can be handled by one pump throughout the operating range but the head is so high that a number of pumps in series is needed. Figure 6.4c describes the fact that with pumps in series, the head developed by all the pumps is added together at the same system flow. Turbine-type pumps are excellent for such high-head, low-flow service. Two or more bowls or stages are connected in series to provide the required head.

Some large HVAC systems exist in which multistage pumps must operate in parallel; in this case, the resulting pump curves are a combination of Fig. 6.4a and c with 200 percent capacity and 200 percent head of one pump impeller.

6.4 Affinity Laws of Pumps

Centrifugal pumps are variable-torque machines, like centrifugal fans, and they obey the same laws as do centrifugal fans. The term *variable torque* indicates that the horsepower required to turn the pump or fan does not vary directly with their speed. It varies with the cube of the speed. Constant-torque pumps, like most positive-displacement pumps, are those in which the horsepower does vary directly with the speed.

The laws that define centrifugal pump performance are called the *affinity laws*. They dictate the changes in pump performance with variations in speed, flow, head, and impeller diameter. The basic affinity laws are as follows.

For a fixed-diameter impeller:

1. The pump capacity varies directly with the speed.

$$\frac{Q_1}{Q_2} = \frac{S_1}{S_2} \tag{6.1}$$

2. The pump head varies as the square of the speed.

$$\frac{h_1}{h_2} = \frac{S_1^2}{S_2^2} \tag{6.2}$$

3. The pump brake horsepower required varies as the cube of the speed:

$$\frac{\text{bhp}_1}{\text{bhp}_2} = \frac{S_1^3}{S_2^3} \tag{6.3}$$

Figure 6.5a describes these three laws graphically. These are the basic laws that must be understood in any attempt to comprehend centrifugal pump performance. It must be remembered that these laws pertain to the pump itself. *They do not describe pump performance when the pump is connected to a system of piping containing both constant and variable head.*

Figure 6.5b describes the variation in flow, head, and power for a variable-speed pump operating with a constant head of 20 ft; data for these curves are provided in Table 6.2 later in this chapter. This constant head can be static head on a cooling tower or constant differential pressure on a secondary chilled or hot water system. It is obvious that there is no comparison between the curves in Fig. 6.5a and b. At 50 percent speed and 885 rev/min, the flow is 36 percent, not 50 percent, the head is 30 percent, and the brake horsepower is 10 percent, not 12.5 percent. This demonstrates the need to evaluate pump performance with the actual system conditions and not depend on the classic affinity laws of Fig. 6.5a.

This is emphasized at this point because it may appear from Fig. 6.1a that pump impellers can be changed just through observation of the basic affinity laws of Fig. 6.5a. Pump impeller diameter should not be changed without a thorough evaluation of the system head curve or area.

Changing the pump impeller diameter changes pump performance, as does pump speed.

For a fixed pump speed:

1. The pump capacity varies directly with the impeller diameter:

$$\frac{Q_1}{Q_2} = \frac{d_1}{d_2} \tag{6.4}$$

2. The pump head varies as the square of the impeller diameter:

$$\frac{h_1}{h_2} = \frac{d_1^2}{d_2^2} \tag{6.5}$$

3. The pump brake horsepower required varies as the cube of the impeller diameter:

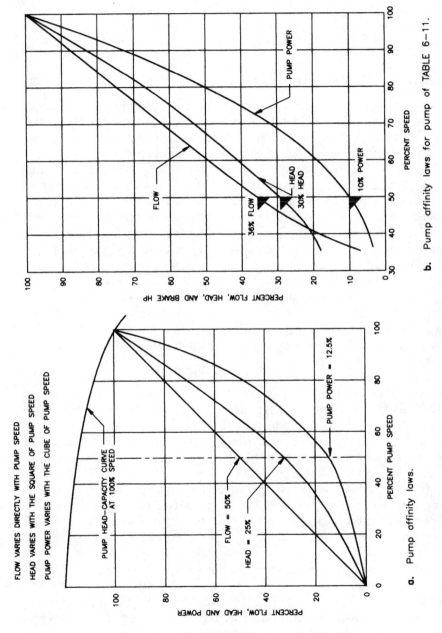

Figure 6.5 Pump affinity laws. *(From the Water Management Manual, Systecon, Inc., West Chester, Ohio, 1992, Fig. 3-17.)*

a. Pump affinity laws.

b. Pump affinity laws for pump of TABLE 6–11.

111

$$\frac{\mathrm{bhp}_1}{\mathrm{bhp}_2} = \frac{d_1^{\ 3}}{d_2^{\ 3}} \tag{6.6}$$

It should be reemphasized that pump impellers should not be changed without a thorough review of the water system characteristics.

6.5 Pump Suction Limitations

Centrifugal pumps operate on water, a liquid stream that can change its state under certain pressure-temperature relationships. Some particular characteristics of water were included in Chap. 1 to help understand these conditions. One of the first rules of thermodynamics is that there is a temperature for every absolute pressure at which water will change its state from a liquid to a gas. A little understood word is *cavitation*. Cavitation is the result of the changing of part of a water stream from a liquid to a gas. It occurs when the temperature of the water reaches the evaporation temperature for the absolute pressure of that stream. The specific gravity of the gas (steam) is much less than that of the liquid water, so the result is "hammering" as dense water and then "light" steam hit the internal parts of a water system. Such hammering can erode the internal parts of the impeller or casing of a pump. Cavitation can damage many parts of a water system; the damage usually occurs first in a pump due to the fact that the suction of a pump may be the point of lowest pressure in the water system.

For example, assume that a hot water system requires 240°F water. From Table 2.4, the vapor pressure for 240°F water is 24.97 psia or 10.27 psig with an atmospheric pressure of 14.7 psi. At least a 10-psi cushion, or 35 psig, should be maintained on this hot water system so that no part of the system approaches the vapor pressure of the 240°F water. If the pressure does drop, cavitation will occur whenever this hot water pressure falls below 24.97 psia.

How can cavitation be avoided in a water system? Simply by ensuring that the water pressure in every part of the water system is greater than the evaporation pressure for any potential temperature of water in that system. Tables 2.3 and 2.4 provide vapor pressures for water at different temperatures. There is so much information available and tools such as the pressure-gradient diagram that there is no excuse for cavitation to occur in a normally operating HVAC water system. Since the pump appears to be the HVAC equipment most vulnerable to cavitation, it is a good place for the designer to start the evaluation process needed to avoid cavitation.

6.5.1 Net positive suction head

Centrifugal pumps are tested very carefully to determine the net positive suction pressure that must be placed on the pump suction at various flows through that pump to eliminate cavitation. This is called the *net positive suction head required (NPSHR) curve* and appears in Figs. 6.1*a* and 6.1*b* as the net positive suction head required by the pump. Most of the centrifugal pumps used in the HVAC field cannot pull water into the pump casing; it must be pushed. The NPSHR curve merely defines the amount of push. If this amount of pressure is maintained on the pump suction, cavitation will not occur in the pump. Figure 6.1*a* provides the type of NPSHR curve found on most pump curves; more precise NPSHR curves are shown in Fig. 6.1*b*. Figure 6.6*a* describes the pressure gradient along the liquid path into a pump.

How do we calculate the net positive suction head needed to avoid cavitation? The actual net positive suction head on the pump suction must always be equal to or greater than the net positive suction head required by the pump. The actual suction pressure is called the NPSHA or *net positive suction head available (NPSHA)*, so

$$NPSHA \geq NPSHR \qquad (6.7)$$

The problem for the uninitiated designer when calculating NPSHA is the attempt to do so in absolute pressures (psia). If the calculations are made in feet of head, they become much easier. This is reasonable, because the NPSHR curve of the pump is always in feet of head. Equation 6.8 is acceptable for HVAC operations at temperatures up to 85°F. The amount of "push" available, or net positive suction pressure available, is simply the atmospheric pressure plus the system pressure minus the vapor pressure of the water minus the friction of the suction pipe. Figure 6.6*b* demonstrates this using the most common point in HVAC systems for calculating NPSHA, namely, a cooling tower. As shown,

$$NPSHA = P_a + P_s - P_{vp} - P_f, \text{ in ft of head} \qquad (6.8)$$

where P_a = atmospheric pressure in feet at the installation altitude. This can be done by reading the value directly from Table 2.1 for water from 32 to 85°F.

P_s = static head of water level above the pump impeller (This is negative if water level is below the impeller.)

P_{vp} = vapor pressure of water, in feet, at operating temperature (see Tables 2.3 and 2.4)

P_f = friction of suction pipe, fittings, and valves, in feet of head

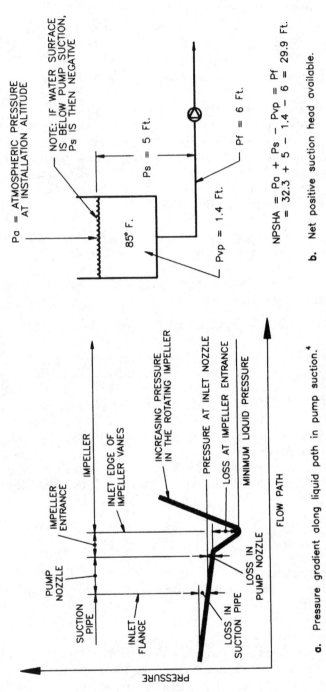

a. Pressure gradient along liquid path in pump suction.[4]

NOTE: IF WATER SURFACE IS BELOW PUMP SUCTION, Ps IS THEN NEGATIVE

Pa = ATMOSPHERIC PRESSURE AT INSTALLATION ALTITUDE

Ps = 5 Ft.

Pf = 6 Ft.

Pvp = 1.4 Ft.

85° F.

$$NPSHA = Pa + Ps - Pvp = Pf$$
$$= 32.3 + 5 - 1.4 - 6 = 29.9 \text{ Ft.}$$

b. Net positive suction head available.

Figure 6.6 NPSH conditions for HVAC pumps. (*From Robert Kern, How to design piping for pump-suction conditions, Chemical Engineering, April 28, 1975, p. 119.*)

For precise calculation of NPSHA, the atmospheric pressure for higher temperatures should be calculated in feet of water at the temperature of the water entering the pump. This is done by converting the atmospheric pressure P_e (Table 2.1) at the installation altitude in pounds per square inch to feet of water at that pressure and temperature. The equation for this is

$$P_a = \frac{144 \cdot P_e}{\gamma} \qquad (6.9)$$

Equation 6.8 thus becomes

$$\text{NPSHA} = \frac{144 \cdot P_e}{\gamma} + P_s - P_{vp} - P_f \qquad (6.10)$$

Assume the following:

1. The maximum cooling tower water temperature is 95°F, so $P_{vp} =$ 1.9 ft of vapor pressure and $\gamma = 62.03$ lb/ft^3.

2. The cooling tower sump is above the pump impeller, and $P_s = 5$ ft.

3. The installation altitude is 1000 ft; $P_e = 14.2$ lb/in^2. Thus

$$P_a = \frac{144 \cdot 14.2}{62.03} = 33.0 \text{ ft}$$

4. The friction loss of the suction pipe, fittings, and valves P_f is 6 ft. Thus

$$\text{NPSHA} = 33.0 + 5 - 1.4 - 6 = 30.6 \text{ ft}$$

Any pump handling this water must have an NPSHR of less than 30.6 ft at any possible flow through the pump.

Warning. *Calculation of NPSHA for a cooling tower should be made at the maximum possible operating water temperature in the tower, not the design water temperature.*

6.5.2   Air entrainment and vortexing

A companion to cavitation is *air entrainment* in the suction of centrifugal pumps. Air can enter the water system at several points, as shown in Fig. 6.7a. The effect of air entrainment on pump performance is described in Fig. 6.7b; it indicates the drastic reduction in both head and capacity when air is present. Every effort must be made to ensure that the water in a pump is free of air.

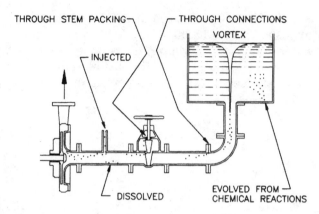

THROUGH STEM PACKING

THROUGH CONNECTIONS

VORTEX

INJECTED

DISSOLVED

EVOLVED FROM
CHEMICAL REACTIONS

a. Centrifugal pumps and entrained—air problems.[5]

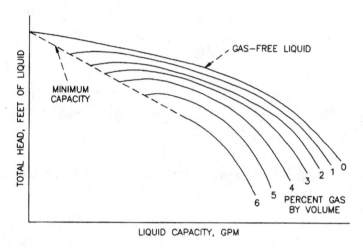

TOTAL HEAD, FEET OF LIQUID

GAS—FREE LIQUID

MINIMUM
CAPACITY

3   2   1   0

4

6   5   PERCENT GAS
BY VOLUME

LIQUID CAPACITY, GPM

b. Effects of various amounts of entrained gas on pump characteristics.

Figure 6.7 Entrainment of air in HVAC pumps. (*From John H. Doolin, Centrifugal pumps and entrained-air problems, Chemical Engineering, January 7, 1963, p. 103.*)

Air entrainment is often confused with cavitation when drawing water from a tank. It is thought that air entrainment cannot happen when a tank of water is 10 to 15 ft deep. Air entrainment can occur easily when water is taken from a free surface of water regardless of the depth of the water. Figure 6.8 describes the air entrainment that can occur if water is not removed properly from an open tank. As shown in Fig. 6.8a, a small whirlpool occurs on the surface, and this deepens into a vortex that will extend down to the water outlet from

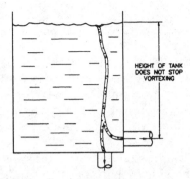

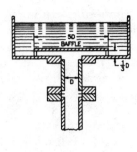

a.   Vortexing in an open tank.

b.   Baffle for bottom connection in an open tank.[7]

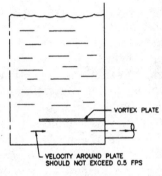

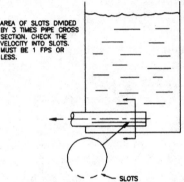

c.   Vortex plate for side exit from tank.[8]

d.   Slotted pipe in vortex breaker.[9]

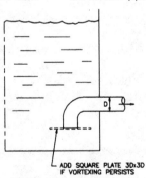

e.   Suction elbow in open tank.

Figure 6.8   Vortexing in HVAC pumps. (*From James H. Ingram, Suction side problems: Gas entrainment, Pump and System Magazine, September 1994, p. 34.*)

the tank, and finally, air will pass into the impeller of the pump, causing hammering similar to cavitation. Tank depth does not deter the development of the vortex.

Cavitation and air entrainment can cause similar noises in a pump. A simple method of eliminating cavitation as a source of noise is to run the pump at full speed and close the manual discharge valve on the pump until only a small flow is passing through the pump. Cavitation should cease, but air noise will persist at the low flow through the pump.

Vortexing can occur easily when water is taken from a tank, particularly if the water outlet is on the side or bottom of the tank, as shown in Fig. 6.8a; a fully developed vortex is shown in this figure. Vortexing can be eliminated by placing an antivortexing plate above a bottom inlet, as shown in Fig. 6.8b; a side inlet can be corrected by placing an antivortexing plate above the inlet (Fig. 6.8c). On shallow cooling tower sumps, if vortexing persists, it may be necessary to install a slotted vortex breaking tube, as shown in Fig. 6.8d. As an alternative, an elbow can be installed on the suction connection of a side inlet and pointed downward, as shown in Fig. 6.8e. If vortexing still occurs, a square plate can be installed on the suction of the elbow. Vortex-breaking devices must be designed to avoid substantial additional friction losses. Normally, the friction loss of an entrance from a tank is equal to about $0.5 \times V^2/2g$, where V is the velocity of the water in the entering pipe. The vortex-breaking device should not increase this entrance loss if possible.

A simple method to determine if vortexing is occurring is to place mats or rafts on the water surface above the inlet from the tank. This prevents the vortex from forming and can be done without draining the tank.

Vortexing should not occur in HVAC tanks and sumps. The precautions are so simple that the design of these tanks and sumps should always accommodate them.

6.5.3   Submergence of pumps in wet pits or open tanks

Pumps installed in a tank, such as vertical turbine pumps, have a specific submergence requirement. The *submergence,* or distance above the inlet bell of such a pump, must be great enough to ensure that the friction loss of the water passing through the bell and entering the pump is made up by the static height of the water over the suction bell. This height is determined by whether the suction of the pump is or is not equipped with a suction strainer. Most manufacturers of axial-flow pumps have adequate data on submergence and

TABLE 6.1   Typical Submergences for Vertical Turbine Pumps (For Operation at 1750 rev/min)

Bowl size, in	Submergence, in	Bowl size, in	Submergence, in
4	7	13	23
6	11	14	30
8	12	15	32
10	16	16	36
11	20	18	36
12	24	20	42

clearance from the bottom of the tank for the water system designer; their recommendations should always be followed. Table 6.1 describes typical submergences for vertical turbine pumps operating at 1750 rev/min; these submergences should not be applied to any specific manufacturer's pumps.

6.6   Internal Forces on Pumps

There are internal forces on centrifugal pumps that concern only the pump designer in the determination of structural parts of the pump and in the selection of pump bearings. Typical of these forces are the lateral or axial thrust that occurs along a pump shaft on a volute-type pump due to an imbalance between the inboard and outboard thrusts. The designer ensures that the pump bearings are capable of withstanding these thrusts throughout the possible operating range of the pump.

Other thrusts that vary greatly with the flow and speed of the pump are of concern to the water system designer. The radial thrust on a volute-type pump varies greatly with the flow through that pump. Figure 6.9a is a general description of a radial thrust curve. This thrust curve is for a single-volute pump at constant speed. It is obvious that this pump should operate as closely as possible to the best efficiency point, where the radial thrust is the lowest. Reducing the speed lowers this radial thrust, as shown in this figure, which explains why variable-speed pumps generally have longer lives than constant-speed pumps. The water system designer should therefore ensure that any constant-speed pumps involved do not operate with either very low or very high flows through them. As will be emphasized throughout this book, pumps should be selected and operated as closely as possible to their best efficiency point. With the knowledge now in existence on pump operation, it is inexcusable to allow HVAC pumps to operate with high radial thrusts on them.

Some vertical turbine pumps have an axial thrust curve that must be recognized in the application of these pumps. Figure 6.9b describes

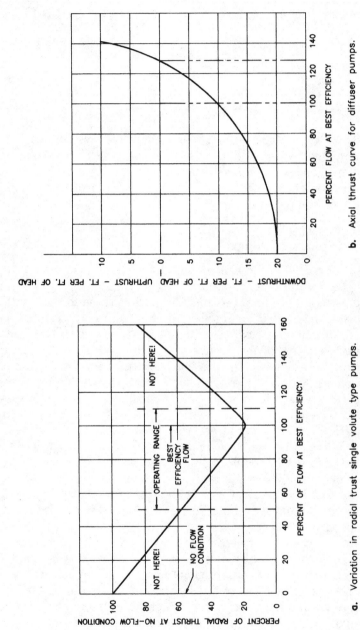

Figure 6.9 Radial and axial thrusts on HVAC pumps. (*From The Water Management Manual, Systecon, Inc., West Chester, Ohio, 1992, Fig. 3-24.*)

a. Variation in radial trust single volute type pumps.

b. Axial thrust curve for diffuser pumps.

a typical axial thrust curve for an enclosed impeller, vertical turbine pump. Such pumps operate normally with a downthrust; this downthrust is equal and opposite to the force of water that is moving upward through the pump. This downthrust is greatest near the no-flow or shutoff condition. It reduces until it is zero at some point at a flow greater than at the best efficiency point. If the flow increases, the downthrust goes to zero and then becomes an upthrust. If flow is increased on high-head pumps, the upthrust can be great enough that it may damage the top drive coupling of the pump motor. If the pump designer utilizes these efficient pumps, care must be taken that the upthrust condition will not occur or that the pump drive train and motor are designed to withstand any upthrust that may occur.

6.7   Pumping Energy

Following is a review of the various energy equations that must be used in determining the energy requirements of a water system and the energy consumed by pumps and their drivers in satisfying those requirements. It is important that the HVAC system designer understand the difference between the energy absorbed by the water and that consumed by the pump or its motor.

6.7.1   Water horsepower

The energy required by a pump, as indicated by the affinity laws, depends on its speed and the diameter of its impeller. The energy imparted to the water by a pump is called the *water horsepower* whp. Its equation is

$$\text{whp} = \frac{Q \cdot h \cdot s}{3960} \tag{6.11}$$

where Q = flow, in gal/min
    h = head, in ft
    s = specific gravity

Most HVAC water systems operate from 32 to 240°F water, where the specific gravity can vary from 1.001 at 32°F to 0.948 at 240°F. For these applications, the specific gravity is generally assumed to be 1.00; systems designed to operate at 240°F must be able to function at startup with water temperatures near 50°F or at specific gravities of around 1.00. Water horsepower for these systems usually ignores the specific gravity and assumes it to be 1.00. This should not be the case with medium- and high-temperature water systems, where the operating temperatures can vary from 250 to 450°F. The specific gravity of 450°F water is 0.825 and should not be ignored in calculating pump operating energy for these systems. Again, pump motor or driver brake horsepower should recognize that the pump may be required to operate with colder water and specific gravities near 1.0.

Following is an example of water horsepower: If a pump is delivering 1000 gal/min at a total head of 100 ft, the water horsepower is

$$\text{whp} = \frac{1000 \times 100}{3960} = 25.25 \text{ whp}$$

6.7.2 Pump brake horsepower

The energy required to operate a pump is determined in brake horsepower bhp; the difference between water horsepower and pump brake horsepower is the energy lost in the pump. The pump brake horsepower equation is therefore

$$\text{bhp} = \frac{\text{whp}}{P\eta} \tag{6.12}$$

where P_η = pump efficiency as a decimal

For example, if the pump in the preceding example is operating at an efficiency of 85 percent,

$$\text{bhp} = \frac{25.25}{0.85} = 29.7 \text{ hp}$$

6.7.3 Pump motor power in kilowatts

The electrical energy, *pump kilowatts,* for a motor-driven pump must take in consideration the efficiency of the motor on constant-speed pumps and the wire-to-shaft efficiency of the motor and variable-speed drive on variable-speed pumps. Again, the difference between the pump brake horsepower and electric power input to the motor or variable-speed drive and motor is the efficiency of these devices. The power must be converted from brake horsepower to kilowatts, so the pump brake horsepower must be multiplied by 0.746. The equation is

$$\text{Pump kW} = \frac{bhp \cdot 0.746}{E\eta} \tag{6.13}$$

where $E\eta$ is the efficiency of the electric motor or the motor and variable-speed drive as a decimal.

For example, if the preceding pump is variable speed with a wire-to-shaft efficiency of 89 percent for the variable-speed drive and high-efficiency motor, then

$$\text{Pump kW} = \frac{29.7 \cdot 0.746}{0.89} = 24.9 \text{ kW}$$

Summarizing Eqs. 6.11 through 6.13,

$$\text{Pump kW} = \frac{Q \cdot h \cdot s}{5308 \cdot P\eta \cdot E\eta} \tag{6.14}$$

As will be explained in Chap. 7, the combined efficiency of an electric motor and a variable-speed drive cannot be just the multiplication of the efficiencies of the motor and the variable-speed drive. It must be determined by the variable-speed drive manufacturer with recognition of the manufacturer and the quality of the electric motor.

Steam turbine–driven pumps also will be reviewed in Chap. 7; the energy required by them is usually based on a water rate in pounds per pump brake horsepower. Equation 6.12 is used to compute the maximum water rate at maximum flow and head.

6.7.4 Actual energy consumed by variable-speed pumps

The energy consumed by variable-speed pumps requires knowledge of the water system head curve or area on which they are operating. These curves and areas will be described in Chap. 9. Table 6.2 de-

TABLE 6.2   System and Pump Variables

System gal/min	Pump head, ft	Pump speed, rev/min	Pump efficiency, %	Pump bhp	Wire-to-shaft efficiency, %	Total kW	Wire-to-water efficiency, %
100	20.9	626	41.1	1.3	58.1	1.7	23.8
200	23.4	674	63.2	1.9	61.4	2.3	38.4
300	27.2	742	74.3	2.8	65.8	3.1	47.9
400	32.2	826	79.7	4.1	70.7	4.3	54.7
500	38.6	918	82.6	5.9	75.4	5.8	59.9
600	46.1	1013	84.1	8.3	79.5	7.8	63.8
700	54.7	1113	85.0	11.4	83.0	10.2	66.8
800	64.5	1217	85.4	15.3	85.9	13.3	69.1
900	75.4	1324	85.6	20.0	88.1	17.0	70.6
1000	87.5	1433	85.7	25.8	89.7	21.4	71.7
1100	100.6	1544	85.8	32.6	90.7	26.8	72.2
1200	114.8	1656	85.8	40.5	91.3	33.1	72.5
1300	120.0	1770	85.7	49.8	91.7	40.5	72.6

scribes the variation in pump efficiency, pump brake horsepower, speed, motor kilowatts, and overall wire-to-water efficiency of the pump, motor, and variable-speed drive with a system head curve that contains a constant head of 20 ft. The pump fitting losses are accounted for in these calculations. The pump head-capacity curve in Fig. 6.1a is used in these calculations.

The computer program for calculating the wire-to-water efficiency for this pump is available from manufacturers of variable-speed pumping systems. The data for the wire-to-shaft efficiency included in this table are secured from the variable-speed drive manufacturer. The overall wire-to-water efficiency of variable-speed pumping systems that have two or more pumps operating in parallel is discussed in detail in Chap. 10.

6.8   Sources of Pump Information

It is obvious that the information included in this chapter on pump performance requires substantial data from the pump manufacturers on the selection and operation of pumps on HVAC water systems. Most of the pump manufacturers active in this market have gone to great lengths to provide technical information on their pumps.

Traditionally, pump catalogs contain detailed information on pump physical dimensions and general arrangement as well as pump head-capacity curves at 1150, 1750, and 3500 rev/min. Data on slower speeds such as 850 and 720 rev/min are provided for larger-capacity pumps. The development of computers has enabled pump manufacturers to provide these data in software form such as diskettes or CD ROM. Some pump manufacturers are furnishing computer selection programs where the required pump head and capacity are inserted and the pump selection is made for the designer. This selection can be on the basis of (1) the most efficient or (2) the most economical in first cost.

6.9   Summary

The increase in cost of energy has thrust efficiency of operation to the forefront. Highly efficient centrifugal pumps are now available at little added cost over less efficient pumps. The first and foremost evaluation of HVAC pumps should therefore be efficiency. Since so many of these pumps are now variable speed, efficiency throughout the operating range must be considered.

Ease of pump maintenance was of major concern in the past. The reduction of pump maintenance due to variable speed has lessened the concern for maintenance capability. Traditionally, pumps have been given much more space for operation than centrifugal fans. Many fans are tucked up in air handling units where repair is quite difficult. With the application of variable speed and greater care in pump operation, the large spaces around HVAC pumps are no longer needed. Most HVAC pumps should operate for years without any serious maintenance.

Along with efficiency and first cost, these pumps should be evaluated for the space they require. As can be seen in Chap. 5, there are many different pump configurations for all the applications in the HVAC industry. The development of vertical pump assemblies has reduced the space needed without sacrificing ease of maintenance. Typical of this is the vertically mounted double-suction pump, which offers space saving as well as ease of maintenance.

The HVAC industry has many capable manufacturers of pumps and pumping systems that can provide efficient pumping for any application encountered in heating ventilating and air conditioning.

6.10 Bibliography

John H. Doolin, Centrifugal pumps and entrained-air problems, *Chemical Engineering,* January 7, 1963.

James H. Ingram, Suction side problems: Gas entrainment, *Pumps and Systems Magazine,* September 1994.

Robert Kern, How to design piping for pump-suction conditions, *Chemical Engineering,* April 28, 1975.

The Water Management Manual, Systecon, Inc., West Chester, Ohio, 1992.

Pump Drivers and Variable-Speed Drives

7.1 Introduction

HVAC pumps are driven by electric motors, steam turbines, and gas- or oil-fueled engines. By far most of the pumps are operated by electric motors. Steam turbine–driven pumps are found on large central chilled water or cogeneration plants where steam is available from boilers or exhaust steam from turbine-driven chillers or electric generators. Engine-driven pumps are found on large pumping installations where a source other than electricity is needed; this can be in areas of power curtailment or where emergency pump operation is required.

7.2 Electric Motors

The following discussion makes reference to two principal organizations in the electrical industry. These are (1) the Institute of Electrical and Electronic Engineers (IEEE) and (2) the National Electrical Manufacturers Association (NEMA).

Most electric motor–driven pumps in the U.S. HVAC market are single- or three-phase, 60 Hz. Table 2.8 provides the motor nameplate voltages and the accompanying distribution system voltages. This table provides the best match of motor voltages to distribution voltages and meets current motor design practice.

Although 50-Hz power is generally not used in the United States, HVAC designers will encounter it occasionally in the United States and when working in foreign countries. Table 2.9 also includes information on 50-Hz power and motors. Standard pump curves are available from the pump manufacturers for operation with 50-Hz motors at induction motor speeds of around 960 and 1450 rev/min, compared with 1150 and 1750 rev/min for such motors operating on 60-Hz power.

Almost all electric motors in the HVAC field are induction-type. Large cooling tower or condenser motors over 1000 hp may be synchronous-type, running at 1200 and 1800 rev/min. Wound rotor and direct-current motors offer few advantages for most of these pump applications.

7.2.1 Electric motor power characteristics

All electric motors are designed to operate with some variation in the electric power supplied to them. Following are some general rules for the variation in power characteristics:

1. Voltage variation must be limited to ±10 percent at rated frequency.

2. Frequency variation must be limited to ±5 percent at rated voltage.

3. Combined variation of voltage and frequency must be limited to the arithmetic sum of 10 percent.

4. On polyphase motors, the voltage imbalance should be kept within 1 percent between phases. Table 7.1 describes the great decrease in motor rating caused by greater imbalances. Balancing of loads between phases in a building becomes a necessity to ensure maximum output from the polyphase electric motors.

TABLE 7.1 Derating Factors Due to Voltage Imbalance for Polyphase Motors

Percent voltage imbalance	Derating factor
0	1.00
0.5	1.00
1.0	1.00
1.5	0.97
2.0	0.95
2.5	0.93
3.0	0.89
3.5	0.85
4.0	0.82
4.5	0.78
5.0	0.76

SOURCE: *AC Motor Selection and Application Guide*, Bulletin GET-6812B, General Electric Company, Fort Wayne, Ind., 1993, p. 3; used with permission.

7.2.2 Motor output ratings

The following discussion pertains to induction-type electric motors, since most of the pump motors in the HVAC industry are of this design. These ratings are for 104°F ambient air and for motors installed between sea level and 3300 ft of elevation. For motors installed at higher elevations:

1. A motor with a service factor of 1.15 can be operated at higher elevations with a service factor of 1.0.

2. A motor can be operated at its normal service factor at higher elevations at lower ambient temperatures. Refer to the manufacturer's nameplate information.

3. Medium- and high-voltage motors should be operated at higher elevations with caution due to a corona effect.

All high-altitude motor operations should be referred to the motor manufacturer for approval.

7.2.3 Motor speed

The speed at which an induction motor will operate depends on the input power frequency and the number of electrical magnetic poles for which the motor is wound. The speed of a pump varies linearly with the frequency; the higher the frequency, the faster the motor runs. This is how the present electronic-type variable-speed drives are designed and why they are therefore called *variable-frequency drives.* Conversely, the more magnetic poles in a motor, the slower it runs. For an induction-type motor, the speed of the rotating magnetic field in the stator is called the *synchronous speed,* and it is determined by the following equation:

$$\text{Synchronous speed, rev/min} = \frac{60 \cdot 2 \cdot \text{frequency}}{\text{number of poles}} \qquad (7.1)$$

For example, a motor operating at 60 Hz with four poles would have a synchronous speed of

$$\frac{60 \cdot 2 \cdot 60}{4} = 1800 \text{ rev/min}$$

In the pump industry, a 1200 rev/min motor is called a *six-pole machine,* an 1800 rev/min motor, a *four-pole machine*; and a 3600 rev/min motor, a *two-pole machine.*

7.2.4 Types of polyphase motors and code letters

NEMA has established four different motor designs and has given them a letter for each design, namely, A, B, C, and D. Each of these four designs has unique speed-torque-slip relationships. The type used on HVAC pump motors is usually design B.

Polyphase motors from 1 to 200 hp have been designated by NEMA with reference to their speed-torque relationships; NEMA has developed these code letters that include percentage of slip. Full-load speed for an induction motor differs from synchronous speed by the percentage of slip. All induction motors are designed to various amounts of slip and are included in the design letters from A to D as described in Table 7.2.

Most HVAC pump motors are NEMA design B with a maximum allowable slip of 5 percent. Motors with a slip of greater than 5 percent are called *high-slip motors* and are designed for loads requiring high starting torques. Most 1200 rev/min, six-pole induction motors have a full-load speed of around 1150 rev/min, while 1800 rev/min, four-pole motors have a full-load speed of near 1750 rev/min.

TABLE 7.2   Comparison of NEMA Designs for Induction Motors

NEMA design letter	Percentage of slip	Starting current	Locked-rotor torque	Breakdown torque	Applications
A	Max. 5%	High to medium	Normal	Normal	Broad application including fans and pumps
B	Max. 5%	Low	High	Normal	Normal starting torque for fans, blowers, and pumps
C	Max. 5%	Low	High	Normal	For equipment with high inertia start such as positive displacement pumps
D	—	Low	Very high	—	Very high inertia starts, choice of slip to match the load
	5 to 8%	—	—	—	Punch presses, etc.
	8 to 13%	—	—	—	Cranes, hoists, etc.

SOURCE: *AC Motor Selection and Application Guide,* Bulletin GET-6812B, General Electric Company, Fort Wayne, Ind., 1993, p. 7; used with permission.

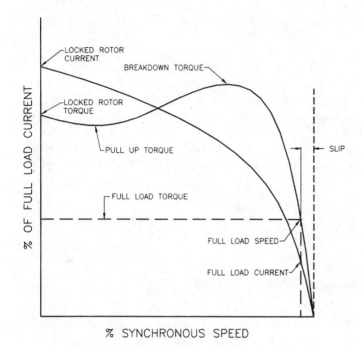

% SYNCHRONOUS SPEED

SPEED VS. TORQUE, CURRENT CURVES

Figure 7.1 Electric motor performance curves. (*From AC Motor Selection and Application Guide, Bulletin GET-6812B, General Electric Company, Fort Wayne, Ind., 1993, p. 5.*)

7.2.5 Torque and horsepower

Torque is the turning force acting through a radius and is rated in pound feet. *Horsepower* is the rate of doing work and is rated in foot pounds per minute. Thus

$$1 \text{ hp} = 33,000 \text{ (ft} \cdot \text{lb)/min} \tag{7.2}$$

There are several torques for an induction motor, and they are described in Fig. 7.1 for torque, speed, and motor current curves. *Locked-rotor torque* is also called *starting torque*. This is the torque that the motor can develop when at rest or zero speed. *Pull-up torque* is the minimum torque developed by the motor as it runs from zero to full-load speed.

Since most centrifugal pumps in the HVAC field are variable-torque machines, where the torque varies as the cube of the speed, the locked-rotor or pull-up torque developed by NEMA design B motors is adequate for most of these pumps. The only torque required of the motors for these pumps is the inertia of the motor rotor and the pump rotating element. This is called the *starting-load inertia* WR^2 for the pump shaft and impeller and WK^2 for the motor rotor and shaft. This inertia is low enough for most centrifugal pumps used in this field that it is of no concern. On very large condenser pumps with large-diameter impellers, it may be necessary to check the inertia, particularly when reduced-voltage types of starters with reduced starting torques are used.

Full-load torque is the torque required to produce the rated horsepower of the motor at full-load speed in pound feet. The equation for this torque is

$$\text{Full-load torque, lb} \cdot \text{ft} = \frac{\text{bhp} \cdot 5250}{\text{max. rev/min}} \qquad (7.3)$$

where bhp = maximum brake horsepower
rev/min = maximum speed

7.2.6 Motor currents

Figure 7.1 describes a typical motor current curve for a NEMA design B motor. The locked-rotor current is the maximum steady-state current that the motor will draw at 0 rev/min with rated voltage and frequency applied to the power terminals. NEMA has developed a set of code letters for various locked-rotor amperes for electric motors. These code letters run from A to V. The code letter for most HVAC pump motors is G, which indicates that the locked-rotor current is approximately 650 percent of the full-load current of the motor. This code letter appears on the nameplate of every NEMA-rated motor. Full-load current is the steady-state current of a motor operating at full-load torque with rated voltage and frequency applied to the power terminals of the motor.

There may be several reasons for reducing the locked-rotor or starting current. The electric utility may require it or the electrical distribution of the building may be better served with reduced starting currents. A number of different devices are available for this, and they are known as *reduced-voltage starters*.

1. *Autotransformer.* This method provides several taps to adjust the starting voltage.

Starting voltage at motor, %	Line current, %	Starting torque, %
80	64	64
65	42	42
50	25	25

With the preceding three taps of 80, 65, or 50 percent, the starting current is limited to approximately 125 to 390 percent of full-load current. This starter is large and expensive.

2. *Primary resistor.* Starting characteristics are fixed; starting current is 390 percent of full-load current. This starter has an added power loss in the resistors.

3. *Part winding.* Starting characteristics are fixed; starting current is 390 percent of full-load current. It requires special motor connections for 460-V service. It is the smallest and least expensive reduced-voltage starter.

4. *Wye delta.* Starting current is approximately 200 percent of full-load current; starting torque is low, only 33 percent of full-load torque. This starter requires special motor design and more complex control than the other reduced-voltage starters.

5. *Solid state.* Solid-state starters are being used on both constant-speed and variable-speed pumps due to their adjustability and size. A typical range of adjustable starting currents is from 100 to 400 percent of full-load current. They may or may not be less expensive than other types of reduced-voltage starters.

6. *Variable-frequency drive.* The variable-frequency drive has proved to be an excellent method for reducing the current inrush. Since the starting torque on most HVAC pumps is limited to inertia torque WR^2, the variable-frequency drive can be used on almost all HVAC pump motors. It is usually more expensive at this date than other reduced-voltage starters; its ability to be programmed into the overall duty cycle of HVAC pumps often results in its selection due to the energy savings achieved through variable speed.

Due to special motor windings, the part winding and wye delta types of reduced-voltage starters cannot be used as standby starters for variable-speed drives at 460-V power supply.

7.2.7 Output horsepower

A motor is designed to produce its rated horsepower with nameplate voltages and frequency applied to its power terminals. Most HVAC pump motors have a service factor that has been developed by NEMA for standard polyphase motors. These service factors are defined as the permissible overload beyond nameplate rating with standard voltage and frequency. Most NEMA design B, open-frame, drip-proof motors have a service factor of 1.15 up to 200 hp and 1.0 for over 200 hp. Totally enclosed, fan-cooled (TEFC) and explosion-proof (class 1, group D) motors may have a service factor of 1.0 hp and no overload or 1.15 hp. The manufacturer or supplier of a motor should verify the actual service factor for a motor. Also, the nameplate for existing motors should be inspected to confirm the allowable service factor.

7.2.8 Power factor

The *power factor* for a three-phase motor recognizes the magnetizing current of the motor. Its equation is

$$\text{Power factor} = \frac{\text{watts applied}}{\sqrt{3} \cdot \text{volts} \cdot \text{amps}} \tag{7.4}$$

Power factor is a characteristic of a polyphase electric motor's operation. Electric utilities may charge penalties for low power factor. Likewise, some governing bodies may establish a minimum power factor.

Power factor can be improved by purchasing motors with high power factor ratings, providing no loss in efficiency results. A more satisfactory method may be the installation of power factor correction capacitors. Total power factor correction programs are within the province of the consulting electrical engineer, but the HVAC designer should have a working knowledge of power factor and its correction.

One advantage for some variable-speed drives, particularly pulse width modulation type, is that the input power factor of the variable-speed drive and the motor as a combination is equal to the power factor of the drive itself. Most of these drives have a power factor close to 95 percent. More on power factor will be discussed later in Sec. 7.3 of this chapter.

7.2.9 Motor efficiency

Motor efficiency is an important design consideration and has been the object of much redesign and rating in the motor industry. The equation is

$$\text{Motor efficiency} = \frac{\text{hp output} \cdot 746}{\text{watts input}} \tag{7.5}$$

IEEE has established and NEMA has adopted Standard 112, *Method B for the Testing of Electrical Motors*. This test establishes uniform methods of testing and rating electric motors and is the basis for rating motors for compliance with the latest government requirements.

The Energy Policy Act of 1992 establishes nominal full-load efficiencies for both open and closed motors. Table 7.3 lists these efficien-

TABLE 7.3   Nominal Full-Load Efficiencies, Polyphase Electric Motors

| Motor hp | Open motors (ODP) | | Closed motors (TEFC or EXP PRF) | |
	1200 rev/min	1800 rev/min	1200 rev/min	1800 rev/min
1	80.0	82.0	80.0	82.5
1.5	84.0	84.0	85.5	84.0
2	85.5	84.0	86.5	84.0
3	86.5	86.5	87.5	87.5
5	87.5	87.5	87.5	87.5
7.5	88.5	88.5	87.5	89.5
10	90.2	89.5	89.5	89.5
15	90.2	91.0	90.2	91.0
20	91.0	91.0	90.2	91.0
25	91.7	91.7	91.7	92.4
30	92.4	92.4	91.7	92.4
40	93.0	93.0	93.0	93.0
50	93.0	93.0	93.0	93.0
60	93.6	93.6	93.6	93.6
75	93.6	94.1	93.6	94.1
100	94.1	94.1	94.1	94.5
125	94.1	94.5	94.1	94.5
150	94.5	95.0	95.0	95.0
200	94.5	95.0	95.0	95.0

SOURCE: *Energy Policy Act of 1993 as It Relates to Motors,* Bulletin GEK-100919, General Electric Company, Fort Wayne, Ind., 1993; used with permission.

cies for up to 200-hp motor sizes. These efficiencies apply to all motors manufactured after October 1997.

All electric motor companies must develop a bell curve of efficiency distribution where the minimum or guaranteed efficiency can only be 10 percent less than the nominal efficiencies shown in Table 7.3. Some manufacturers list this minimum efficiency on their motors as the guaranteed efficiency. Motors manufactured prior to October 1997 should comply with the efficiencies listed in NEMA Bulletin G-1-1993.

7.2.10   Motor construction

Most horizontally mounted and some vertically mounted electric motors for HVAC pumps are manufactured in three different enclosures, namely, drip-proof; totally enclosed, fan cooled; and explosion-proof. Particular atmospheres that are dusty or contain specific chemicals may require special motor enclosures.

Drip-proof motors are designed for cooling by ambient air and are called *open-frame* on horizontal motors and *weather-protected* (WP1) on vertical motors. Open-frame drip-proof motors are designed for relatively clean indoor applications, whereas weather-protected motors can be used outdoors and are so constructed to minimize the entrance of rain, snow, and airborne particles. Drip-proof motors are adequate for most HVAC equipment room installations.

Totally enclosed, fan-cooled motors are cooled by an external fan mounted on the motor shaft. This enclosure is available for applications where the motor is wetted periodically. It should not be used for routine motor applications in this industry. Usually, it does not have any higher efficiency than an open-frame drip-proof motor.

Explosion-proof class 1, group D construction is for applications where the ambient air may contain combustibles such as gasoline, petroleum, naphtha, or natural gas. There are other types of explosion-proof constructions for other hazardous atmospheres. Most HVAC atmospheres do not contain any combustibles. If a specific hazardous material is encountered, the insuring agency should verify the correct motor enclosure. These motors should not be used with variable-frequency drives without approval of the motor manufacturer. They may not be certified for this use.

Vertical motors are available in two physical constructions, vertical hollow shaft (VHS) and vertical solid shaft (VSS). The hollow-shaft construction is usually applied to vertical turbine pumps due to the ability to adjust the lateral setting on these pumps without disassembling the motor. This motor has a top drive coupling with an adjustment nut that provides this adjustment.

Most motors for HVAC pumps are single speed. Although two-speed motors are available for these applications, the variable-speed drive has all but eliminated the use of multispeed motors for these pumps. The added cost of the special two-speed motor and two-speed starter is at this time still less than the cost of a variable-speed drive. However, the complex control and the energy loss due to only two speeds, i.e., two pump head-capacity curves, usually result in the selection of a variable-speed drive instead of a two-speed motor. Another disadvantage for two-speed motors on large HVAC water systems is the hydraulic shock that is produced when the motor changes from one speed to the other.

7.2.11 Motor sizing for HVAC pumps

The important factors on most HVAC installations in the proper sizing of an electric motor are long life and efficient operation. These factors dictate that the motor should not operate beyond its nameplate rating. For example, assume that a pump is selected for 2000 gal/min at 125 ft of head and a 75-hp motor is fully loaded at the pump's selection point. If the pump can carry out to (operate at) capacities greater than

2000 gal/min, the brake horsepower required will exceed 75 hp, and the motor will be overloaded. For long life and efficient operation, a 100-hp motor should be selected for this pump. It is recommended that the specification for electric motors for pumps include the statement that the motor will not be overloaded beyond its nameplate rating when the pump operates at any point on its head-capacity curve.

7.3   Variable-Speed Drives for Electric Motors

7.3.1   History

Up to about 1970, variable-speed drives in the HVAC industry usually meant eddy-current or fluid couplings between a fixed-speed induction motor and a variable-speed pump. The coupling was controlled electrically or hydraulically to allow a variable slip between the motor speed and that of the coupled load. These drives established a high order of reliability and were quite satisfactory for pump loads in which the torque requirements dropped off rapidly as the speed was reduced. In such applications, the losses inherent in the coupling between the motor and the pump were offset by the reduction in losses due to overpressuring caused by a constant-speed pump. For new installations, however, this arrangement has largely been superseded by variable-frequency drives. On applications that have environments hostile to variable-frequency drives, mechanical variable-speed drives should still be used. This includes dusty or corrosive atmospheres and high ambient temperatures where it is impossible to provide adequate cooling for the variable-frequency drive.

The advantages of variable-frequency drives (VFDs) for fans, pumps, and chillers have been known for many years. They permit the use of simple, reliable, and inexpensive induction motors yet provide the operating economies of variable-speed. Unfortunately, motor generator sets, thyratrons, and ignitrons, the only methods of obtaining a variable-frequency source, were too expensive for all but the most critical applications.

The invention of the thyristor (SCR) in the mid-1960s changed the picture dramatically. Here was a device that could control power at the megawatt level yet was both economical and reliable.

Variable-speed drives soon appeared for direct-current (dc) motors and shortly thereafter for alternating-current (ac) induction motors. The early variable-frequency thyristor drives, which were mostly voltage source inverters, were sometimes less than ideal in their characteristics; they opened up a huge new field of application in the HVAC industry. Now, drive designs have matured, but there are still important new developments that solve some of the major application problems on the electrical side.

7.3.2 Types of variable-frequency drives

For many years the variable-frequency drive field has been dominated by six-pulse voltage-source and current-source thyristor inverters (Fig. 7.2). Block diagrams of these drives are shown in Fig. 7.2a. Both generate the output ac voltage by alternately switching between three pairs of thyristors. Capacitor banks are used to force the load current to switch from one set to the other. In the voltage-source inverter, a set of six diodes in a rectifier is used to charge a filter capacitor bank to a dc voltage equal to the peak output voltage of the load. The filter capacitor bank serves to isolate the inverter from the ac line. This filter capacitor bank is, in addition to the capacitors, used for switching.

The current-source inverter replaces the filter capacitor bank with a large inductor that serves the same purpose of isolation and filtering but makes the drive more tolerant of line and load disturbances. It also permits regenerating energy from the load to the line, an important advantage when high-inertia loads must be brought to rest quickly.

Both these drives, in their basic form, generate a six-step output waveform, voltage for the voltage source, and current for the current source. Typical waveforms are shown in Fig. 7.2b. The magnitude of the voltage is directly proportional to the load frequency so that the volts-per-hertz ratio remains constant. The motor operates at a constant flux level and, except for ventilation considerations, is able to operate at constant torque. Harmonics in the output current cause additional heating, however, and this generally results in a derating of 10 to 15 percent in horsepower for standard motors.

PWM Drives.   The development of large power transistors in the 1980s spawned a new type of variable-frequency drive (Fig. 7.3). The acronym *PWM* stands for *pulse-width modulation,* a totally different technique from the six-step unit for obtaining voltage control and a variable output frequency. Whereas the six-step voltage and current source drives vary the amplitude of the switched voltage, PWM drives vary the output voltage by repetitively connecting and disconnecting a fixed voltage at rapid intervals. The ratio of "on" to "off" periods determines the voltage magnitude. Figure 7.3a illustrates this process in generating a 320-V rms, 40-Hz sine-wave approximation from a 675-V constant-potential bus. This is an actual requirement for a PWM drive on a 480-V, 60-Hz motor. The switching frequency is about 1 kHz in this example.

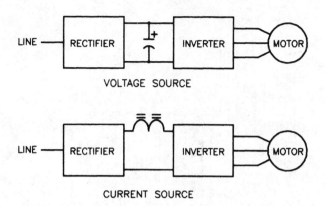

a. Block diagrams for six pulse drives.

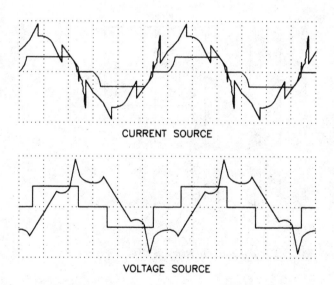

b. Six—step waveforms at the motor.

Figure 7.2 Current-source and voltage source drives. (*From Keith H. Sneker, private communication, Halmar Robicon Group, Pittsburgh, PA., 1994.*)

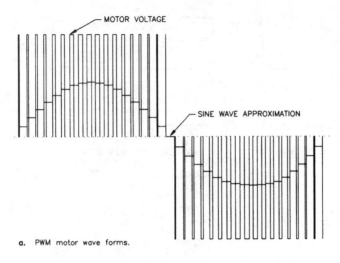

a. PWM motor wave forms.

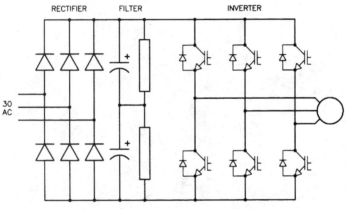

b. Simplified schematic of PWM VFD.

Figure 7.3   Pulse-width-modulating (PWM) drives. (*From Keith H. Sneker, private communication, Halmar Robicon Group, Pittsburgh, PA., 1994.*)

At the beginning and end of the 40-Hz sine wave, the switching is mostly "off," but the duty cycle rises to two-thirds at the peak to generate the required 453-V peak for 320-V rms. In this fashion, the output voltage frequency and magnitude are controlled for the motor. More sophisticated switching techniques are used in practice, but this example illustrates the basic approach. Motor current is nearly sinusoidal.

A PWM drive consists of a diode rectifier, a filter capacitor bank, and a set of six switching transistors. A simplified schematic is shown in Fig. 7.3b. The rectifier diodes charge the filter capacitors to a constant potential for approximately 675 V dc. The transistors are controlled so that one transistor in each phase is always conducting, and a path for current exists through a transistor or its inverse diode.

The diode rectifier and filter capacitor act to greatly reduce the harmonic effects on the power system. Although the low-frequency harmonic currents remain, line notching is virtually eliminated. The interference potential is much less, but elimination of the low-frequency current harmonics still requires corrective measures at increased cost. Power factor is also much better for the PWM drive than for the six-step units. It is always 90 percent or better and is nearly independent of motor speed.

Multipulse input circuits.   An input transformer or autotransformer can be used to power two or three input rectifiers to a drive. Phase-shifted voltages cancel the low-frequency current harmonics and greatly reduce the line-current distortion. Block diagrams of 12- and 18-pulse systems are shown in Fig. 7.4. The disadvantage of multipulse systems is an increase in cost over the simpler systems. In drives over 500 hp, however, multipulse operation is usually the arrangement of choice. In general, a 12-pulse drive is sufficient, and the added cost and complexity of an 18-pulse drive are seldom warranted.

Clean power variable-speed drives.   A dramatic series of technical developments has recently yielded variable-frequency drives that will meet the requirements of IEEE Standard 519-1992 with no filters or multipulse circuitry. All these new PWM drives are based on fast-switching power transistors. Most use an active filter that detects the current distortion and injects corrective currents to cancel the harmonics. Some manufacturers, however, employ an ingenious switching algorithm that eliminates the harmonic currents in the first place. This arrangement requires transistors in place of the rectifier

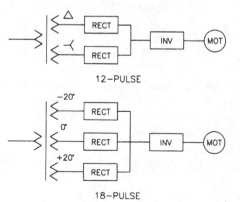

Figure 7.4   Twelve- and eighteen-pulse drives. (*From Keith H. Sneker, private communication, Halmar Robicon Group, Pittsburgh, PA., 1994.*)

diodes, but it needs no active filter. It also permits regeneration for rapid deceleration of an overhauling load. These clean power drives are currently available through 1000 hp.

7.3.3 Harmonics and variable-frequency drives

A problem common to all types of variable-frequency drives is harmonics on the ac power line. Harmonic currents generated by the converter thyristors cause distortion of the power-line voltage (the infamous *notching*), and this distortion may affect equipment in the building or in neighboring facilities. The problems may become acute when power factor correction capacitors are installed. The capacitors resonate with the power-line inductance and may amplify harmonic currents to many times their original values. This can result in failed capacitors, interference with data-processing equipment, overvoltages, and other undesirable effects. If capacitors are used on variable-frequency drives, the motors should be installed as closely as possible to those drives.

Harmonics were addressed by the IEEE, which, in 1981, issued Standard 519-1981 establishing limits on the allowable voltage distortion on a feeder common to several facilities. In most cases, the total harmonic voltage distortion was limited to 5 percent. The use of a current-source or voltage-source thyristor inverter often required the installation of a high-pass filter to control the distortion. In addition to increasing the cost of the installation, the filter introduced losses that reduced operating efficiency. Nonetheless, these filters did the required job, and thousands are presently in service.

In 1992, an IEEE committee revised the standard and issued Standard 519-1992, *IEEE Recommended Practices and Requirements for Harmonic Control in Electrical Power System*. This document spelled out, for the first time, the allowable levels of harmonic currents injected into the utility system. The limit values depend on how "stiff" the supply system is and on the particular harmonic involved. In general, the limits are much more severe and difficult to meet than were the previous limits on voltage distortion only.

Harmonic distortion is of great concern to electric utilities and building or plant operators. The projected harmonic distortion can be computed for a new or existing installation of variable-frequency drives. Figure 7.5 is a form that can be used by the variable-frequency drive manufacturer to estimate the harmonic distortion for most installations. The following information is required:

1. Size in kilovoltamperes and percentage impedance of the distribution transformer

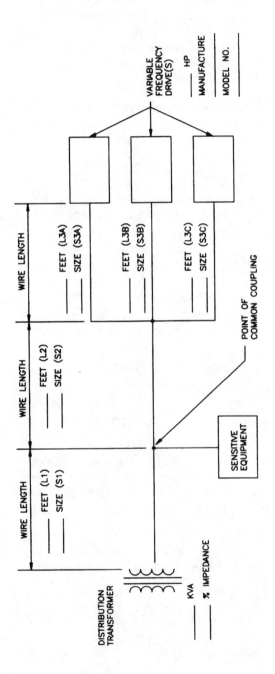

POWER DISTRIBUTION ANALYSIS PERTAINING TO IEEE-519 STANDARDS

THE FOLLOWING INFORMATION IS REQUIRED FOR THE ANALYSIS OF THE POWER DISTRIBUTION SYSTEM.

1.) DISTRIBUTION TRANSFORMER SIZE IN KVA AND PRECENT OF IMPEDANCE.
2.) LENGTH (L1) AND SIZE (AWG) OF CONDUCTOR (WIRE) FROM DISTRIBUTION TRANSFORMER TO THE POINT OF COMMON COUPLING (PCC).
3.) TYPE OF SENSITIVE EQUIPMENT I.E. COMPUTERS, MEDICAL EQUIPMENT ECT.
4.) LENGTH (L2) AND (L3), SIZE (AWG) OF CONDUCTOR (WIRE) FROM THE PCC TO THE VFD OR VFDS IF A MULTIPLE INSTALLATION.

Figure 7.5 Harmonic distortion analysis. *(From Halmar Robicon Group, Pittsburgh, Pa., 1990.)*

2. For current source and voltage source drives, the wire sizes and lengths between any sensitive equipment and the distribution transformer as well as wire sizes and lengths between any sensitive equipment and the variable-speed drive installation (Wire sizes and distances between variable-speed drives should be secured where more than one is installed.)

Most public utilities or electrical designers will specify a maximum allowable harmonic voltage distortion in percent for a specific installation. Typically, this can be 3 or 5 percent.

7.3.4 Advantages of variable-frequency drives

Variable-frequency drives are available for nearly any HVAC application and have been the preferred means of varying the speed of a pump. They have become the drive of choice for new applications for many reasons, such as

1. First cost is lower in most sizes.
2. All are air-cooled. Larger slip-type drives require water cooling.
3. Wire-to-shaft efficiency is much higher than any slip-type drive. For typical wire-to-shaft efficiencies for variable-frequency drives, see Table 10.2.
4. It is very easy to integrate drive-control software into the software of pumping systems or building management systems.

7.3.5 Application of variable-speed drives

The application of variable-frequency drives requires some care to ensure proper operation and reasonably useful life. The manufacturer's installation instructions should be reviewed carefully. Following are some of the more pertinent concerns for installation. Contemporary variable-speed drives are very reliable and will provide years of uninterrupted service if installed and operated properly.

1. *Ventilation.* On variable-torque applications such as centrifugal pumps, the heat expended by the variable-speed drive can be computed easily.

$$\text{btu/h} = \text{max. kW of drive} \cdot 3412 \cdot (1 - E_\eta) \qquad (7.6)$$

where 3412 = thermal equivalent of a kW in btu/h
$\quad\quad E_\eta$ = efficiency of the drive as a fraction at maximum speed

If the energy consumption of the drive is 100 kW and the efficiency is 95 percent, the heat expended would be $100 \cdot 3412 \cdot (1 - 0.95)$, or 17,060 btu/h.

2. *Cleanliness.* Most commercial variable-frequency drives operate very well in normal HVAC equipment rooms. Industrial applications that are dusty may require special enclosures to keep the dust away from the internals of the variable-speed drive. NEMA 12 enclosures with internal air-conditioners may be required. The air-conditioners are usually air-cooled, so provisions must be made to keep clean the condenser coil on any air-conditioner.

Some terribly dusty or dirty applications should be equipped with a water-cooled drive instead of a variable-frequency drive. The water-cooled drive could be a slip-type drive using an eddy-current or fluid coupling.

Sometimes a variable-frequency drive can be cooled with clean outside air through ducts that eliminate the dirt or dust problem. Adequate duct size or auxiliary fans must be furnished so that there is no external static on the ventilating fans inside the drive.

3. *Chemical attack.* Variable-frequency drives contain many copper or copper-alloy parts that are susceptible to attack by acids or other sulfur-bearing compounds. Again, most HVAC installations do not have such atmospheres, but some industrial cooling and heating systems do. For example, very small quantities of hydrogen sulfide in the air can have a deleterious effect on variable-speed drives. When these conditions exist, special construction must be used such as NEMA 12 enclosures with internal air-conditioners or ducted fresh air as described above for dusty atmospheres (see Fig. 7.6a). The condenser coils for these air-conditioners must be coated with a compound, such as heresite, that resists attack from the hydrogen sulfide.

4. *Maximum temperature.* The maximum ambient temperature for most variable-speed drives is 104°F (40°C); for higher ambient temperatures, internal air-conditioners or ducted cool air should be considered for these applications.

5. *Location.* Variable-speed drives should be located in dry areas where they cannot be wetted by either surface water or water from overhead pipes. When there is no other location than under water or steam pipes, the cabinets should be equipped with drip shields. *Never locate variable-speed drives outdoors in the open where they are exposed to sunlight!* If it is necessary to install them outdoors, locate them under a sun shield that will protect them from direct sunlight.

6. *Power supply.* Variable-frequency drives are like any other commercial power equipment in that they are capable of withstanding the acceptable variations of commercial power that were defined earlier

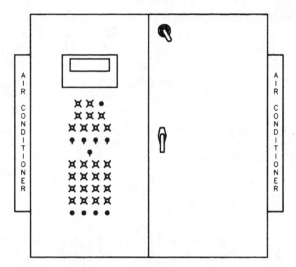

a. Enclosure with air conditioners.

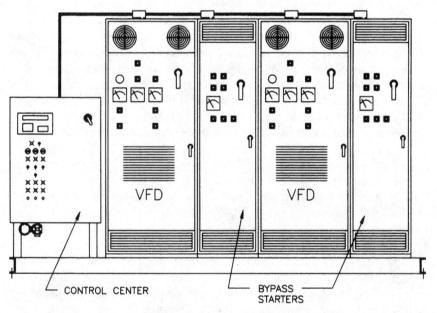

CONTROL CENTER   BYPASS
                                   STARTERS

b. Typical variable frequency drive assembly with control center and bypass starters, NEMA 1 construction.

Figure 7.6 Enclosures for variable-frequency drives.

in this chapter. Likewise, they may not be able to operate properly with harmonic distortions in the power supply that exceed the acceptable percentages established by the IEEE.

7. *Number of drives.* The simplest arrangement for HVAC pumps is one drive for each pump that is to be variable speed. More than one pump can be operated on a single drive. Usually, the cost of transition equipment to add motors to an operating drive is so expensive that it is cheaper to purchase multiple drives instead. Also, a multiplicity of drives provides better standby capability. Almost without exception in the HVAC industry, a variable-speed drive is furnished for every pump that is to be variable speed.

7.3.6 Variable-speed drive accessories and requirements

Following are popular accessories and requirements for variable-frequency drives:

1. *Certification.* Variable-frequency drives are often part of pumping systems that bear the label of various approving agencies such as Underwriters Laboratories (UL), Electric Testing Laboratories (ETL), or in Canada, the Canadian Standards Association (CSA). Care must be taken in evaluating these drives and their accessories when compliance is required with such standards.

   For example, some drive manufacturers have a certain approval on their basic drives, but they do not have such approval on the drive's accessories or enclosures. It is imperative that such approval be ascertained to avoid expensive field approvals or changes.

2. *Bypass starters*

   a. *Configuration.* Bypass starters are often used to keep a pump in operation during failure of the variable-speed drive. The designer must determine how critical the standby operation of a pump is during drive failure. The designer should not just indiscriminately put a bypass starter on every drive.

   The designer also must realize that the use of bypass starters forces the pump to full speed when operating with the bypass starter. On high-pump-head applications, it may be necessary for the operator to adjust the pump discharge manual valve to alleviate some of the pressure on the water system.

   Facilities with critical operations such as hospitals, computer centers, and research facilities should have consideration for these starters. It is doubtful that commercial buildings or educational facilities require them. It is the responsibility of the designer to make this decision.

Bypass starters must be designed properly. First, the decision must be made as to whether across-the-line starting is acceptable or whether some form of reduced-voltage starting is required. When this decision is under consideration, if reduced-voltage starting is required, the type used must have the same motor wiring as that required by the variable-speed drive. Usually, this eliminates part-winding or wye delta types of starters on 460-V power service. The solid-state starter is proving to be a very acceptable means of accomplishing reduced-voltage starting.

The arrangement of the bypass starter is critical to ensure that adequate safety is provided to the operator or service personnel. Figure 7.6*b* describes a bypass starter arrangement that provides some safety for operators. Although it is recommended that the variable-speed drive be serviced with both the drive and bypass starter depowered, it is recognized that this is impossible on some facilities. Only qualified electricians should service this equipment. As indicated above, it is imperative that the bypass starter, its enclosure, and the arrangement be approved by the same standard as the basic drive.

b. *Starting.* Transferring pump operation from a variable-speed drive to a bypass starter can be accomplished two ways, manually and automatically by transferring the pump motor from the drive to the bypass starter and returning the pump to full speed.

(1) *Manual transition.* Under this procedure, the operator recognizes a failure because the pump is usually stopped on drive failure. The power is transferred manually from the drive to the bypass starter, and the pump is returned to full speed. This procedure has several advantages: (1) the operator can inspect the equipment and ensure that it is acceptable to return the pump to duty, and (2) the operator can adjust the discharge valve on the pump to avoid the imposition of excessive water pressures on the system. *This is the recommended procedure for utilizing standby starters with variable-speed drives.* The use of multiple pumps in parallel provides the best method of handling variable-speed drive failure.

(2) *Automatic transition with stopped pump.* If the water system cannot accept a stopped pump until an operator can transfer the pump to a bypass starter, the pump motor can be transferred automatically. The pump-control system, upon sensing a drive failure, transfers the pump motor to

its standby starter and starts the pump. This has disadvantages, since there is no visual inspection during the transition, and there may be a possibility that the pump operating at full speed may overpressure the water system.

3. *Drive enclosures.* Standard NEMA 1 enclosures are adequate for most indoor HVAC installations of variable-speed drives (see Fig. 7.6*b*). Figure 7.6*b* describes a typical variable-speed drive assembly with control center and bypass starters. There is seldom any need for closed, nonventilated enclosures such as NEMA 3, 3R, or 4. Special ambient conditions such as the presence of water may require these enclosures. For hazardous locations, NEMA 7 or 9 enclosures may be required. Local and insurance codes will dictate the use of these special types of enclosures.

If there is a need for a nonventilated enclosure for use with internal air-conditioners or ducted cooling air, the Nema 12 type is usually the best enclosure (see Fig. 7.6*a*).

4. *Instrumentation.* All drives should have at least the following instrumentation.

   a. Ammeter for supply power
   b. Percent speed meter
   c. Hand-off-automatic switch
   d. Manual speed potentiometer
   e. Common fault alarm

   Some drives have a number of diagnostic indications and procedures that replace the common fault alarm. Others provide additional information digitally.

5. *Control.* Variable-speed drives can be furnished with internal microprocessors for controlling the speed of pumps. However, pump speed control is only part of the pump-control algorithm. Total pump control is so dependent on the water system characteristics that it is often included with its software in a pumping system control center (see Fig. 7.6*b*).

   Contemporary software is so flexible that the pump-control center can be interfaced with the rest of the building management system through data-gathering panels or protocols such as BACnet that is being developed by ASHRAE. Practically, there is seldom any need for any special software or interfacing at the point of installation.

Variable-frequency drives have become so standard for the HVAC pumping industry that there is no reason why there should not be a reliable variable-frequency drive with a minimum of service for any HVAC variable-speed pumping system.

7.4   Steam Turbine Drives for HVAC Pumps

On a central energy plant with high-pressure boilers, there may be high-pressure steam available for operating HVAC pumps. These turbine-operated pumps could be for boiler feed service as well as chilled and hot water system pumps. Steam turbine drives on pumps may be economical where large chillers are driven by steam turbines. The overall heat cycle for the boiler plant may indicate economies if the pumps are turbine-driven as well, taking exhaust steam from the chiller turbines. This is a detailed economic evaluation that must be made by the boiler room designer.

On installations where high-pressure steam is available but the chillers or other major boiler house equipment is not turbine-driven, it may be economical to equip certain pumps with turbines for use during periods that have high electrical demand charges. Typical of this are systems using storage chilled water to reduce the peak power during these periods of high demand charges.

Steam turbines usually operate with horizontal double-suction pumps. Steam turbine selection is based on the maximum brake horsepower required by the pump. Other parameters that must be determined are minimum supply steam pressure and maximum exhaust pressure.

Turbines are rated in brake horsepower capability at a certain water rate. Water rate is the pounds of steam per hour consumed when the turbine is operating at a particular speed and brake horsepower.

Turbine speed control consists of a supply steam pressure regulator that receives an analog signal from the pumping system controller. Turbine speed control is much like that for variable-frequency drives.

7.5   Engine-Driven Pumps

Engine-driven pumps are not common today in the HVAC field. They are used for emergency backup in event of power failure on critical installations or for peak shaving where high demand charges occur during a specific period of the day. Their value is contingent on the relative costs of electric power, fuel oil, and natural gas. Natural gas–driven engines may become more common as the peak cooling loads increase on electric power distribution systems. They can be an economic alternative to ice or chilled water storage on some installations.

Engines for this purpose normally operate on natural gas, while some may operate on no. 2 fuel oil. All parts of the fuel storage or delivery installation must meet the requirements of an approving agency such as Underwriters Laboratories or the attending insuring agency.

Noise abatement may be a problem with the use of these engines. Also, the exhaust effluent of these engines must meet local environmental codes, particularly oil-fired engines with the higher carbon percentage in the fuel.

7.6 Summary

The selection of motors and drives for HVAC pumps is critical for the realization of a cost-effective and efficient installation. It is obvious that there are a number of calculations and decisions that must be made by the designer to actualize such a pumping installation. Care should be taken in the selection of electric motors, ensuring that the best type, rating, and enclosure have been selected for each application. Too often the motor is just specified as a three-phase induction motor.

7.7 Bibliography

AC Motor Selection and Application Guide, Bulletin GET-6812B, General Electric Company, Fort Wayne Ind., 1993.

Energy Policy Act of 1993 as It Relates to Motors, Bulletin GEK-100919, General Electric Company, Fort Wayne, Ind., 1993.

Keith H. Sucker, PE, *Private Communication,* Halmar Robicon Group, Pittsburgh, PA, 1994.

Pivate Communication, Halmar Robicon Group, Pittsburgh, PA, 1990.

The HVAC World

Chapter

8

The Use of Water in HVAC Systems

8.1 Efficiency of Operation

The use of water in HVAC systems has been much like the ordinary use of water in our civilization. Water has been plentiful and cheap, so why not use it? Such has been the situation in HVAC water systems. Now the cost of energy to move water through these systems is no longer less than a penny per kilowatthour, and water itself should be conserved. Conservation must be practiced, and we have the tools to achieve the efficient use of pumping energy in HVAC water systems. The development of digital electronics has opened the door to the achievement of higher efficiency in the movement of water through these systems. This added efficiency is most pronounced in hot water, chilled water, and condenser water systems in the HVAC field. The elimination of energy-consuming mechanical devices, the use of variable-speed pumps, and the wiser control of water temperatures to increase the efficiency of boilers and chillers have all brought conservation of energy in the HVAC field.

The major users of water and pumping energy in the HVAC field are chilled, condenser, hot, and boiler feed water. This chapter reviews the design considerations that are common to all these systems. Later chapters will be devoted to the specific design requirements for each of the water systems. Energy consumption is a major concern for all these systems.

For hot water systems, the development of the condensing-type boiler has brought a much higher level of boiler efficiency, over 90 percent compared with the traditional 80 percent for noncondensing boilers. This means also that hot water systems need not be designed with higher water temperatures to avoid condensing. Noncondensing boilers cannot operate with water at temperatures below 140°F. The development of the condensing boiler has enabled the designer to select great differential temperatures between the supply and return temperatures. This has reduced appreciably the water flow and, therefore, the pump flow and horsepower.

Likewise, on chilled water systems, it is recognized that chillers operate at lower kilowatt per ton rates with higher chilled water temperatures. In recognition of these facts, HVAC water systems should be designed for lower hot water temperatures and higher chilled water temperatures. Reset of system water temperature at boilers or chillers should be used rather than zone reset in most cases. Zone reset should use efficient pumping such as distributed pumping, which will be described in Chaps. 15 and 20 on chilled and hot water distribution systems.

Condenser water systems are realizing greater efficiencies through the use of variable-speed pumps. The types of controls that are available today for removing heat from the chillers have allowed the use of these variable-speed condenser pumps. The consumption of energy for boiler feedwater is relatively less than that for the other major HVAC water systems.

8.1.1 Determination of useful energy

The advent of digital electronics, as discussed in Chap. 1, provides the means to determine rapidly the efficient use of pumping energy for all HVAC water systems. It remains for us to develop what is efficient use and what is not. Following are some guidelines that will help us in this determination of the value of various practices that consume pumping energy.

Useful consumption of pumping energy

1. Pipe friction required to transport water

2. Friction in fittings to connect the pipe

3. Friction in heating and cooling coils

4. Properly sized pumps and motors

5. Properly selected variable-speed drives

6. Friction in properly evaluated chillers and boilers.

Inefficient use of energy

1. Temperature-control valves

2. Balance valves, manual or automatic

3. Pressure-reducing or pressure-regulating valves

4. Most crossover bridges

5. Any other mechanical device that regulates water flow

6. Constant-speed pump overpressure

There may be some question about the preceding categories of efficient and inefficient uses of pump energy, particularly temperature-control valves on heating and cooling coils. Any device that forces water into a certain path or circuit must be considered inefficient even if it is useful, as are these temperature-control valves for heating and cooling coils. A perfect system would be one that did not need a control valve to regulate the water flow through a coil. Typical of this would be a water system with a variable-speed pump for each coil and no control valve. The objective here is to establish a means of determining the overall efficiency of a water system. Likewise, it should be our objective to reduce pumping energy by eliminating, wherever possible, the devices listed above as inefficient users of energy.

8.1.2 Calculation of system efficiency

The overall efficiency of a water system is like any other equation for efficiency—the useful energy divided by the energy input. The *useful energy* is the friction of the piping, fittings, and heating and cooling coils; their sizing obviously affects the amount of useful energy. The engineer must, as indicated in Chap. 3, balance friction against first cost of the pipe to achieve an economical answer for the water system under design.

The efficient energy for an HVAC water system K_e can be calculated in kilowatts as follows:

$$K_e = \frac{\text{useful frictions (ft)} \cdot \text{system flow (gal/min)} \cdot 0.746}{3960}$$

$$K_e = \frac{\text{useful frictions (ft)} \cdot \text{system flow (gal/min)}}{5308} \tag{8.1}$$

The energy consumed by such a system K_i can be the energy input to the system if it is known, or it can be calculated (in kilowatts) as follows:

$$K_i = \frac{\text{total system head (ft)} \cdot \text{system flow (gal/min)}}{5308 \cdot P_\eta \cdot E_\eta} \tag{8.2}$$

The efficiency of a water system WS_η utilizing electric motor–driven pumps can be computed by dividing the useful kilowatts by the kilowatt input to the pump variable-speed drives or motors K_i:

$$WS_\eta = \frac{K_e}{K_i} \cdot 100\% \qquad (8.3)$$

Using as an example a chilled water system with variable-speed pumping, if the friction of the piping, fittings, and cooling coils on this system with a capacity of 1000 gal/min amounted to 75 ft of head, the useful energy would be

$$K_e = \frac{1000 \cdot 75}{5308} = 14.1 \text{ kW}$$

To calculate K_i, assume that the friction loss through control valves and other appurtenances amounts to 20 ft; the total pump head would then be 75 + 20 for a total of 95 ft. If the equipment efficiencies were 83 percent for the pumps and 90 percent for the wire-to-shaft efficiency of the motor and variable-speed drive, the input energy K_i would be

$$K_i = \frac{1000 \cdot 95}{5308 \cdot 0.83 \cdot 0.90} = 24.0 \text{ kW}$$

The efficiency of the system WS_η would be

$$WS_\eta = \frac{14.1}{24.0} \cdot 100 = 58.8 \text{ percent}$$

The preceding calculations are for the system at full flow and with no overpressure generated by the pumps. It is apparent that for this equation to be meaningful, similar calculations must be made under part-load conditions.

For example, assume that the system is operating at 50 percent load, where the system flow is 500 gal/min and the useful system head has dropped to 40 ft. The useful kilowatts, therefore, are

$$K_i = \frac{500 \cdot 40}{5308} = 3.8 \text{ kW}$$

Also assume that a constant-speed pump is used on the preceding application at 50 percent load. At this point, the pump has moved up its curve and is operating at 500 gal/min, 105 ft of head, and an efficiency of 78 percent. Since no variable-speed drive is involved, E_η is the efficiency of the electric motor, which would be near 91 percent. K_i, therefore, becomes

$$K_i = \frac{500 \cdot 105}{5308 \cdot 0.78 \cdot 0.91} = 13.9 \text{ kW}$$

The overall efficiency of the water system becomes

$$WS_\eta = \frac{3.8 \cdot 100}{13.9} = 27.3 \text{ percent}$$

This demonstrates that only slightly more than one-fourth of the energy applied to the pump motors is being used for efficient transportation of the water through the system. As dramatic as this may seem, there are water systems utilizing constant-speed pumps and mechanical devices to overcome the pump overpressure where less than 10 percent of the energy applied is used to move the needed water through the system at moderate and low loads on the system.

These equations for energy consumed, energy applied, and system efficiency are for the water system in total. Similar evaluations for energy consumption must be made for all parts of a water system to ensure that maximum system efficiency is achieved. In Chap. 6, wire-to-water efficiency for pumping systems was addressed. Examples of this efficiency will be found throughout this book.

The use of Eq. 8.3 to evaluate the efficiency of pumping for an HVAC system may be cumbersome and the answer difficult to determine. Also, the use of small pipe and high friction losses may provide a relatively high efficiency for a poorly designed system. The use of Eq. 8.3 is more relative than absolute in comparing different piping system designs with the same level of pipe friction.

Energy consumption of water distribution. There are other equations for chilled water and hot water systems that are useful to determine the effectiveness of pumping. They are easier to compute than Eq. 8.3 and provide absolute values. For example, for chilled water,

$$\text{kW/100 tons} = \frac{0.452 \cdot H}{P_\eta \cdot E_\eta \cdot \Delta T - {}^\circ F} \tag{8.4}$$

where H = system head
P_η = pump efficiency
E_η = motor efficiency or wire-to-shaft efficiency of a variable-speed drive and motor for variable-speed pumps
ΔT = system temperature difference

For example, if the system head H is 100 ft, the pump efficiency 82 percent, the wire-to-shaft efficiency 89 percent, and the temperature differential 12°F, then

$$\text{kW/100 tons} = \frac{0.452 \cdot 100}{0.82 \cdot 0.89 \cdot 12} = 5.16 \text{ kW/100 tons}$$

If flow and watt transmitters are measuring system flow and kilowatts of the pumps for the chilled water, an alternate equation (Eq. 8.5) utilizes values measured from the actual system. This equation enables the operators of the water system to measure continuously the energy consumed in distributing the chilled water.

$$\text{kW/100 tons} = \frac{2400 \cdot \sum \text{pump kW}}{\text{gal/min} \cdot \Delta T - °F} \qquad (8.5)$$

For example, if secondary pumps are pumping 1000 gal/min at a system temperature difference of 12°F and are consuming 25.8 kW, then

$$\text{kW/100 tons} = \frac{2400 \cdot 25.8}{1000 \cdot 12} = 5.16 \text{ kW/100 tons}$$

for hot water,

$$\text{kW/1000 mbh} = \frac{23.48 \cdot h}{P_\eta \cdot E_\eta \cdot \Delta T - °F \cdot \gamma} \qquad (8.6)$$

where γ is the specific weight of the hot water at operating temperature. See Tables 2.3 and 2.4.

Like chilled water, if watt transmitters are available to measure the energy input to the hot water pumps, the following equation can be used:

$$\text{kW/1000 mbh} = \frac{124,700 \cdot \sum \text{pump kW}}{\text{gal/min} \cdot \Delta T - °F \cdot \gamma} \qquad (8.7)$$

For example, if the secondary hot water pumps are pumping 500 gal/min of 180°F water at a system temperature difference of 40°F and are consuming 10.2 kW, then

$$\text{kW/1000 mbh} = \frac{124,700 \cdot 10.2}{500 \cdot 40 \cdot 60.57} = 1.05 \text{ kW/1000 mbh}$$

where 60.57 is the specific weight of water at 180°F.

Examples of these energy rates in kilowatts per 100 tons or kilowatts per 1000 mbh will be included in various chapters of this book.

8.1.3 Energy lost to mechanical flow-control devices

In this day of concern over energy conservation, as we begin the design of an HVAC water system, it is imperative that we reevaluate our standard practices to see where we are wasting energy. With computer-aided design, it is much easier to develop part-load information and a closer evaluation of diversity on hot and chilled water systems. This provides the basis for more efficient piping designs that do not need mechanical devices to circulate the water throughout the system. For example, on a recent evaluation of a Midwestern university, a chilled water pumping system was in operation with balance valves on the pump discharges. There was a 40-lb/in^2 pressure drop across these balance valves, and each year around 900,000 kWh was wasted by them.

As discussed in Chap. 2, the thermal equivalent of a brake horsepower is 2544 Btu/h, and that for a kilowatthour is 3412 Btu/h. This energy must be accounted for in the calculation of heating and cooling loads. Motor horsepower is therefore a plus for heating loads and is a deduction when computing the total heating load for a building. It is an added load on chilled water systems and must be included in the total cooling load on a building. It is important that we watch for energy wasters, particularly on chilled water systems. Following is an evaluation of such mechanical devices that are used on hot and chilled water systems. Appreciable energy is used to circulate water to all the terminal units that are on such a system. The HVAC industry is full of various devices that regulate the flow of water to achieve a desired flow distribution of hot or chilled water in a building.

Typical energy consumers

1. Balance valves, manual and automatic
2. Pressure-regulating valves
3. Pressure-reducing valves
4. Three-way temperature-control valves
5. Most crossover bridges

The most popular of the preceding are balance valves; many of them may be installed in a single building. Usually, a manual-type balance valve has two ports that are used to measure the pressure loss across the valve. Each manufacturer publishes a table for the valve that will provide the flow in gallons per minute at the measured pressure loss.

8.1.4 Energy losses for an element of a water system

All the calculations and formulas in this chapter are based on electric motor–driven pumps. Following are calculations that can be used to determine how much energy is being consumed by any element in or part of a water system such as a single balance valve on a hot or chilled water system. This procedure also can be used for pressure-reducing and pressure-regulating valves, but it may be difficult to accurately determine the flow through these valves.

The basic formula for computing pump horsepower also computes the energy loss HP_v for part of a system such as a balance valve. This formula is as follows, where Q_v is the flow in gallons per minute through the part of a system being evaluated or for a balance valve; likewise, H_v is the loss in feet of head for that part of a system or for a balance valve.

$$hp_v = \frac{Q_v \cdot H_v}{3960 \cdot P_\eta} \tag{8.8}$$

To convert to kilowatts,

$$kW_v = \frac{hp_v \cdot 0.746}{E_\eta} \tag{8.9}$$

or

$$kW_v = \frac{Q_v \cdot H_v}{5308 \cdot P_\eta \cdot E_\eta} \tag{8.10}$$

where P_η = pump efficiency
E_η = motor efficiency for constant-speed pumps and wire-to-shaft efficiency for variable-speed pumps

This formula (8.10) computes the energy lost through part of a system such as a balance valve by inserting the flow and head difference recorded for each balance valve. The following discussion will be based on balance valves.

A quick formula for estimating this energy loss without knowing the pump and motor efficiencies is as follows: Assuming that the average pump efficiency is 75 percent and the average motor efficiency is 90 percent, Eq. 8.5 can be reduced to the following:

$$\text{Lost } kW_v = \frac{Q_v \cdot H_v}{3600} \tag{8.11}$$

If the pressure drop H_v is indicated in psig, multiply it by 2.31.

The annual lost energy is

$$\text{Annual kWh} = \text{kW}_v \cdot \text{annual hours} \qquad (8.12)$$

Although Eq. 8.11 is an approximation, it is accurate enough to determine generally how much energy is being lost in a building due to balance valves. If the loss is appreciable, a further study should be made using Eqs. 8.8 and 8.9 with actual pump and motor efficiencies.

Serious studies of the energy consumption of condenser, chilled, and hot water systems are now under way. Out of this work has come evidence of the great loss that can be incurred by balance valves on cooling coils in particular. A specific study was conducted by a balancing contractor and supervised by a consulting engineer that showed a pronounced reduction in energy consumption by opening wide existing balance valves on a variable-volume, variable-speed chilled water system in a high school. Table 8.1 is a summary of these data.

The increase in system flow with the balance valves opened was caused by the addition of the auditorium air-handling units. Some significant system changes were achieved, namely, reduction in pump speed and discharge pressure and an increase in the differential temperature.

Although these are very specific data for a single installation, the over 30 percent reduction in pumping energy demonstrates that the use of mechanical devices such as balance valves and any kind of pressure-regulating valve should be reviewed carefully to determine their effect energywise on an HVAC water system.

TABLE 8.1 Energy Data for an Existing High School

Item	Balance valves set traditionally	Balance valves opened completely	Percent change
Flow, gal/min	700	750	+7
Cooling load, tons	215.8	278.1	+29
Supply temp., °F	45.0	45.4	—
Return temp., °F	52.4	54.3	+4
Differential temp., °F	7.4	8.9	+20
Secondary pump discharge pressure, psig	67	58	−13
Pump speed, rev/min	1232	1032	−16
Pumping system kW	18	16	−11
Secondary pumping, kW/ton	0.083	0.058	−30

SOURCE: From Ben L. Kincaid and Andrew Spradley, Removing Manual Balancing in a High School, 1995.

8.1.5 Effect of energy consumption on chillers and boilers

On chilled and hot water systems, it must be remembered that all the energy lost through a flow-regulating device such as a balance valve results in heat in the water. Therefore, on chilled water applications, this heating effect must be taken into consideration as an additional load. On hot water systems, it is beneficial and should be deducted from the energy lost. This is one reason why the energy lost in balance valves on heating systems was seldom questioned.

Following is a procedure for checking the total energy lost to balance valves on a particular chilled water system. When the balance valves are checked or set for a desired flow, the flow and pressure drops should be recorded for each valve and the preceding equations used to compute the energy loss for each valve. All the balance-valve losses can be totaled for an annual energy loss. Table 8.2 can be used for this computation.

This annual energy loss in kilowatthours can be used to determine the additional annual energy consumption of the chillers due to the balance valves:

$$\text{Added chiller kWh} = \frac{\text{annual kWh}_v \cdot 3412 \cdot \text{average kW/ton}}{12,000}$$

$$= \frac{\text{annual kWh}_v \cdot \text{average kW/ton}}{3.52} \qquad (8.13)$$

where total annual kWh_v is for the balance valves, and average kilowatts per ton is the annual kilowatthour consumption of the chillers divided by the annual ton-hours of cooling produced by the chillers.

For example, if the annual loss due to balance valves is 100,000 kWh and the average kilowatthours per ton-hour is 0.80, then

$$\text{Added chiller kWh} = \frac{100,000 \cdot 0.80}{3.52}$$

$$= 22,727 \text{ kWh}$$

The total annual loss for the balance valves is therefore 100,000 + 22,727 kWh, or 122,727 kWh.

The energy loss for the balance valves on hot water systems must recognize the heating effect of the energy consumed in these valves. This is done as follows for gas fired boilers. The equivalent fuel per kilowatt EF/kW in cubic feet is calculated first, and then the net deduction in percent is developed:

TABLE 8.2   Energy Calculations for Balance Valves on Constant-Volume Systems

Building:_____   Date: _____

Valve number or location	Flow, gal/min	Head loss in feet	Energy loss, kW

Total kW for building:_____

Annual hours:_____

Annual energy loss = _____ Total kW × _____ hours

= _____ kWh

NOTE: See text for calculation of added chiller energy consumption due to balance valves.

$$\frac{EF}{kW} = \frac{3412 \ \text{Btu/kW}}{B_\eta \cdot 1000 \ \text{Btu/ft}^3} = \text{ft}^3 \ \text{gas/kW} \qquad (8.14)$$

where B_η is boiler efficiency as a decimal. The heating value of natural gas is assumed to be 1000 Btu per cubic foot. Thus Heating effect H_{ef} in percent is

$$H_{ef} = \frac{EF/kW \cdot \text{fuel cost (cents/ft}^3) \cdot 100}{\text{overall power cost (cents/kW)}} = \% \ \text{reduction} \quad (8.15)$$

Using these equations, with a natural gas cost of $4 per 1000 ft³, or 0.4 cents/ft³, overall power cost of 6 cents/kW, and a boiler efficiency of 80 percent,

$$EF/kW = \frac{3412}{0.8 \cdot 1000} = 4.265 \ \text{ft}^3/\text{kW}$$

$$H_{ef} = \frac{4.265 \cdot 0.4 \cdot 100}{6} = 28.4 \ \text{percent reduction}$$

If the annual loss to balance valves is 100,000 kW, recognizing the heating value of this energy is achieved by multiplying 100,000(1−0.284), which gives a net energy loss of 71,600 kW/year.

On hot and chilled water systems, the advent of variable-speed pumping and digital control has made possible the conversion of constant-volume systems to variable-volume systems. This is accomplished through the replacement of three-way control valves by two-way valves and adjusting of the balance valves to the full-open position. The procedure included herein for computing energy losses of balance valves can be used to help determine the energy savings that could be achieved by installing variable-speed pumps with proper control, replacing the three-way valves with two-way valves, and opening the balance valves.

8.2   Efficient Use of Water in HVAC Systems

The preceding discussion points the way to efficient use of water and pumping energy in HVAC systems. The remainder of this chapter will be used to develop the load range for a water system and to describe the various uses of water in these systems.

8.2.1   Load range for a water system

All of this chapter has been devoted to an analysis of energy at design load or maximum water flow. Unfortunately, in many instances on actual water systems this is the last part of the analysis. It is of utmost importance that the minimum load as well as the maximum load be determined for a water system to achieve an accurate consumption of energy. Part-load calculation is a subject of considerable study by the American Society of Heating, Refrigerating, and Air Conditioning Engineers (ASHRAE). Its *Handbook on Fundamentals* describes the "bin" method for computing part-load system flows. Using the formula for pipe friction, the system heads can be calculated for these part-load conditions. Likewise, with proper use of the affinity laws, the energy consumption of the pumps can be computed from minimum to maximum load. In all the system evaluations that will be discussed in this book, *minimum load will always be included when computing pumping system performance and energy consumption.*

8.2.2 Energy consumption and water uses in HVAC

The use of the energy consumers as described above develops a possibility that the water system could have been designed differently; this would have eliminated the need for their use and reduced the energy consumption in the system. It must be remembered that many existing systems were designed at a time when energy was lower in cost and the use of these mechanical devices was the most economical procedure at that time. The following discussion will review the use of water in hot and chilled water systems. A similar evaluation will be made for condenser water in Chap. 11.

8.3 Hot and Chilled Water Systems

There are three zones of different activity in a hot or chilled water system. The first zone is energy generation, the second zone is transportation of that energy, and the third zone is use of the energy (Fig. 8.1).

Zone 1: Energy generation. Energy for hot and chilled water systems is generated in boilers and chillers. The efficiency of the boilers and chillers may be affected appreciably by how the water is circulated through them. The development of bypasses and special piping arrangements has been achieved to enhance their efficiency.

The efficiency of boilers and chillers should not be compromised in an effort to save pumping energy. Here lies one of the major responsibilities of the water system designer, who must evaluate various connections and pumping arrangements to acquire the optimal configuration for a particular system.

Zone 2: Transportation of energy. Once the hot or chilled water has been generated, it must be moved out to the heating or cooling coils. Com-

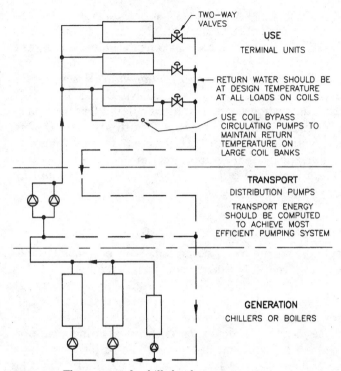

Figure 8.1   Three zones of a chilled or hot water system.

bining the pumps for this transportation with the pumps for the boilers or chillers may or may not improve the efficiency of the entire
water system.

Often, the most efficient system of delivering the water to the end
uses is to develop the transportation pumps strictly for the purpose of
distributing the water and not to generate the energy. A number of
different pumping and piping arrangements are provided herein to
assist the designer in selecting the most efficient system for a particular water system.

Zone 3: Use of the energy.   Water in hot and chilled water systems is
used by hot and chilled water coils, heat exchangers, and process
equipment. Most of these energy users are equipped with a control
valve to regulate the flow of water through them.

Some coils are operated without control valves and are called "wild"
coils. They are often equipped with face and bypass dampers that control the use of energy by them. The water flows continuously through
the coil, and the air flows around the coil on low heating or cooling

loads. Wild coils should not be used on systems with high pumping heads and broad load ranges due to the pumping energy wasted by them. Some designers use these coils on outside makeup air to prevent freezing in the coil. There are other methods to prevent freezing besides wild coils that control the flow and therefore conserve pumping energy. The other problem with these uncontrolled coils is the return of the water unused to the boilers and chillers. This raises the return water temperature on hot water systems and lowers the return water temperature on chilled water systems.

Much will be made of different coil connections to develop the efficiencies or inefficiencies of the various systems that are used in HVAC systems today. It is obvious that the three-way valve on a heating or cooling coil is a wasteful device like the wild coil with a face and bypass damper. Water is bypassed around the coil and returned to the boiler or chiller unused.

Sizing of coils from a standpoint of water friction loss is becoming a very detailed task. From a standpoint of pumping energy, if the coil is sized with too great a friction drop, energy is wasted. If the coil is sized with a low velocity in the tubes and low friction loss, laminar flow develops and excessive water flow results. The designer must balance the selection of coils between high water pressure loss and laminar flow. The flow of water in the coil should not pass into the laminar range at any known load on the coil.

How well we use hot or chilled water will determine the efficiency of operation of the entire water system. Hot and chilled water systems are described together here, because there is so much similarity between hot water and chilled water coils in their use and in their connections.

Hot and chilled water coil connections must be studied in detail to ensure that the supply water is being used efficiently. There are a number of ways to connect these coils, and many are being advocated in the HVAC industry. Some of them are being sold to solve a particular system problem.

The system problems that are being addressed with these connections are

1. Laminar flow in coils

2. Dirty coils—water side, air side, or both

3. Improperly sized coil control valves or valve actuators

8.3.1   Three-way valves for hot and chilled water coils

When three-way valves were used predominantly on these coils, the systems were constant flow, and there was little concern for energy

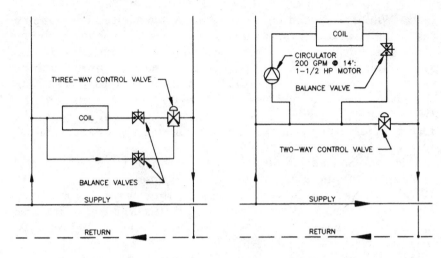

a. Coil with three—way control valve.

b. Coil with circulator and two—way control valve.

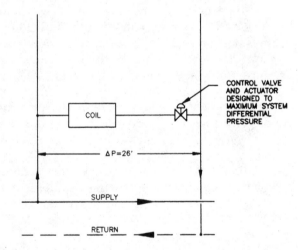

c. Contemporary coil with two—way valve.

Figure 8.2 Coil connections for hot and chilled water systems.

savings. The three-way control valve wastes energy by bypassing the supply water around the coil, as shown in Figure 8.2a. On variable-volume chilled water systems, this has a negative effect on the chillers, since the return temperature to the chillers is reduced. The chillers reach rated water flow long before they achieve design load in tons. This forces more than one chiller to operate when the cooling load is less than the capacity of one chiller.

8.3.2 Two-way valve with circulator

Recognizing the deficiency of the three-way valve, efforts have been made to control the return water temperature by replacing the three-way valve with a two-way valve using a circulator on the coil (see Fig. 8.2b). Laminar flow was eliminated with this arrangement, but dirty coils could still pass water through the control valve below design temperature on chilled water systems. The significant defect of this arrangement is the continuous operation of the coil pump whenever the building chilled water system is in operation. Also, the wire-to-water efficiency of many coil pumps is low, around 35 to 45 percent, when compared with most secondary pumps that have similar efficiencies of 60 to 70 percent. Further, the added piping and wiring raises the first cost and increases the total friction head of the system.

8.3.3 Contemporary two-way valve connection

The ability to evaluate more accurately the pressure drops that could occur across control valves has resulted in the selection of valves better fitted to a specific application. This has culminated in use of the two-way valve directly connected to its coil and the supply and return headers (see Fig. 8.2c). With proper calculation of maximum possible pressure drop across the valve, the valve and its actuator can be sized to operate under these pressures without damage or lifting of the valve head off the valve seat.

8.3.4 Energy evaluations for three different coil connections

Following are energy analyses using these three coil and control-valve connections. This energy audit should demonstrate the need to eliminate three-way valves and coil circulators on most variable-volume systems. Circulators in coil bypasses do have specific uses on existing systems, and these will be demonstrated in this chapter.

It is difficult to locate the right place in this book for the following energy discussion. This is so important that it must be addressed before specific installations are reviewed. Chapter 9 will develop the primary/secondary system that is used in the following three figures. Also, the remote differential transmitter shown in these figures for pump speed control is described in Chap. 10.

Energy consumption in HVAC water systems is very dependent on the proper use of water in these systems. The following coil connections impinge heavily on energy consumption in hot and chilled water systems. The great variation in energy use will be demonstrated for the three systems using different coil and control-valve arrangements. Similar energy evaluations will be made for condenser water in Chap. 11.

Three-way valve system. Figure 8.3*a* describes a secondary system with 10 air-handling units, each with a requirement of 200 gal/min, for a total design flow of 2000 gal/min. Three-way valves are installed on the air handlers so that the flow is constant regardless of the load. The coil, control-valve, and piping loss is 26 ft, the pump fitting loss is 8 ft, and the header losses are 50 ft, for a total pump head of 76 ft. If the pumps have an efficiency of 86 percent and the motors have an efficiency of 92 percent, the total electrical consumption at any load is 36.2 kW. Equations 6.12, 6.13, and 6.14 are used for this calculation.

Two-way valve system with coil circulators. This system is similar to the three-way valve system above and is described in Fig. 8.3*b*. Each coil is fitted with a 1½-hp circulator that has a pump duty of 200 gal/min at 16 ft. The head of 16 ft results from a coil loss of 10 ft, pump fittings of 2 ft, balance valve of 2 ft, and pipe friction and fittings of 2 ft. The pump has an efficiency of 67 percent, and the pump motor has an efficiency of 84 percent if it is of the high-efficiency type. This pump runs continuously whenever the coil is in operation, and its energy consumption is 1.1 kW.

The secondary pumps have a total head of 72 ft, consisting of 8 ft of loss in the pump fittings, 50 ft in the system headers, and 14 ft of differential pressure across the coil, and its control valve and piping. Each pump has an energy consumption of 17.5 kW at design flow with a pump efficiency of 86 percent and a wire-to-shaft efficiency of 90 percent. Adding the circulator pump kilowatts to those of the secondary pumps produces a total kilowatts of $10 \times 1.1 + 2 \times 17.5$ for a total energy consumption of 46.0 kW at full load with a system flow of 2000 gal/min.

Contemporary two-way valve system. The same secondary system, when fitted with two-way valves, is much simpler and has a lower overall energy use (see Fig. 8.3*c*). With the two-way valves, the secondary pump head is now 86 ft, consisting 50 ft of loss in the system headers, 8 ft of loss in the pump fittings, and 26 ft of differential pressure across the coil, its control valve, and the connecting piping. The total energy consumption for each pump is 21.1 kW with a pump efficiency of 85 percent and a wire-to-shaft efficiency of 90 percent for the variable-speed drive and motor. The energy consumption at design load is 2×21.1, or 42.2 kW.

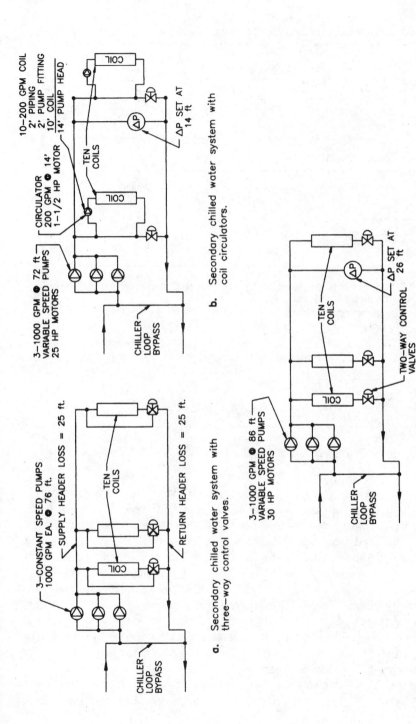

a. Secondary chilled water system with three-way control valves.

3—CONSTANT SPEED PUMPS
1000 GPM EA. @ 76 ft.

SUPPLY HEADER LOSS = 25 ft.

TEN COILS

COIL

RETURN HEADER LOSS = 25 ft.

CHILLER LOOP BYPASS

b. Secondary chilled water system with coil circulators.

10–200 GPM COIL
2' PIPING
2' PUMP FITTING
10' COIL
14' PUMP HEAD

COIL

CIRCULATOR
200 GPM @ 14'
1–1/2 HP MOTOR

TEN COILS

COIL

ΔP SET AT 14 ft

3—1000 GPM @ 72 ft
VARIABLE SPEED PUMPS
25 HP MOTORS

CHILLER LOOP BYPASS

c. Contempory system with two-way control valves.

TEN COILS

COIL

TWO-WAY CONTROL VALVES

ΔP SET AT 26 ft

3—1000 GPM @ 86 ft
VARIABLE SPEED PUMPS
30 HP MOTORS

CHILLER LOOP BYPASS

Figure 8.3 Water systems with the three coil connections.

173

TABLE 8.3   Comparison of Pump Energy Consumptions (kW)

Percent load	Three-way valve (Fig. 8.3a)	Two-way valve with circulator (Fig. 8.3b)	Contemporary two-way valve (Fig. 8.3c)
25	36.2	13.9	4.4
50	36.2	18.9	11.1
100	36.2	46.0	42.2

TABLE 8.4   Chiller Effect (kW)

Percent load	Three-way valve (Fig. 8.3a)	Two-way valve with circulator (Fig. 8.3b)	Contemporary two-way valve (Fig. 8.3c)
25	8.2	3.2	1.0
50	8.2	4.3	2.5
100	8.2	10.5	9.6

TABLE 8.5   Total Energy Consumption (in kW) for the Three Coil/Valve Arrangements on a Chilled Water System

	Table 8.3 + Table 8.4		
Percent load	Three-way valve (Fig. 8.3a)	Two-way valve with circulator (Fig. 8.3b)	Contemporary two-way valve (Fig. 8.3c)
25	44.4	17.1	5.4
50	44.4	23.2	13.6
100	44.4	56.5	51.8

8.4   Energy Comparison of the Three Systems

Table 8.3 is a comparison of the three coil and valve arrangements on a system that has a system load range from 25 to 100 percent of maximum load. Data are provided at 25, 50, and 100 percent system load. The wire-to-water efficiencies of the variable-speed drives and motors were computed from the data shown in Table 8.7. The pump affinity laws (see Chap. 6) were used to compute the efficiencies of the secondary pumps at reduced loads. Table 8.3 provides the energy consumption of the pumps, while Table 8.4 lists the kilowatt effect on any chillers involved, Table 8.5 gives the net kilowatt effect on a chilled water system, and Table 8.6 produces the same for a hot water system.

For the system as chilled water, assuming that the chiller kilowatts per ton is 0.80 and using Eq. 8.10, Table 8.4 provides the chiller effect for the three different valve and coil arrangements.

Adding Tables 8.3 and 8.4 together into Table 8.5 gives the total energy effect that the three coil and valve arrangements will have on a chilled water system. It is obvious from this table that by far the two-

TABLE 8.6 Net Energy Consumptions (in kW)for the Three Coil/Valve
Arrangements on a Hot Water System

	71.6 percent of Table 8.3		
Percent load	Three-way valve (Fig. 8.3a)	Two-way valve with circulator (Fig. 8.3b)	Contemporary two-way valve (Fig. 8.3c)
25	25.9	10.0	3.2
50	25.9	13.5	7.9
100	25.9	32.9	30.2

way valve without a circulator is the most efficient arrangement of cooling coils and their control valves.

Assuming the same capacities for a hot water system, Table 8.5 must be adjusted to accommodate the heating effect of the electrical energy. The calculation of net energy effects on hot water coils are developed in Eqs. 8.11 and 8.12. If as above for balance valves, the chiller kilowatts per ton is 0.6 and the cost of natural gas is $4 per thousand cubic feet, or 0.4 cents/ft^3, then

$$EF/\text{kW} = 4.265 \text{ ft}^3/\text{kW}$$

and

$$H_{ef} = 28.4 \text{ percent reduction due to the heating effect}$$
$$\text{of the electrical energy}$$

Table 8.6 provides the net energy consumed for the various coil and control valve arrangements by deducting 28.4 percent of the kilowatt figures from Table 8.3.

Tables 8.3 through 8.6 demonstrate the great variation of energy caused by the various coil and control-valve arrangements. It must be remembered that these tables were computed using specific chiller and boiler efficiencies. A comparison of actual coil and valve arrangements requires the determination of these efficiencies.

Table 8.7 lists wire-to-shaft efficiencies used in the above calculations was provided by a manufacturer of variable-frequency drives.

It should be pointed out that systems that are constantly loaded can utilize three-way valves with some energy savings. Most HVAC water systems, however, have variable loads due to building occupancy, lighting and machine loads, and variable outdoor air conditions.

These calculations should put to rest the use of three-way valves or circulators on most coils on variable-volume hot and chilled water systems. There are specific uses of circulators on coils to eliminate laminar flow and freezing conditions; Figure 8.4 describes this use. It

TABLE 8.7   Wire-to-Shaft Efficiencies for Variable-Speed Calculations

Percent speed	25-hp motor	30-hp motor
40	58	61
50	70	72
60	79	80
70	84	84
80	87	87
90	89	89
100	90	90

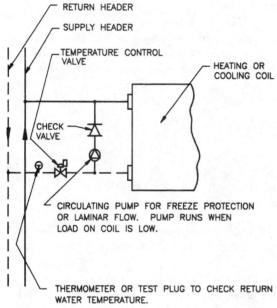

Figure 8.4 Heating or cooling coil pump for freeze protection or laminar flow. (*From The Water Management Manual, Systecon, Inc., West Chester, Ohio, 1992, Fig. 4.3.*)

should be noted that the circulator is in the bypass, so it does not run except when there are low loads on the coil. Usually, its motor is one-third to one-quarter the size of a circulator installed on the supply connection to the coil, as shown in Fig. 8.3. For elimination of freezing possibilities, the circulator can be programmed to run whenever the outdoor temperature is below 32°F and/or the load on the coil is at some discreet level such as 70 percent.

8.5   Categorization of HVAC Water Systems

HVAC water systems can be categorized by whether they are open or closed. Any classification will create some confusion, but these categories should be adequate for evaluating pumps on all HVAC water systems.

Open systems have no external pressures placed on them, and usually, they include an open tank holding a reservoir of water. Water makeup is made to these systems at the reservoir or tank. Following is a listing of open-type systems and the chapter of this book that describes them.

Cooling tower pumps	Chap. 11
Pumps for process cooling	Chap. 12
Open thermal storage systems	Chap. 13

Closed systems utilize external pressure to maintain the desired operating pressure. This is accomplished through the use of makeup water that is fed through a pressure regulator. Since they are closed systems, they have an expansion tank to accommodate thermal expansion and contraction caused by changes in system water temperature. Following is a similar listing for closed systems described in this book.

Chillers and their pumps	Chap. 14
Chilled water systems	Chap. 15
Closed condenser water systems	Chap. 16
Closed thermal storage systems	Chap. 17
District heating and cooling	Chap. 18
Hot water boilers	Chap. 19
Low-temperature hot water heating	Chap. 20
Medium- and high-temperature hot water	Chap. 21

8.6   Suggested Design Rules

Generally, water consumption is not a concern in most HVAC systems, since most systems are circulatory. *The use of energy is important!* If the recommendations and equations developed herein are used, efficient pumping procedures will result for these water systems. Following are some additional recommendations.

1. The prospective water system should be designed to the specific requirements of the owner, utilizing the following principles to achieve the most efficient system possible within the first cost budget of the project.

2. The water system should be configured to distribute the water efficiently with a minimum use of energy-wasting devices. These devices are listed here:
   a. Three-way temperature-control valves
   b. Balancing valves, manual or automatic
   c. Pressure-reducing or pressure-regulating valves
   d. Some crossover bridges
3. The piping should be designed without
   a. Reducing flanges or threaded reducing couplings
   b. Bullhead connections (e.g., two streams connected to the run connections of a tee with the discharge on the branch of the tee)
4. The friction for the piping should be calculated for all pipe runs, fittings, and valves.
5. Distribution pumps should be selected for maximum efficiency at the design condition and within the economic constraints of the project.
6. Distribution pumps should be added and subtracted to avoid operation of pumps at points of high thrust and poor efficiency. Pump sequencing should achieve maximum possible system efficiency.
7. Cooling and heating coils should be selected with a high enough water velocity in the tubes to avoid laminar flow throughout the normal load range imposed on the coils.
8. Coil control valves and their actuators should be sized to ensure that they can operate at all loads on the system without lifting the valve head off the valve seat.

It is obvious from the preceding that the design of HVAC water systems is not a simple task. Much analysis must be done, balancing many cost factors against operating costs to achieve the economically feasible design for each installation. Reiterating, the use of computers enhances this effort and eliminates much of the drudgery of this design work.

8.7 Eliminating Energy Waste During Commissioning

Every effort should be made in the design, testing, and initial operation of HVAC water systems to eliminate the waste of energy in such systems. Typical of this is the testing and balancing procedures that are used to commission a new chilled or hot water system. In the past, a calibrated balance valve was used to check the flow through an individual coil. This balance valve was adjusted until the design flow was achieved through that coil. This is shown in Fig. 8.5a; Fig.

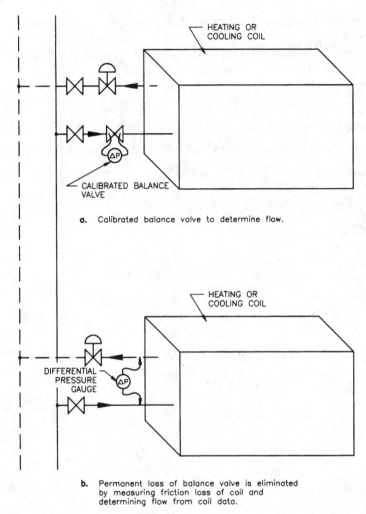

a. Calibrated balance valve to determine flow.

b. Permanent loss of balance valve is eliminated
    by measuring friction loss of coil and
    determining flow from coil data.

Figure 8.5 Testing for coil flow.

8.5*b* describes the correct method for checking the flow through such coils. The differential pressure across the coil in Fig. 8.5*b* is checked, and the flow is verified from technical data on that coil. This is just as accurate as the balance-valve method; it may provide greater accuracy because a much larger differential signal is achieved by measuring the pressure drop across the coil. The permanent loss across the calibrated balance valve is eliminated. The flow through the coil during commissioning should be controlled by its control valve, not any other device. This should check the coil flow and the control-valve operation at the same time.

Every traditional practice during initial operation of a water system such as this coil testing must be evaluated to determine if any energy waste is occurring.

8.8   Bibliography

Ben L. Kincaid and Andrew Spradley, *Removing Manual Balancing in a High School.* 1995

The Water Management Manual, SYSTECON, Inc., West Chester, Ohio, 1992.

Configuring an HVAC Water System

9.1 Introduction

This chapter presents a comprehensive evaluation of the configuration of HVAC water systems. It may appear to be beyond the scope of a book on HVAC pumps; however, the economical application of these pumps depends totally on proper system design. An understanding of these principles of good water system configuration should result in the economical selection of pumps and in efficient pump operation. The continuing theme of this book is the intelligent application of HVAC pumps to eliminate the mechanical devices used in the past to overcome the overpressure of these water systems that was caused by improperly sized and misapplied pumps.

Reviewing the basic equation for pumping energy,

$$\text{Pump kW} = \frac{Q \cdot H \cdot 0.746}{3960 \cdot P_\eta \cdot E_\eta}$$

$$= \frac{Q \cdot H}{5308 \cdot P_\eta \cdot E_\eta} \qquad (6.14)$$

the energy required by a pump rises with increases in system flow or head and with decreases in pump efficiency and motor or motor and variable-speed drive efficiency. It is therefore incumbent on the HVAC designer to develop a water system with a minimum of flow and head for a particular duty. This chapter addresses the piping for these water systems in an attempt to secure the optimal design that recognizes the preceding equation. This chapter embraces hot and chilled water systems, whereas the energy economies of condenser water systems are included in Chaps. 11 and 16.

9.2 Selection of Temperature Differential

The first decision a designer of a chilled or hot water system must make is the selection of the temperature differential. *Temperature differential* is the difference between the supply water and the return water temperatures. A number of conditions must be recognized before making the final selection of temperature differential:

1. An increase in temperature differential decreases water flow and therefore saves pumping energy.

2. An increase in temperature differential may increase the cost of coils that must operate with a higher mean temperature difference.

3. Higher temperature differentials increase the possibilities of loss of temperature difference in coils due to dirt on the air side and chemical deposits on the water side of them.

4. Laminar flow on the water side due to lower velocities at low loads on a coil is always a concern of the water system designer. The possibility of laminar flow is greater with higher temperature differences.

The quick promise of energy savings with higher temperature differences can be offset later by problems in system operation due to failure to have enough pumping capacity after the system has aged. Only experienced designers should entertain water temperature differences in excess of 12°F on chilled water and 40°F on hot water systems. A careful balance between energy savings and first cost should be made by the designer. *There is no one temperature difference for all chilled or hot water systems.*

Following are two formulas for calculating the water flow, in gallons per minute, for hot and chilled water systems:

$$\text{Hot water gal/min} = \frac{\text{system Btu/h}}{500 \cdot \text{temperature difference}} \qquad (9.1)$$

$$\text{Chilled water gal/min} = \frac{\text{system load (tons)} \cdot 24}{\text{temperature difference}} \qquad (9.2)$$

These are general formulas based on a specific gravity of 1.0 for water. If it is desired to secure more exact water flows for hot water, then specific gravity and specific heat of water at the average temperature of the system should be considered; Eq. 2.4 should be used. Computer programs for load calculations should insert the specific gravity automatically for the water temperature at all parts of the water system.

The actual temperature difference that is selected for a specific installation is determined by the cost of the coils for various temperature differences and the effect that higher differences may have on the operating cost of the chiller or boiler. Reducing the leaving water temperature of a chiller increases its energy consumption and may offset the savings of higher temperature differences; likewise, elevating the leaving water temperature of a boiler decreases its efficiency and can have the same effect on a hot water system. These are the decisions that must be made by the designer for each application.

A second factor that the designer must take into consideration is laminar flow in coils, whether they are for heating or cooling. Laminar flow reduces the heat-transfer rate and increases the flow through a coil. If laminar flow occurs in a coil, the design temperature difference is not maintained, and a greater system flow in gallons per minute will be required, as indicated in the preceding formula. Laminar flow should not occur in a coil at any point in its load range. Many systems are operating inefficiently today because of coils that were selected at too low a friction loss through them at design load; therefore, at reduced loads and flows, they are operating with laminar flow. Coil velocities of 2 to 3 ft/s in the tubes under design conditions help eliminate the possibility of laminar flow. The actual coil velocity used should be the recommendation of the coil manufacturer and the decision of the designer.

It is obvious that higher temperature differences and lower system flows increase the chances of laminar flow in heat-transfer equipment. Laminar flow also creates problems for the pumping system, since the system flows are higher than design flow.

9.3 Modeling a Water System for System Head and Area

The first task confronting the designer of a water system is to compute the water flow and pump head required by that system after the temperature differential is determined. It can be a daunting task to compute these values by hand from minimum to maximum load on the system.

As described in Chap. 1, the advent of high-speed computers and special software has eliminated the drudgery of such water system analysis. This enables the engineer to evaluate a system rapidly and determine system flows and head losses under variable loads on the water system. The designer can evaluate the building under various load conditions, develop a better understanding of the energy consumptions of the mechanical systems, and achieve an accurate estimate of the diversity of the heating, ventilating, and cooling loads.

The significant fact that has been found about building loads is that very few multiple-load chilled or hot water systems are ever uniformly loaded. Analysis of many systems has demonstrated a decided variation in the water loading on the heating or cooling coils. Some coils on a system will be fully loaded, while others will have very little load on them. This fact must be taken into consideration during the system analysis, as well as during the piping and control design. Computer-assisted evaluation allows the designer to study the many load variations that can occur on the prospective water system.

Chapter 3 provided the means to calculate pipe friction in HVAC piping systems. This chapter will use those data to demonstrate how the head changes in a water system as the flow varies in that system. It should be remembered that most hot and chilled water systems are actually a number of small systems using common supply pipes. Each heating or cooling coil is a water system in itself, since each coil has a different total friction head. These friction loads for a specific coil consist of the supply main loss, branch main loss, and loss through the coil itself with its piping and control-valve losses. If the designer recognizes this individuality of every coil and seeks to design the water system with recognition of this fact, a simpler and more efficient system will result.

Traditionally, the pump head required for a water system has been shown as a curve that results when the system head in feet is plotted against the system flow in gallons per minute. Therefore, this curve has been named the *system head curve*. Through many years of work with HVAC water systems, it has been demonstrated that the head requirements for many water systems cannot be represented by this simple parabolic curve. Instead, the head varies through a broad area that may be difficult to calculate. Before tackling this knotty problem, the elements of a system head curve or area should be reviewed.

9.3.1 System head curve components

A system head curve consists of plotting the flow in the system, in gallons per minute, horizontally and the system head vertically, in feet of head. The system head must be broken down into variable head and constant head. *Variable head* is the friction head of the water distribution system. *Constant head* can be divided into either static head or constant friction head. *Static head* is simply raising water from one level to another. Typical of this is the height of a cooling tower. *Constant friction head* can be the loss across a heating or cooling coil, its control valve, and connecting piping, as shown in Fig. 9.1*a*. This constant head is typified by the differential pressure maintained across a heating or cooling coil and its appurtenances at a constant value by a differential pressure transmitter. This is the signal used to control the system to which the coil is connected. Constant friction head occurs in many systems with two-way temperature-control valves on the coils; it is not found in old systems that used three-way temperature-control valves. Constant friction head is plotted vertically similar to static head of a cooling tower. A typical system head curve is that for a variable-volume system with two-way temperature control valves or a system head curve for a cooling tower installation.

As described in Part 1 on pipe design, the formulas for pipe friction reveal that pipe friction varies as parabolic curves with exponents of from 1.85 or 2.00. The total system head curve is achieved by adding the static or constant friction head to the system friction head; the following equation can be used for calculation of the uniform system head curve for a hot or chilled water system:

$$H_a = H_2 + \left(\frac{Q_a}{Q_1}\right)^{1.90} \cdot (H_1 - H_2) \text{ ft of head} \qquad (9.3)$$

where H_1 = total head in feet on the system at the design flow Q_1, in gallons per minute

H_2 = constant head on the system in feet

Q_a = flow at any point on the curve between points 1 and 2

H_a = head at any point on the curve between points 1 and 2 at a flow of Q_a

1.90 = the exponent that has been found to follow the Darcy-Weisbach equation closer than the normally accepted 2.0

The use of this equation can be demonstrated by the following example. Assume that

1. The maximum system flow is 1000 gal/min at 100 ft of head. This is the maximum point on the uniform system head curve and is identified as Q_1 and H_1 in the equation.

2. The maximum pressure loss in a coil with its control valve and branch piping is 20 ft. This is the zero point on the system head curve with no flow in the system. This point is H_2 in the equation.

3. The preceding two points set the ends of the uniform system head curve.

Table 9.1 provides the system head curve points from a minimum flow of 100 gal/min to the maximum of 1000 gal/min for this example.

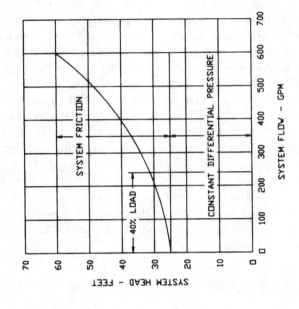

b. Uniform system head curve for model building of A.

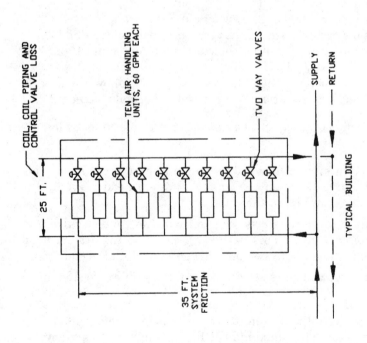

a. Model building for system head area evaluation.

Figure 9.1 Model building for system head evaluation.

**TABLE 9.1   System Head Curve
Coordinates**

System flow, gal/min	System head, ft
100	21.0
200	23.8
300	28.1
400	34.0
500	41.4
600	50.3
700	60.6
800	72.4
900	85.5
1000	100.0

Reiterating, this is the uniform system head curve for an HVAC system where all heating or cooling loads are loaded to the same percentage of design load.

A closed chilled or hot water system with three-way valves has only system friction, and such a system does not have a system head curve because it operates at only one point, maximum system flow and head. In the preceding example, the system always operates at 1000 gal/min and 100 ft of head when the coils are equipped with three-way control valves. This is a beginning indication of the terrible waste of energy in a constant-volume system equipped with three-way control valves on the heating or cooling coils; no matter what the load is on the system, the pumps continue to operate at full system flow.

9.3.2   System head areas

The preceding example and data describe systems with uniform flow in all heating or cooling coils. Obviously, this does not exist in most actual buildings, since some coils will be loaded while others will not have any load on them. If the building has windows, as the sun moves around the building during the day, the loads on the various coils will change even with constant outdoor temperature and internal load. This is also demonstrated by the term *diversity,* which was defined in Chap. 1.

Recognizing that systems are not loaded uniformly and that diversity does exist, we must evaluate chilled and hot water systems to determine how to compute and display graphically the actual head on them. Figure 9.1a describes a typical building with 10 air-handling units on different floors; this will be our model building to illustrate a typical system head area. The uniform system head curve for this building is shown in Fig. 9.1b. Assume that a 40 percent uniform load occurs on each of the 10 air-handling units; this is detailed graphically in Fig. 9.2a. Now, let the 40 percent load shift so that only the 4

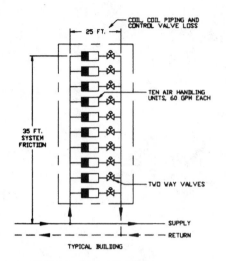

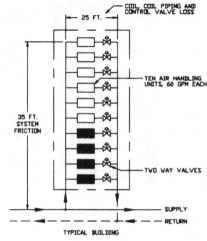

a. Uniformly loaded building, 40% load on each air handling unit.

b. Non-uniformly loaded building, air handling units close to pumping source fully loaded.

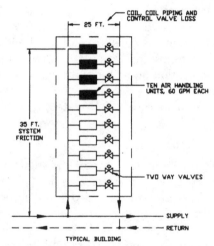

c. Non-uniformly loaded building, air handling units far from pumping source fully loaded.

Figure 9.2   Model building loading.

bottom coils are fully loaded and the top 6 coils have no load on them (see Fig. 9.2*b*). Since these 4 coils are nearer to the pumps than all 10 coils, the system friction will be less than that for all 10 coils uniformly loaded at 40 percent each. Next, transfer all the 40 percent load to the upper 4 coils (see Fig. 9.2*c*). Now the load is farther than all the 10 coils uniformly loaded, and the system friction head will be greater

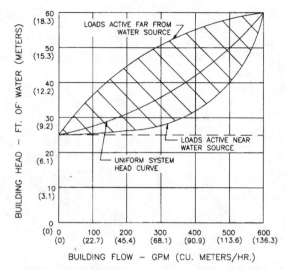

Figure 9.3 System head area.

than that for the uniformly loaded condition. This procedure can be done for various loads on the building, from 10 to 90 percent load. The result of these moves of the load on the building is shown in Fig. 9.3; this is a system head area, and it exists for almost any building with more than one heating or cooling coil. This is one of the most important figures in this entire book, since it makes the designer aware of how complex true system evaluation can be and that a HVAC water system cannot be represented by a simple system head curve.

This is elementary manual modeling of a water system. With computers, such system head areas can be developed easily. Without a computer, a simple procedure that works practically is to adjust the distribution friction at 50 percent water flow, i.e., 36 ft in our model building at 50 percent load. This variable or distribution friction can be multiplied by 75 percent for the lower curve and then by 150 percent for the upper curve. Drawing curves similar to Fig. 9.3 through these points will generate an approximate system head area. This system head area will give designers a rough idea of what the system head area will be, and it will enable them to predict pump performance within this system head area. A further discussion of the operation of pumps with the system head area will be presented in Chap. 15.

Campus-type installations with a number of buildings, as shown in Fig. 9.4a, create another dimension in system head areas and system modeling. As demonstrated in this figure, there are buildings near the central energy plant and buildings far from it. This figure has two of

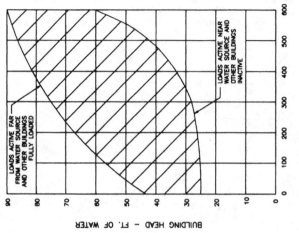

b. System head area caused by non-uniform flow in building "B" and other buildings.

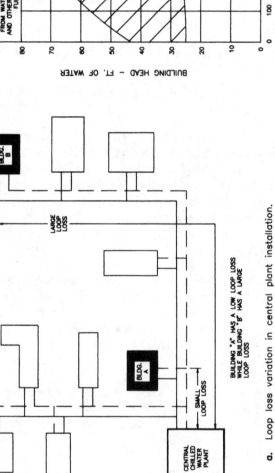

a. Loop loss variation in central plant installation.

Figure 9.4 Campus type chilled or hot water system.

our model buildings, building A near the central plant and building B far from the central plant. It is obvious that the campus loop loss will be greater for building B than for building A. If the loop loss is 30 ft in the distribution mains out and back for building B with all the buildings fully loaded, this loss will be determined by how active are all the intermediate buildings. There are two ultimate conditions that must be checked: first, with all the intermediate buildings fully loaded, and second, with all these buildings with no load. Using the preceding procedure for determining system head produces the system head area of Fig. 9.4b. The upper curve is for the first condition, where all the intermediate buildings are fully loaded, and the bottom curve is for the second condition, where these buildings have no load. This figure demonstrates the great variation that can occur in the head required for a campus building located far from the central energy plant. Here again, computer modeling can produce system head areas for every building on a campus under a number of different load conditions.

Figure 9.4b is a typical head variation for many campus buildings. Most college and university campuses now utilize many energy-wasting devices such as balance valves, pressure-reducing valves, and crossover bridges to overcome this friction variation. This system head area demonstrates how broad are the actual head requirements of a campus-type system. For example, at 50 percent load on the building, the pump head can vary from 27 to 74 ft. With constant-speed pumps, all this head difference must be destroyed by mechanical devices to maintain adequate flow to all parts of the building. With the correct pumping arrangement and variable-speed pumps, all this wasted head can be eliminated. With proper control, the variable-speed pump will always operate at the needed system flow and head without any overpressure or wasted energy. A careful review of this figure will be made in Chap. 15, as well as how pump performance is affected by such broad variations in system head.

9.3.3 Pumping system losses

All the foregoing discussion has been for water system flow and head evaluation. Losses through pumping systems themselves are often ignored or included with the system head losses. It has been learned that pumping system losses can be significant and should be analyzed separately from the water system losses. Figure 9.5a describes an end view of an actual pumping system consisting of five pumps, each with a capacity of 700 gal/min, for a total system capacity of 3500 gal/min. The individual loss for each fitting is shown along with its K factor. The total loss for the pump fittings is 6.9 ft. Recognizing Hydraulic Institute's estimate that the losses in pipe fittings can vary by as

SYSTEM FLOW: 3500 GPM
PUMP FLOW: 700 GPM
VELOCITY HEAD IN 6" PIPE: 0.94 FT
TOTAL SYSTEM LOSS = 6.9 x 1.2 = 8 FT

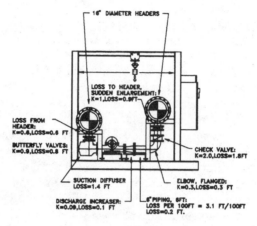

a. Friction losses for a pumping system.

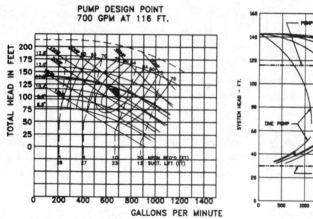

b. Pump head—capacity curve for system of A.

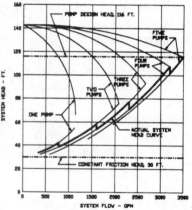

c. System head curve adjusted for pump fitting losses.

Figure 9.5 Friction losses for a pumping system.

much as 10 to 35 percent, these losses have been multiplied by plus 20 percent to achieve a reasonable loss for the fittings. The losses for this system would therefore be 8 ft with 700 gal/min flowing through an individual pump. Figure 9.5*b* is the pump curve for this system; each pump could operate at a maximum flow of 1100 gal/min if improperly controlled. At 1100 gal/min, the pump fitting loss becomes $(1100/700)^{1.90} \times 8$, or 18.9 ft. A loss of only 8 ft may not raise much concern, but a loss of close to 19 ft should be of concern to anyone making a serious energy evaluation of this water system.

This system had a differential pressure of 30 ft across the largest coil and its control valve and a system loss of 78 ft. With the estimated 8 ft of loss throughout the pump fittings, the total system loss becomes 116 ft. Figure 9.5*c* describes the uniform system head curve for this system, which has been adjusted to recognize the variable loss through the pump fittings. Again, no longer is the uniform system curve a smooth curve, but rather it is one of scalloped shape. It is difficult to select the most efficient pump add points without a computer program similar to the wire-to-water efficiency program that will be described in Chap. 10. In this example, the pump add/subtract points in Fig. 9.5*c* were set arbitrarily at 800 gal/min for each pump. This is not necessarily the most efficient point to change the number of operating pumps. This curve demonstrates conclusively that the old tradition of just running one pump until it cannot pump any more is a wasteful practice.

It is obvious that the system head curves and areas described earlier in this chapter are affected by this variation in pump fitting losses. How do designers incorporate pump fitting losses into their calculations for a proposed system? It is difficult unless a computer program is available for system friction loss calculation that includes such fitting loss calculations. The *wire-to-water evaluation* that will be reviewed in various chapters accounts for pump fitting losses. This evaluation separates the pump fitting losses from system losses to achieve a reasonable procedure for estimating system performance and sequencing of pumps on multiple-pump systems.

System flow and head evaluation is a very important matter, since efficient pump selection depends on it. The preceding discussion demonstrates that it can be a complicated subject. Any time devoted to such an evaluation is well worth its cost.

One important factor that emerges from true system evaluation is the indication of losses that are caused by energy wasters such as balance valves, pressure-reducing valves, etc. Their losses become pronounced under this type of careful scrutiny. It should be noted that most of the HVAC water systems that are reviewed in this book seldom use these devices.

The advent of computer software has enabled the designer to achieve better part-load evaluation as well as a closer estimate of the true diversity of the system under examination. Many different hypothetical loads can be inserted as data in the computer by the designer, who can then select the design that most closely fits the economic parameters of the proposed system.

9.4 Static Pressure

The *static pressure* of an HVAC system is the system pressure that avoids (1) drying out the tops of the water system or (2) imposing too high a system pressure on the water system. Basically, the static pressure for most systems consists of the height of the building plus a cushion at the top of the building. Usually this cushion is around 5 to 10 lb/in^2 or 10 to 20 ft. With a 200-ft-tall building and a cushion of 20 ft on top of the building, the static pressure would be 200 + 20, or 220 ft minimum (95.2 psig at 60°F). When the water system is totally within a building, the actual elevation above sea level of the building can be ignored.

Such is not the case with a campus-type installation consisting of several buildings. With this application, actual elevations of the buildings must be included. The elevations of the tops of all the buildings must be recorded as well as the elevation of the expansion tank at the operating level in the central plant. Static pressure is the elevation at the top of the highest building minus the elevation at the operating level of the central energy plant plus the required cushion at the top of the tallest building. That is,

$$\text{Static pressure} = \frac{Z_b + Z_c - Z_p}{(144/\gamma)} \text{ psig} \qquad (9.4)$$

where Z_b = elevation of the top of the tallest building
  Z_c = elevation of the operating floor of the central energy plant
  Z_p = cushion required on top of that building in ft of head
  γ = specific weight of the water at the operating temperature

For example, assume that the top of the tallest building is 650 ft, the desired cushion on that building is 10 ft, the operating level of the central energy plant is 420 ft, and the system is a hot water system operating at 180°F (60.57 lb/ft^3). Thus

$$\text{Static pressure} = \frac{650 + 10 - 420}{(144/60.57)} = 101 \text{ psig}$$

Distributed pumping, which relies on static pressure for movement of the water in the system, requires special calculation of the static pressure. The remote building whose height, cushion, and distribution loss are the greatest determines the static pressure. Equation 9.4 becomes

$$\text{Static pressure} = \frac{Z_b + Z_c - Z_p + H_d}{(144/\gamma)} \tag{9.5}$$

where H_d = distribution loss in feet H_2O from central plant to the building that determines the static pressure

If the building in the preceding example were the building that determines the static pressure for the system and the friction loss between it and the central energy plant was 50 ft at design flow, then the static pressure would be

$$\frac{650 + 10 - 420 + 50}{(144/60.57)} = 122 \text{ psig}$$

This will be explained further in this book with other design criteria for this important method of water distribution.

9.5   Three Zones of HVAC Water Systems

Almost all HVAC water systems are of the loop type; this means that the water is returned to its source such as a chiller, cooling tower, or boiler. In rare cases, groundwater is used for cooling and is dumped into a sewer or stream after use.

Most of these water systems consist of three zones, namely, (1) an energy source, (2) energy transportation, and (3) energy use as described in Fig. 8.1 for chilled or hot water. Cooling towers are similar in arrangement. Chapter 8 discussed the use of water, or the third zone. It is important to understand that each of these three zones must be designed individually for optimal use of energy. Therefore, these zones often are combined together with resulting poor energy generation, transportation, or use. Each zone must be evaluated separately; the first and second zones will be reviewed in the chapters on chillers and boilers. True energy conservation in HVAC systems begins with careful analysis of energy consumption in boilers, chillers, or cooling towers. Chapters 11, 14, and 19 of this book will study this analysis of energy consumption.

Many times the second zone, energy transportation, is mixed with energy generation; this can result in poor energy generation and transportation if care is not used in evaluating the total energy consumption of the system. Efficient energy transportation will be discussed in Chaps. 15, 16, 18, and 20 on cooling and heating systems.

Finally, the use of water in cooling and heating coils or heat exchangers must be efficient or pumping costs will be excessive. As indicated in Chap. 8, the indiscriminate use of balance valves, three-way control valves, and improper heating and cooling coil connections results in poor system efficiency. If these three zones are remembered during design and each treated as efficiently as possible, this effort should result in a good pumping and piping system.

9.6   Piping Configurations

A number of different piping configurations are used in HVAC water systems; only the more popular types will be discussed here. These are

1. Open or closed systems and how water is returned to the source, namely, direct or reverse return

2. Number of pipes to transport the water through the system

9.6.1   Three types of systems

There are three basic system arrangements in HVAC piping: (1) direct return systems for loop piping, (2) reverse return systems for loop piping, and (3) open piping systems for open tanks such as cooling towers and energy-storage systems. Figure 9.6a and b describes direct and reverse return piping, and Fig. 9.7a describes an open piping system utilizing a cooling tower.

At one time, reverse return piping was always used to balance the friction to all terminal units, particularly when they were equipped with three-way temperature-control valves. Today, with the great emphasis on energy savings, three-way temperature-control valves have very few logical applications. Reverse return piping can be useful on low-rise buildings, where it may be the most economical arrangement; seldom should it be used on high-rise structures owing to the vertical weight of the extra piping. One advantage for reverse return piping is the reduction in maximum pressure drop across the temperature-control valves. Figure 9.6c is a pressure-gradient diagram for a reverse return system that demonstrates this fact. This diagram demonstrates why reverse return piping should be used on constant-volume systems. With the same pressure drop across each cooling coil, its control valve, and piping, balancing the system is much easier than with direct return. This diagram also demonstrates why reverse return piping is no longer needed with contemporary control valves on variable-volume systems.

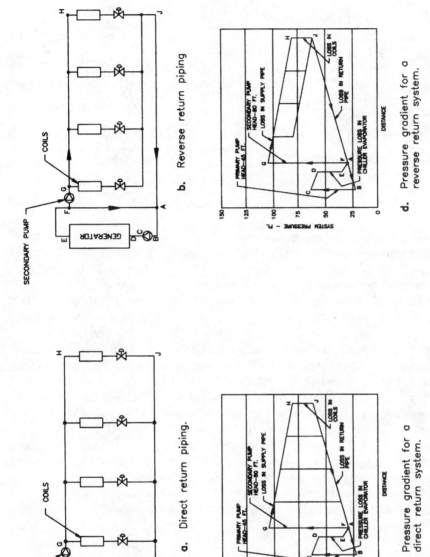

a. Direct return piping.

b. Reverse return piping

c. Pressure gradient for a direct return system.

d. Pressure gradient for a reverse return system.

Figure 9.6   Direct and reverse return systems.

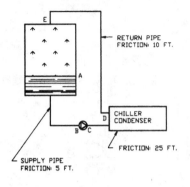

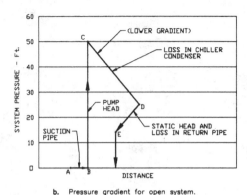

a. Open piping system.

b. Pressure gradient for open system.

Figure 9.7 Open water systems.

Direct return piping is the most economical arrangement for most contemporary buildings utilizing the energy savings of variable-volume systems. It requires a minimum of piping and usually has less pipe friction than an equivalent reverse return system. The pressure-gradient diagram for a direct return system is shown in Fig. 9.6. As seen in this diagram, the full pump head can be exerted across the temperature-control valves. This is not a problem for the designer of contemporary control valves as long as the maximum head that can be imposed across the control valves is recognized. The maximum head is not imposed on the valves except during full-load conditions.

Open piping systems usually are encountered with cooling towers or open energy-storage tanks; Figure 9.7a describes a typical open system with a cooling tower serving the condenser of a chiller. The pressure-gradient diagram for this system is shown in Fig. 9.7b. Pressure-gradient diagrams are seldom needed for elementary cooling tower applications because the operating pressures are relatively low. On some complex cooling tower applications and process cooling operations they are valuable for determining pump head and system arrangement.

9.6.2 Number of pipes in a loop system

There are three different numbers of pipes found in HVAC systems. These are

Two-pipe: Two-pipe systems are used where there is only heating or cooling; other systems utilize two pipes where there is both heating and cooling but not simultaneously.

Three-pipe: This is an old configuration that is seldom used today. There are two supply pipes and one return pipe; one supply pipe is for heating, and the other is for cooling. There is a common return pipe for both the heating and cooling systems. The possibilities of energy waste are obvious with improper control.

Four-pipe: This is the preferred piping arrangement for buildings that have simultaneous heating and cooling loads. Most buildings in the temperate zones are of this configuration and have simultaneous heating and cooling loads, particularly in the spring and fall. All the systems discussed in this book will be of the four-pipe variety, two pipes for cooling and two pipes for heating.

9.7 Location of Expansion Tanks

Expansion tanks should be located at the point where the system is to be stabilized, not necessarily at the suction of the pumping system. With the advent of variable-speed pumps and their digital control, the operation of a pump is not dependent on its suction pressure being constant. In the past, the expansion tank was always located at the pump suction and took its air from the air separator. This expansion tank was of the plain type without an internal bladder that separates the air from the water. The result was absorption of air by the water and waterlogging of the tank.

This nuisance was eliminated by the bladder-type expansion tank. This expansion tank is located away from the air separator, and this enables the designer to locate the expansion tank at the desired point of pressure regulation in the water system, not just at the point of installation of the air removal equipment.

On low-rise buildings, the expansion tank for both hot and chilled water systems can be located at the suction of the pumps. On high-rise buildings, the expansion tank and water makeup equipment should be located at the top of the building to ensure continuous pressure at that point in the system.

9.8 Elimination of Air in HVAC Systems

In the past, air elimination from HVAC water systems depended on mechanical devices such as air vents and mechanical separators. The development of better chemical treatment for these systems has resulted in the use of chemicals for complete removal of air. The mechanical separator did not remove any of the dissolved air from a water system. The so-called pot feeder is an economical chemical feeder can accept a passivating chemical such as sodium sulfite for removing all the oxygen from an HVAC water system. It is much less costly than a mechanical separator for many sizes of pipe.

On chilled water systems, mechanical separators do very little air separation except at the initial fill of the water into the system. The reason for this is the fact that the temperature of the makeup water is equal to or greater than the chilled water operating temperature. For example, the makeup water may be at 60°F and then reduced to 40 to 50°F. The solubility of air in the water is actually increased. Table 2.5 demonstrates this.

Chilled water systems should be equipped with manual air vents at the high points of the system. These air vents should be equipped with air-collection chambers below them. The chemical feeder should be installed at the central energy plant for complete treatment of the entire water system. Larger systems might be equipped with automatic chemical feeders. Chemical feeders offer a better and in many cases less expensive method of removing air from these systems than depending on a mechanical separator, which cannot eliminate all the oxygen from a water system. *Reiterating, mechanical separators cannot remove dissolved air!*

Hot water systems can use mechanical separators to remove some of the air, as is shown in Figs. 3.6*a* and 3.7*a*. Seldom should they be located on the discharge from boilers. Little additional air removal is achieved on most installations, and the location on the discharge pipe from boilers may be cumbersome. The use of dip tubes on boilers, manual air vents, and a chemical feeder may produce a more economical and better method of removing air from hot water systems than large mechanical separators.

9.9 Control of Return Water Temperature

As we began this chapter with system temperature difference, so we will leave it. *Return water temperature is one of the most important operating values for a chilled or hot water system.* It tells the operator just how good a job the control system and coils are doing in converting energy from the chillers or boilers to the air or water systems that are cooling or heating the building. This is such a basic criterion that it should be addressed early in the design of a chilled or hot water system. Individual coils should be equipped with thermometers or insertion plugs on their return connections, as described in Fig. 9.8*a*; zone or building returns should be equipped with temperature transmitters, as also shown in this figure. This information should be displayed and recorded at the central data-acquisition point for the entire system.

There is a great argument ongoing in the HVAC industry about how to control return water temperature in loop-type systems such as chilled and hot water systems. The proper method of controlling return

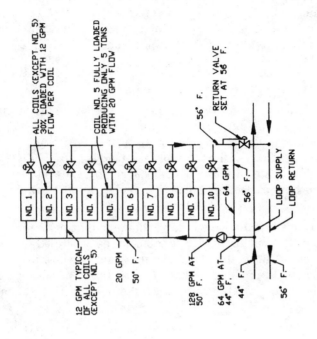

b. Unsatisfactory use of return temperature valves.

a. Proper piping and instrumentation for return temperature control.

Figure 9.8 Return water temperature indication and control.

temperature is through the correct selection of control valves and heating and cooling coils. The use of return temperature-control valves, as shown in Fig. 9.8b, is a quick way to solve many system operating problems, but it is fraught with its own problems. For example,

1. The return temperature-control valve adds head loss to the entire system and is therefore an energy waster.

2. There is a danger that warm return water will be bypassed through the crossover bridge back to the supply side of a coil that is fully loaded and needs design supply temperature at that moment. The result is that the coil will not supply the load required of it. For example, in Fig. 9.8b, the fully loaded coil no. 5 needs 44°F water to produce 10 tons of cooling; since it is receiving 50°F water, it can produce only around 5 tons of cooling.

3. If certain coils are exhibiting laminar flow, circulating pumps can be installed on the coils to avert this problem, as described in Fig. 8.4. This is an excellent method for controlling laminar flow in a heating coil. It was stated in Chap. 8 that, in most cases, pumps or circulators should not be piped in series with the coil but in the bypass, as shown in Fig. 8.4.

4. Return temperature-control valves obliterate the true return water temperature from the system and prevent the operating and maintenance people from understanding which coils or control valves are creating problems. *Return temperature is the operator's guide to efficient use of water in HVAC systems.*

5. Return water temperature control lulls maintenance people into believing that there is no problem with any of their coils when some of those coils may be very dirty on the air or water side.

   In conclusion, one of the designer's most important tasks is the selection of a sound temperature differential that will provide maximum possible system efficiency. The second step in this process is to ensure that that differential is maintained after the system is commissioned.

9.10 Bibliography

James B. Rishel, *Variable Water Volume Is Hydro-Electronics,* SYSTECON, Inc., West Chester, Ohio, 1982.
James B. Rishel, *The Water Management Manual,* SYSTECON, Inc., West Chester, Ohio, 1992.

10

Basics of Pump
Application for HVAC Systems

10.1 Introduction

The configuration of a hot or chilled water system was emphasized in Chap. 9 to achieve the desired heating or cooling with a minimum of flow and head. This chapter undertakes evaluation of the pumps themselves to achieve optimal energy consumption for the pumps. It was necessary for the readers to have a basic understanding of HVAC water systems before initiating this application of pumps.

Before embarking on the actual application of pumps to HVAC water systems, there should be a discussion of when to use constant-speed pumps and when to use variable-speed pumps. With the rapid reduction in the cost of variable-speed drives, there is almost no limit, in terms of motor size, for the use of variable-speed pumps. Very low head pumps, i.e., less than 50 ft, can still be constant-speed pumps when applied to small systems. Otherwise, the rule should be *constant-speed pumps for constant-volume systems and variable-speed pumps for variable-volume systems.*

If a water system has a large variation in its flow, from less than 50 to 100 percent of design, it is probably a candidate for variable speed. Likewise, if the flow rarely changes and there is little variation in temperature or viscosity, this is a constant-speed pump application.

As was pointed out elsewhere, variable-speed pumping offers great savings in maintenance as well as in energy. Further, variable-speed pumps, if controlled properly, operate at the system's flow and head requirements, not necessarily at the design conditions. They eliminate overpressuring, which causes operational problems as well as loss of pumping energy.

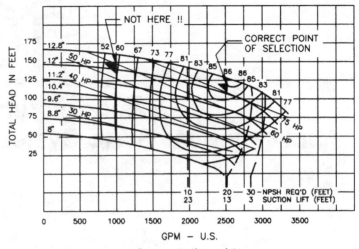

a. Pump selection points.

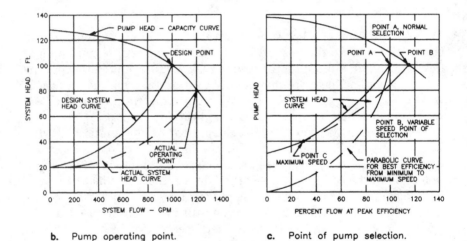

b. Pump operating point.   c. Point of pump selection.

Figure 10.1   Points of pump selection and operation.

10.2   Point of Selection

With the advent of variable-speed pumping, the point at which a centrifugal pump is selected has become more complicated. The basic rule that has been offered in the industry and in this book is to select the pump as closely as possible to its best efficiency point. This rule must be expanded to achieve efficient performance from both constant- and variable-speed pumps. Figure 10.1 will be used to demonstrate some of the points that should be brought out for selecting a pump for a particular duty.

10.2.1   Selecting constant-speed pumps

The pump head-capacity curve shown in Fig. 10.1a is for an excellent pump for a pump duty of 2500 gal/min and 120 ft of head. The peak efficiency for this pump is 86 percent. For constant-speed operation, this would be satisfactory for system flows between 2000 and 2700 gal/min. To operate this pump at lesser or greater flows would produce poorer efficiencies, and the radial thrust would increase so that there could be increased wear within the pump. One pump company urges that constant-speed pumps operate at no greater flow range than ±25 percent of the flow at the best efficiency point. Normally, pumps for constant-speed operation would be selected just to the left of the best efficiency point, or at 2200 to 2500 gal/min for the preceding pump.

A dangerous point to operate a pump is at 1000 gal/min and 130 ft of head, as shown in Fig. 10.1a. Often, inexperienced pump representatives who do not have a smaller pump at higher heads will offer this pump improperly. This is an unacceptable selection because of the poor efficiency and the high radial thrust existing at this point. Additional wear may occur due to hydraulic imbalance within the pump. When a pump shaft breaks, it may be an indication that the pump is running too closely to the no-flow or shutoff head.

Sometimes, HVAC water systems are designed with more estimated pump head than actually exists in the system. The result is that the pump does not operate at the design point but at a point farther out on its curve. This is shown in Fig. 10.1b. This pump was selected for a capacity of 1000 gal/min at a head of 100 ft. When new, the system had much less head requirements, and this resulted in the pump being operated at 1200 gal/min at 80 ft of head.

Another example of this so-called carryout of a pump's operation is given in Fig. 10.2 for the model building of Chap. 9 when located at the end of a campus chilled water loop. This figure uses the system head area of Fig. 9.4a. If a constant-speed pump is selected at the design condition of 300 gal/min at 90 ft of head for this broad system head area, the pump can operate at capacities as high as 540 gal/min. At this point, the pump would operate noisily, at a poor efficiency, and with a high radial thrust.

In an attempt to avoid poor pump operation, and recognizing the inability to compute system head accurately, the usual practice with constant-speed pumps has been to add a pump head conservation factor and then pick the pump to the left of the best efficiency point to

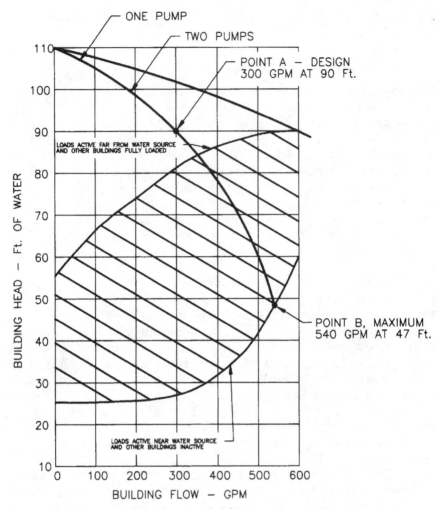

Figure 10.2 Pump operation with system head area.

ensure that the pump will operate without damage at the higher flows and lower heads. The other practice has been to install a balancing valve on the pump discharge, either manual or automatic, and add enough friction to prevent some of the operation at higher flows. One of the advantages of variable-speed pumps is that this "carryout" condition is eliminated by their controls without all the inefficiencies of pump head conservation factors and balancing valves.

10.2.2 Selecting variable-speed pumps

Selection of a variable-speed pump is more complicated than that for a constant-speed pump because the efficiencies of the motor and variable-speed drive must be evaluated along with the efficiency of the pump itself. The point of selection is less critical because the pump controls should not allow the pump to "carry out," as was the case with the constant-speed pump in Fig. 10.2.

The variable-speed pump should be selected slightly to the right of the best efficiency point where possible. Figure 10.1c describes this point. If this were a constant-speed pump, the pump would be picked at or just to the left of point A. As a variable-speed pump, it should be selected at point B or to the right of the best efficiency point. As the pump speed is reduced to minimum speed point C, the pump passes closer to the parabolic curve for best efficiency, from maximum to minimum speed. This demonstrates graphically what may need to be done with a computer program to make the best possible selection. Several pump selections should be made at various points, letting the computer make the wire-to-water efficiency run at each point. Computation of wire-to-water efficiency will be described in detail later in this chapter. The "carryout" condition described above for the constant-speed pump is eliminated by the variable-speed pump and its controls, so it can be selected to the right of the best efficiency point without fear of improper pump operation.

10.2.3 Special pump speeds

Traditionally, pumps have been selected for the HVAC industry at 1750 or 1150 rev/min, and in large pumps, speeds of 720 and 850 rev/min are not uncommon. The advent of the variable-frequency drive has brought a new dimension to pump selection that is often overlooked. Pumps now can be selected at speeds other than these standard induction motor speeds due to the flexibility of the speed output of the variable-speed drive.

Pump speeds as high as 5 to 10 percent greater than electric motor induction speed can be used to increase the overall performance of a pump. Most variable-speed drive manufacturers allow this speed increase on their drives. The nameplate rating in amperes must not be exceeded, and the available torque for the drive at such maximum speeds must be greater than that required by the pump at any point on its head-capacity curve. Likewise, the electric motor must be selected for the maximum horsepower required by the pump at the maximum speed. This enables the water system designer to secure more efficient selection of variable-speed pumps.

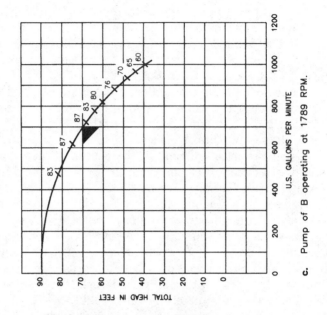

c. Pump of B operating at 1789 RPM.

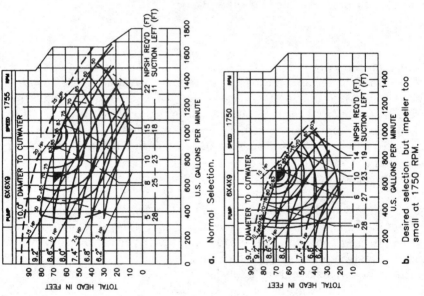

a. Normal Selection.

b. Desired selection but impeller too small at 1750 RPM.

Figure 10.3 Special pump speeds.

Following is an example of this pump selection procedure. Assume pump duty at 700 gal/min at 70 ft of head. There is one selection at 1750 rev/min: a 6 × 6 × 9 pump with a 25-hp motor and efficiency of 72 percent at the design point (Fig. 10.3a). A desirable pump would be the 6 × 4 × 9 pump, but it does not quite achieve the head capacity required at 1750 rev/min (Fig. 10.3b). An alternate selection would be the same 6 × 4 × 9 pump with a 20-hp motor operating at 1789 rev/min and at 87 percent efficiency at the design point (Fig. 10.3c).

The alternate selection was determined by computing the capacity of the smaller pump at a higher speed using the pump affinity laws. The pump speed was calculated at a point where the required condition of 700 gal/min at 70 ft was achieved by the smaller pump. Following is the method of computing the desired speed. From the affinity laws:

$$\frac{Q_1^{\,2}}{h_1} = \frac{Q_2^{\,2}}{h_2}$$

where Q_1 and h_1 are the desired conditions of 700 gal/min and 70 ft. Q_2 and h_2 are the equivalent points on the known pump curve.

By trial and error, the equivalent flow Q_2 can be computed and is 685 gal/min (see Eq. 10.1 developed in Sec. 10.2.5 on the use of the affinity laws for pumps). The required speed S_1 of the pump therefore is

$$S_1 = \frac{700}{685} \cdot 1750 = 1789 \text{ rev/min}$$

Figure 10.3c describes the actual operating curve for the 6 × 4 × 9 pump operating at 1789 rev/min. It is obvious that a 20-hp motor and variable-speed drive will be adequate because the maximum brake horsepower at any point on this curve is 16.8 bhp. The motor and drive manufacturers can certify that their equipment is acceptable for operation at this speed and brake horsepower.

An examination of Figs. 10.3b and c reveals that the use of this technique has increased the pump efficiency from 72.5 to 87 percent and has reduced the nonoverloading motor horsepower from 25 to 20 hp. This demonstrates the added dimension that variable-speed drives can provide in the selection of pumps. Engineers should not be concerned about the operation of a pump near its maximum impeller diameter; pump manufacturers will certify their pumps within the impeller diameters listed in their catalogs. Other engineers are reluctant to select pumps at maximum impeller diameter in anticipation of future increases in pumping requirements. This procedure of using speeds as high as 1210 or 1840 rev/min provides an additional load factor. Reiterating, in all this work, the maximum pump horsepower must be calculated to ensure that the motor and the variable-speed drive are not overloaded. Likewise, it is wise to verify that the pump manufacturer approves the operation of the pump at the proposed speed.

10.2.4 Proper use of affinity laws

It must be emphasized that the affinity laws apply to pump performance alone. They cannot be used directly for calculating pump performance with an actual water system. The reason for this is the fact that a pump changes its point of operation on its head-capacity curve as the system flow and head change. This point of operation on the pump head-capacity curve can be developed through the use of the affinity laws as follows:

Referring to the affinity laws (Eqs. 6.1 and 6.2),

$$\frac{Q_1}{Q_2} = \frac{D_1}{D_2} \quad \text{and} \quad \frac{H_1}{H_2} = \frac{D_1^2}{D_2^2}$$

and substituting Eq. 6.1 in Eq. 6.2 and transposing, we get

$$\frac{Q_1^2}{Q_2^2} = \frac{H_1}{H_2} \quad \text{or} \quad Q_1 = \sqrt{Q_2^2/H_2 \cdot H_1} \qquad (10.1)$$

If Q_2 and H_2 are the desired flow and head, the point of operation on the pump head-capacity curve, Q_1 and H_1, can be computed by Eq. 10.1. This is accomplished by trial and error, solving for Q_1 by inserting various heads H_1 until the resulting values for Q_1 and H_1 land on the pump curve.

Referring to Fig. 10.4a, which includes the system head curve of Fig. 9.1b for the model building of Chap. 9, assume that it is desired to determine the speed and efficiency of the pump when operating at 500 gal/min and 50 ft of head, point 2, or Q_2 and H_2. Inserting these values in Eq. 10.1, the solution of the equivalent pump flow Q_1 can be achieved by selecting values for H_1 in this equation until Q_1 and H_1 land on the pump curve. The equivalent point in this case is point 1, or 561 gal/min at 63 ft. The pump speed would be $Q_1 \div Q_2 \cdot 1750$ rev/min, or $561 \div 600 \cdot 1750$, which equals 1636 rev/min. The efficiency of a variable-speed pump would be 83 percent when running at point 2, which is the efficiency of the pump when operating at point 1 on the 1750-rev/min curve.

If this were a constant-speed pump and it was desired to determine the impeller diameter that would produce 500 gal/min at 50 ft, Eq. 6.1 would apply, and the impeller diameter would be $561 \div 600 \cdot 10$ in, or 9.35 in. The pump efficiency would be 75 percent at this point. Only by these equations can pump performance or impeller diameter be determined. If one were to try to determine pump impeller diame-

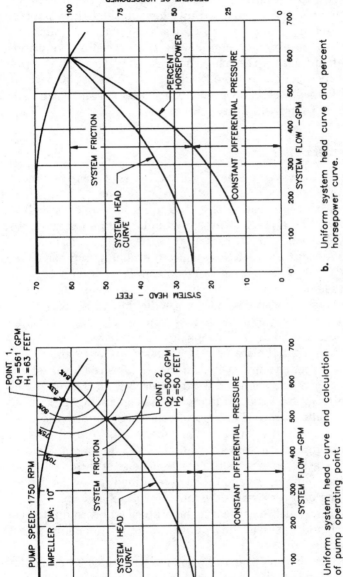

a. Uniform system head curve and calculation of pump operating point.

b. Uniform system head curve and percent horsepower curve.

Figure 10.4 Application of affinity curves.

211

ter by just dividing the two flows, 500 and 600 gal/min, an undercut impeller would result, namely, $500 \div 600 \cdot 10 = 8.33$ in.

The preceding examples demonstrate the *proper* and *improper* use that can be made of the pump affinity laws in the field when trying to change the pump diameter for a specific application. Similarly, the calculation of pump energy must use the preceding equations, not just the affinity laws themselves. Figure 10.4b includes the true pump horsepower for the same system head curve. At 50 percent flow, the pump horsepower is 35 percent, not 12.5 percent, of full flow as would appear from the affinity law curves.

10.3   Number of Pumps

Most HVAC water systems have a broad range of operation due to the variation in heating, cooling, and ventilating loads. As was discussed in Chap. 8, it is imperative that the minimum load as well as the maximum load on a HVAC water system be computed before any additional evaluation be made of that system.

Most HVAC systems also have more than one pump per water system due to (1) the broad range of loads and (2) to provide some redundancy in case of pump failure. Thus, HVAC systems can range from one pump on small systems with a specific flow range and little need for standby pump capability up to a number of pumps operating in parallel to accommodate load range and reliability.

It is obvious that the water system designer must make decisions early in the system design to ensure that the desired levels of efficiency and reliability are achieved for a water system. Let us review the two factors that affect the number of pumps operating in parallel.

"Fear of pump failure" has affected the determination of the number of pumps when it should not be considered. Pumps have a reliability comparable with that of other HVAC equipment. Seldom are two fans installed in parallel because of fear of failure. The consequences of system failure should be determined to develop the redundancy needed on every water system. Hospitals and institutions and critical facilities such as research laboratories and computer centers obviously should have standby pump capability to ensure that there is continuous service.

The question on critical installations is how much standby capacity should be provided? The two most frequent selections are two 100 percent capacity or three 50 percent capacity pumps. The three 50 percent capacity pumps are usually the best selection because they offer three chances to total failure and the pumps and motors are smaller. For example, assume a chilled water system for a hospital has a maximum requirement of 2500 gal/min at 100 ft of head and a system head curve as shown in Fig. 10.5a. This curve shows a constant friction head of 20 ft that is maintained across all the cooling coils and control valves. Two pumps at 100 percent capacity would require two 75-hp motors, while three pumps at 50 percent capacity each would require three 40-hp motors. Figure 10.5b provides the pump head-capacity curve for the pumps of the two-pump system, while Fig. 10.5c does so for the three-pump system. The total horsepower for the two pumps would be 150 hp compared with 120 hp for the three pumps. If space is available for the three pumps, it would appear that the three pumps would be more acceptable for this application; a final decision should be made after a wire-to-water evaluation is made. This will be discussed later in this chapter.

Noncritical installations such as some office and commercial buildings do not require as much redundancy or standby as the buildings mentioned above. For example, if a commercial building had a system head curve as shown above for a hospital (Fig. 10.5a), only two 50 percent pumps could be installed, since one pump would have a maximum capacity of 1750 gal/min, or 70 percent of the total system capacity. It would be a decision for the water system designer or the owner of the building as to whether two or three pumps would be installed. With the reliability and reduced wear of variable-speed pumps, the two-pump installation would offer a desirable arrangement for this office building.

10.4 Mixing Constant- and Variable-Speed Pumps

It should be understood that pumps in parallel *must always operate at the same speed!* There may be some unusual, sophisticated cases where parallel pumps are operated at different speeds, but only experienced pump designers should make evaluations for such a proposed operation. Also, it is better on most HVAC operations to use pumps of the same size when operating them in parallel. This ensures that the operating pumps are producing the same flow, and this simplifies evaluation of the efficiency of operation of the pumps. Small jockey pumps that operate the system on low loads are not justified on most systems. Variable-speed pumps should be controlled so that pumps operating in parallel never have over a 1 percent difference in actual operating speeds.

Mixing of constant- and variable-speed pumps is encountered in the HVAC field, *often with disastrous results.* Following is a discussion of the problems that can occur when a constant-speed pump is operated in parallel with a variable-speed pump (use Fig. 10.6a to follow this discussion):

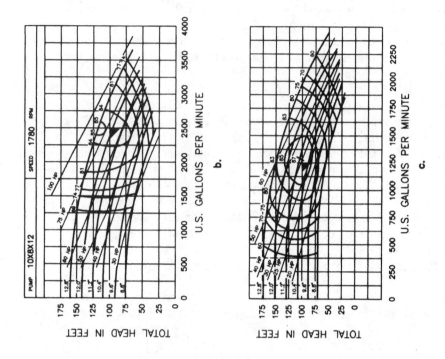

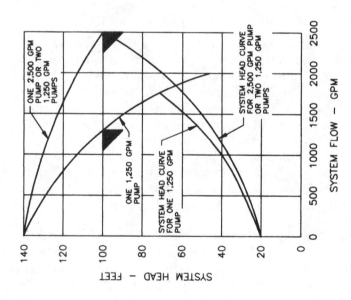

a. System head curves for a hospital requiring a 2,500 GPM chilled water system.

Figure 10.5 System and pump curves for a hospital.

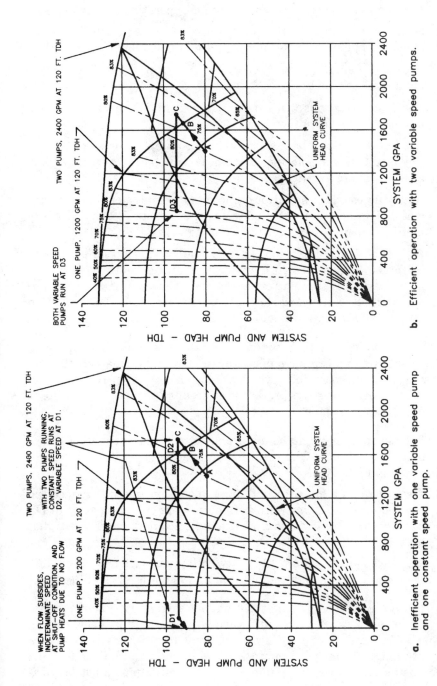

Figure 10.6 Operating variable- and constant speed pumps together.

a. Inefficient operation with one variable speed pump and one constant speed pump.

b. Efficient operation with two variable speed pumps.

215

1. Assume that one variable-speed pump is operating at point A.

2. Next, assume that the flow in the system increases until the variable-speed pump operates at full speed at point B.

3. Let the system load now increase to point C, where one pump cannot handle the load.

4. The system controls will turn on the next pump in an attempt to maintain the system flow.

5. Let the next pump be a constant-speed pump of the same size as the operating variable-speed pump. It will operate at point D_2 because it cannot vary its speed.

6. Since the constant-speed pump is handling most of the load, the variable-speed pump is forced to operate at point D_1. This is a very inefficient point with the pump running with a high radial thrust.

7. If the system load subsides slightly, the variable-speed pump will be forced to run at zero flow, with resulting heating inside the pump casing. If the controls are of the wrong configuration or are set improperly, the variable-speed pump may run at this condition until it is destroyed by heat.

This demonstrates that mixing a variable-speed pump with a constant-speed pump can have disastrous results. This unacceptable operation can be contrasted with the proper operation of two variable-speed pumps. Using the preceding procedure, assume that steps 1 through 4 are the same. Then

5. Let the standby pump be a variable-speed pump instead of a constant-speed pump. The resulting satisfactory operation is demonstrated in Fig. 10.6b.

6. The standby pump will ramp up, and the controls will reduce the speed of the lead pump until both pumps are operating at point D_3. Both pumps are operating at a high efficiency and a low radial thrust.

7. If the system load subsides slightly, both pumps will slow down, and the controls can stop the standby pump when it is efficient to do so.

It is not recommended that a pumping system with four or more pumps consist of constant- and variable-speed pumps operating in parallel. Usually, more energy is consumed with a mixture of constant- and variable-speed pumps. If it is imperative that some of the pumps be constant-speed pumps, the two lead pumps must be variable-speed pumps. Since one of these pumps may be out of service, it

TABLE 10.1   Number of Variable- and Constant-Speed Pumps in Parallel
Operation

Number of pumps	Variable-speed pumps	Constant-speed pumps
1	1	None
2	2	None
3	3	None
4	3	1
5	3	2
6	3	3

is recommended that at least three pumps be variable-speed pumps. Systems with four or more pumps can consist of constant-speed pumps, as demonstrated in Table 10.1, but this forces all the normal operation on the variable-speed pumps. Also, if the load does increase often to where the fourth pump is required, higher operating efficiency is achieved by making it a variable-speed pump.

A careful evaluation of the constant-speed pumps must be made when they are used, including calculating where they will operate on their head-capacity curves, to ensure that they do not operate at an inefficient or high radial thrust point.

10.5 Efficiency of a Pumping System

The efficiency of a pumping system is more than just the efficiency of the individual pumps. There are three different energy losses in a constant-speed electric motor–driven pumping system:

1. Loss in the motor

2. Loss in the pump

3. Friction loss in the pump and header fittings

The variable-speed pumping system has a fourth loss, that of the variable-speed drive.

There may be some concern as to why the pump fitting and header losses are included in the efficiency evaluation for a pumping system. These losses must be computed separately because they are part of the pumping system, not part of the water distribution system. Also, on multiple-pump systems, they must be separated from the water system loss because they do not vary with the total flow on the system. Rather, they vary with the flow through the pump to which they are connected. The flow rate through these fittings affects appreciably the overall efficiency of such a pumping system. The friction loss for these fittings varies closely with the square of the flow through them. The points at which pumps are programmed on and off on a multiple-pump system have a definite impact on the efficiency of the overall pumping system; these points are affected by the variation in friction in the pump fittings and valves.

There can be 6 to 10 specific losses in the fittings and valves on a pumping system:

1. Entrance loss from suction header

2. Suction shutoff valve

3. Suction strainer

4. Suction piping

5. Pump suction reducer

6. Pump discharge increaser

7. Check valve

8. Discharge shutoff valve

9. Discharge piping

10. Exit loss into discharge header

Wire-to-water efficiency is an old term that has been used for many years in the pumping industry to determine the overall efficiency of a pump and motor combination. This efficiency was used first for constant-speed pumps of considerable size, where large motors were involved. When these pumps were equipped with variable-speed drives, the wire-to-water efficiency represented the overall efficiency of drives, motors, and pumps. Wire-to-water efficiency has now been applied to smaller pumping systems because the computer capability is now available to accomplish the tedious calculations.

The wire-to-water efficiency of a proposed pumping system can be estimated using the following procedure:

1. A computer program can be developed to calculate the overall wire-to-water efficiency of a pumping system from minimum to maximum system flow. This program utilizes a standard curve-tracing technique through the calculation of binomial equations for the following curves:
   a. Pump head-capacity curve
   b. Pump efficiency curve
   c. Motor efficiency or wire-to-shaft efficiency curve for the variable-speed drive and motor
   d. System head curve
   e. Pump fitting loss curve

2. These equations are then used to compute system performance using the pump energy equations of Chap. 6; values computed are

a. System flow and head
b. Water horsepower
c. Pump gallons per minute
d. Pump fitting loss
e. Pump brake horsepower
f. Pump revolutions per minute
g. Pump efficiency
h. System brake horsepower
i. Wire-to-shaft efficiency (variable-speed drive and motor efficiencies combined)
j. Input kilowatts
k. Wire-to-water efficiency

For variable-speed pumping systems using variable-frequency drives, the wire-to-shaft efficiency for the motor and variable-speed drive listed above is a single efficiency developed by the drive manufacturer from calculations and tests of various motor and drive combinations. *It cannot be just a multiplication of the motor and drive efficiencies.* All variable-frequency drives notch or disturb the sine wave, which affects the efficiency of the motor.

Typical wire-to-shaft efficiencies for high-efficiency motors and variable-speed drives from 10 through 200 hp are shown in Table 10.2. Actual wire-to-shaft efficiencies for a specific application should be secured from the variable-speed drive manufacturer after receipt of information on the proposed pump motor.

10.5.1 Wire-to-water efficiency for two pumps

An example of this wire-to-water efficiency program is given in Table 10.3. This table is for the two-pump system for the hospital project discussed earlier, and the pump curve is shown in Fig. 10.5b. Only one pump runs at a time.

TABLE 10.2 Typical Wire-to-Water Efficiencies for Variable-Speed Drives with High-Efficiency Motors

Percent full load	Motor horsepower					
	15	50	75	100	150	200
0	0	0	0	0	0	0
40	50.2	63.0	61.5	59.2	67.2	67.0
60	73.2	81.3	80.5	79.6	84.4	84.7
80	84.4	88.3	88.3	88.6	90.9	90.6
90	86.5	89.8	90.9	91.0	91.8	91.8
100	87.7	90.9	91.4	92.2	92.9	93.1

TABLE 10.3   Wire-to-Water Efficiency Calculations for Two 100 Percent Capacity Pumps

System gal/min	System head, ft	Water hp	Fitting loss, ft	Pump head, ft	Pump rev/min	Pump efficiency	Input kW	W/W efficiency
500	23.6	3.0	0.4	24.0	723	67.0	4.8	45.9
625	25.4	4.0	0.6	26.0	771	73.1	5.8	51.2
750	27.5	5.2	0.9	28.4	827	77.0	7.1	55.2
1000	32.9	8.0	1.6	34.5	947	81.3	10.2	60.6
1250	39.4	12.4	2.5	41.9	1071	83.3	14.5	63.9
1500	47.2	17.9	3.6	50.8	1204	84.2	20.2	65.9
1625	51.5	21.2	4.2	55.8	1273	84.4	23.7	66.7
1750	56.2	24.8	4.9	61.1	1343	84.5	27.5	67.2
1875	61.1	28.9	5.6	66.7	1414	84.5	31.9	67.7
2000	66.3	33.5	6.4	72.7	1486	84.6	36.8	68.0
2125	71.8	38.5	7.2	79.0	1559	84.5	42.2	68.1
2250	77.6	44.1	8.1	85.7	1632	84.5	48.3	68.1
2375	83.7	50.2	9.0	92.7	1706	84.5	55.0	68.1
2500	90.0	56.8	10.0	100.0	1780	84.4	62.6	67.7

NOTE: Pump and wire-to-water efficiencies are in percent.

10.5.2   Wire-to-water efficiency for three pumps

The wire-to-water efficiency for the three-pump system for the hospital discussed earlier is listed in Table 10.4, and the pump curve is described in Fig. 10.5c. One or two pumps run depending on the wire-to-water efficiency.

This table demonstrates that, in this case, the three-pump system provides a higher wire-to-water efficiency throughout the load range than does the two-pump system of 2500-gal/min pumps. The actual pump selection has some bearing on these results, so this should not imply that in every case it would be more efficient to use three instead of two pumps. Such a wire-to-water efficiency program should be run on every application where substantial motor horsepowers are involved. A significant fact that can be secured from this particular calculation is that two pumps should be operated from 1000 gal/min upward even though one pump could hand the water system up to 1685 gal/min.

10.5.3   Wire-to-water efficiency for constant-speed pumps

Wire-to-water efficiency can be computed for both constant- and variable-speed pumping systems. One of the unique advantages of this program is the ability to compute the wire-to-water efficiency for a

TABLE 10.4   Wire-to Water Efficiencies for Three 50 Percent System Capacity Pumps

System gal/min	System head, ft	Water hp	Fitting loss, ft	Pump head, ft	Pump rev/min	Pump efficiency	Input kW	W/W efficiency
One pump running
500	23.6	3.0	1.6	25.2	845	84.6	3.3	66.3*
625	25.0	4.0	2.5	27.9	927	86.8	4.4	67.7*
750	27.0	5.2	3.6	31.1	1012	87.3	5.8	67.2*
875	30.0	6.6	4.9	34.9	1104	86.8	7.5	65.7*
1000	32.9	8.3	6.4	39.3	1201	85.9	9.7	63.7
1125	36.0	10.2	8.1	44.1	1302	85.0	12.4	61.7
1250	39.4	12.4	10.0	49.4	1406	84.0	15.5	59.9
1375	43.2	15.0	12.1	55.3	1511	83.0	19.2	58.2
1500	47.2	17.9	14.4	61.6	1619	82.1	23.6	56.6
1625	51.5	21.2	16.9	68.4	1728	81.3	28.7	55.1
1685	53.7	22.9	18.2	71.9	1780	81.0	31.4	54.3
Two pumps running
875	30.0	6.6	1.2	31.3	924	78.2	7.7	64.4
1000	32.9	8.3	1.6	34.5	978	80.7	9.3	66.7*
1125	36.0	10.2	2.0	38.0	1035	82.5	11.2	68.2*
1375	43.2	15.0	3.0	46.2	1158	84.7	16.0	69.9*
1500	47.2	17.9	3.6	50.8	1222	85.4	19.0	70.4*
1625	51.5	21.2	4.2	55.8	1289	85.9	22.3	70.6*
1750	56.2	24.8	4.9	61.1	1356	86.3	26.2	70.8*
1875	61.1	28.9	5.6	66.7	1425	86.5	30.5	70.9*
2000	66.3	33.5	6.4	72.7	1494	86.7	35.2	70.9*
2125	71.8	38.5	7.2	79.0	1565	86.0	40.5	70.9*
2250	77.6	44.1	8.1	85.7	1636	87.0	46.4	70.9*
2375	83.7	50.2	9.0	92.7	1707	87.1	52.9	70.7*
2500	90.0	56.8	10.0	100.0	1779	87.1	60.3	70.3*

NOTE: Pump and wire-to-water efficiencies are in percent.
*Denotes most efficient operation and correct number of pumps to operate at each system gallons per minute.

pumping system with constant- and variable-speed pumping. Table 10.5 describes the wire-to-water efficiency of the two-pump constant-speed system. The pump curve of Fig. 10.5*b* applies in this case, and only one pump runs at a time.

Figure 10.7 provides the data of Tables 10.3, 10.4, and 10.5 as curves plotted with system gallons per minute. This figure demonstrates the advantage of the three-pump variable-speed system energy-wise over the two-pump variable-speed system. It also shows the dramatic energy savings of either variable-speed system over the constant-speed system. Not only do the variable-speed systems save en-

TABLE 10.5   Wire-to-Water Efficiency Calculations for a Constant-Speed Pumping System

System gal/min	System head, ft	Water hp	Fitting loss, ft	Pump head, ft	Pump rev/min	Pump efficiency	Input kW	W/W efficiency
500	23.6	3.0	0.4	145.0	1780	34.0	44.8	5.0
625	25.4	4.0	0.6	144.0	1780	41.0	46.0	6.5
750	27.5	5.2	0.9	142.9	1780	47.5	47.2	8.3
875	30.0	6.6	1.2	141.5	1780	53.4	48.3	10.2
1000	32.9	8.3	1.6	139.9	1780	58.7	49.5	12.5
1125	36.0	10.2	2.0	138.1	1780	63.4	50.6	15.1
1250	39.4	12.4	2.5	136.1	1780	67.6	51.7	18.0
1375	43.2	15.0	3.0	133.8	1780	71.3	52.7	21.2
1500	47.2	17.9	3.6	131.2	1780	74.5	53.7	24.8
1625	51.5	21.2	4.2	128.3	1780	77.2	54.7	28.8
1750	56.2	24.8	4.9	125.1	1780	79.5	55.6	33.3
1875	61.1	28.9	5.6	121.6	1780	81.3	56.5	38.2
2000	66.3	33.5	6.4	117.8	1780	82.6	57.3	43.6
2125	71.8	38.5	7.0	113.8	1780	83:6	58.1	49.5
2250	77.6	44.1	8.1	109.5	1780	84.2	58.7	56.1
2375	83.7	50.2	9.0	104.9	1780	84.4	59.1	63.3
2500	90.0	56.8	10.0	100.1	1780	84.4	59.5	71.3

NOTE: Pump and wire-to-water efficiencies are in percent.

ergy, but they also reduce appreciably the overpressure on the water system and the radial thrust on the pumps themselves.

It should be noted that these tables are based on standard water at 50°F. If wire-to-water efficiencies are computed for liquids of other densities or viscosities, the head inserted in the tables should be corrected for the actual density and viscosity.

10.6   Wire-to-Water Efficiency Indication

Wire-to-water efficiency indication for an actual pumping system can be developed easily through the recording of the system flow, net pump head, and pumping system kilowatts. This instrumentation for wire-to-water efficiency is shown in Fig. 10.8a and includes a system flowmeter, a differential pressure transmitter across the pumping system headers, and a watt transmitter.

Wire-to-water efficiency is computed by dividing the energy applied to the water (water horsepower) by the power input to motor or variable-speed drive. Using the equations of Chap. 6 for pumping energy, the equation for wire-to-water efficiency WWE is

$$WWE\ (\%) = \frac{\text{Water horsepower (whp)} \cdot 0.746 \cdot 100}{\text{kW}}$$

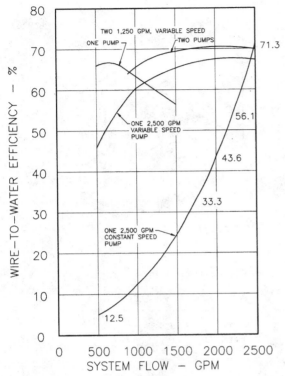

Figure 10.7 Wire-to-water efficiency curves for Tables
10.3, 10.4, and 10.5.

$$= \frac{Q \,(\text{gal/min}) \cdot H \,(\text{ft})}{53.08 \cdot \text{kW}} \qquad (10.2)$$

where Q = flow through the system in U.S. gallons per minute
  H = net pumping system head in ft measured by the differen-
    tial pressure transmitter connected across the headers of
    the pumping system
  kW = power input to the pumping system (The watt transmitter
    must be located ahead of any variable-frequency drives.)

Equation 10.2 demonstrates that the acquisition of the three values
of flow, head, and kilowatts can produce continuously the wire-to-
water efficiency of a pumping system. No allowance for changes in
specific gravity or viscosity need be made, since the kilowatt input
changes automatically with these conditions. Actual wire-to-water efficiencies from a test program on a four-pump system are listed in

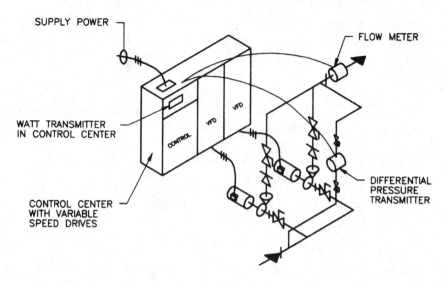

a. Wire—to—water efficiency instrumentation for pump station.

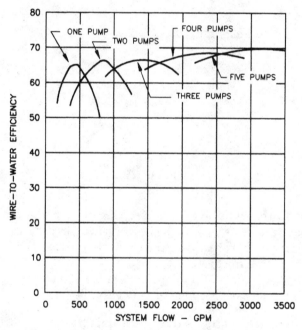

b. Typical wire—to—water efficiency curves.

Figure 10.8 Wire-to-water efficiency instrumentation and typical wire-to-water efficiency curves for a five-pump system.

TABLE 10.6   Wire-to-Water Efficiencies from an Actual Test Program

No. of pumps	System flow, gal/min	System head, ft	Pump rev/min	Input, kW	Wire-to-water efficiency, %
1	2475	28.3	1170	34.6	47.4
2	2475	28.3	852	30.7	53.4*
2	3960	42.7	1123	62.6	58.4
3	3960	42.7	977	62.4	58.6*
3	4950	55.4	1124	91.7	62.5
3	7425	97.0	1518	211	67.0
3	7425	97.0	1394	210	67.3*
4	9900	153.6	1759	404	71.7

*Transition points from one to two, two to three, and three to four pumps for peak wire-to-water efficiency.

Table 10.6. This test is typical of those used to search for peak wire-to-water efficiency for a four pump system.

Wire-to-water efficiency can provide a record of pumping performance that enables the operator to compare the present operation with that in the past when the system was new and properly adjusted. This procedure allows the operator to determine if a particular pump is operating inefficiently; the operator can cycle various pumps and record the differences in efficiency. The pump that is operating inefficiently can be taken out of service and repaired or cleaned. Wire-to-water efficiency indication is a must for large or medium-sized pumps in areas that have high electrical rates.

Wire-to-water efficiencies for a pumping system can be plotted against system flow and used for system evaluation as well as control, which will be described later in this chapter. Figure 10.8b describes the wire-to-water efficiency curves for the five-pump system that was described in Fig. 10.8. This type of information can be secured directly from a computer capable of curve plotting.

10.7   Pump Control

Since variable-speed pumps have entered the HVAC market, there are two aspects to pump control: (1) pump start-stop procedures and (2) pump speed control. These procedures must be developed for both constant- and variable-speed pumping systems.

Pump start-stop procedures consist of a number of control techniques:

1. With system activation or shutdown

2. By system demands such as flow or pressure

3. By system energy evaluation such as wire-to-water efficiency

4. By emergencies caused by pump failure

5. Duty sharing such as alternation

Pumping systems can be interfaced easily so that they can be started or stopped by the building management system. When a chilled or hot water system is actuated, any pumping system involved can be started automatically at that time.

10.7.1 Sequencing of pumps on system demands of flow or pressure

Multiple-pump systems require care in the addition and subtraction of pumps in order to achieve maximum pumping efficiency from minimum to maximum system flow. Items 2 and 3 above list the procedures used for accomplishing this.

Which procedure is used to sequence multiple pumps can have a great bearing on the efficiency of the pumping system. There are several methods of sequencing pumps, and these are

1. *Maximum flow sequencing.* This old method let one pump run until it could no longer handle the system requirements; the controls then started the next pump in the pumping sequence. This has proved to be wasteful in energy because the pump operates far to the right of the design point at a poor efficiency. Also, the pump then operates at a point of high radial or axial thrust with greater wear on bearings, sleeves, and mechanical seals. Maximum flow should be used as a backup control system to maintain system flow on failure of more efficient control systems, such as best efficiency or wire-to-water efficiency control. Arbitrarily selecting a percentage of speed of the operating pumps such as 80 percent to add and subtract pumps does not achieve peak efficiency; this is an attempt to control pumps without evaluating all the system characteristics that determine operating efficiency.

2. *Best efficiency control.* Recognizing the problems with maximum flow sequencing, this procedure was developed to eliminate the preceding inefficient and high-thrust operation with maximum flow sequencing. It has been called *best efficiency control* because the pumps are sequenced so that they operate as closely as possible to their best efficiency point. This is achieved by evaluating the system head for the proposed system and selecting pump sequencing so that the pumps do not "carry out" but operate near their best efficiency point. This is demonstrated in Fig. 10.9, a system with two pumps operating at design load. With maximum flow sequencing, one pump could operate until 167 percent capacity of one pump was reached; at

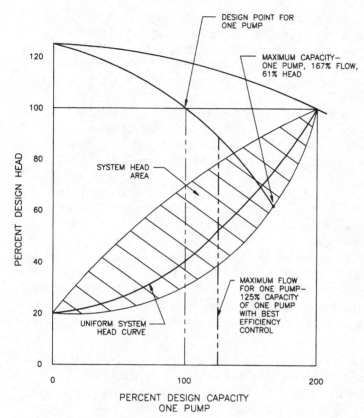

DESIGN POINT FOR
ONE PUMP

MAXIMUM CAPACITY–
ONE PUMP, 167% FLOW,
61% HEAD

SYSTEM HEAD
AREA

MAXIMUM FLOW
FOR ONE PUMP–
125% CAPACITY
OF ONE PUMP
WITH BEST
EFFICIENCY
CONTROL

UNIFORM SYSTEM
HEAD CURVE

PERCENT DESIGN HEAD

PERCENT DESIGN CAPACITY
ONE PUMP

Figure 10.9 Best efficiency control.

this point, the pump efficiency would be very low compared with that at design condition, and the pump would operate noisily. With best efficiency control, the second pump would be added at 125 percent capacity of one pump, and the two pumps would operate at a much higher efficiency if they are variable-speed pumps. Best efficiency control does require a system flowmeter if flow is used as the means for sequencing. Unfortunately, many systems do not have accurate flowmeters that can be used for pump sequencing.

If the pumps are variable speed on systems without flow meters, pump speed can be used to start or stop the second pump. The *wire-to-water efficiency calculation* is used to compute the speeds at which the second pump should be added or subtracted. These computed speeds can be checked under actual operation to ensure that the pumps are being sequenced properly, particularly if a watt transmitter is available to indicate the energy consumption by the pumps.

3. *Wire-to-water efficiency control.* Field experience with best effi-ciency control has proved that pumps still were not being sequenced efficiently under all system conditions. This led to further evaluation of pumping system operation and its refinement by wire-to-water effi-ciency control, which utilizes the procedures listed above for wire-to-water indication.

The wire-to-water efficiency determination can be used to select ap-proximately the points at which the pumps should be added or sub-tracted. The transition points can be selected from the wire-to-water efficiency tables and inserted into the pump-control program. The ac-tual pump operation can be checked by wire-to-water efficiency indi-cation, and the set points can be changed in the field until the pumps are operating on the most efficient program. How the transition points are secured is shown in Table 10.6, data from an actual test for wire-to-water efficiency.

The value of wire-to-water efficiency control is shown dramatically in Table 10.4. If one pump were allowed to operate as high as 1625 gal/min, the input kilowatts would be 28.7 and the wire-to-water effi-ciency 55.1 percent compared with two pumps operating at 1625 gal/min with an input kilowatts of 22.3 and a wire-to-water efficiency of 70.6 percent.

All this discussion on the efficient control of pumps demonstrates the great capability now available to HVAC water system designers through the use of digital electronics in the evaluation of a water sys-tem and the application of pumps.

10.7.2 Emergency backup on pump failure

Equipment failure is always troublesome, and it should be met in a way that will cause the least amount of trouble for the water system. Pump failure must be handled so that the system is not momentarily out of water. The control system must interrogate operating pumps so that a standby pump is started when pump failure is sensed, not when the water system reacts to the failure of a pump. This may seem to be a minor detail to be included in a pump handbook, but it can be a terrible operating problem if the standby pump control pro-cedure is not executed properly. *Reiterating, pump failure should be sensed immediately, and the standby pump started then, not waiting for system pressures to indicate the failure.*

10.7.3 Alternation of operating pumps

Over the years of automatic control, almost every conceivable method of alternating the lead or operating pumps has been used in the pump industry. There are systems called "first on, first off" or "last on, first off." There are duty cycle or run time cycles. All these alternation methods were designed to provide equal wear on all the pumps.

Two facts have resulted from these procedures. First, equal wear is not necessarily the best procedure. As in several cases with large pumps, all the pumps wore out during the same month and year, causing an emergency repair program because all the pumps failed together. Second, automatic alternation in itself allows the operator to neglect the pumps.

A third factor has arisen, and that is the reduction in wear of HVAC pumps. The use of variable-speed pumps and factory-assembled pumping systems where pumps are programmed correctly has reduced the incidence of pump failure. This has resulted in the elimination of most of the preceding alternation cycles. The simplest and cheapest procedure is manual operation by the operator. Equipping the pump control center with an elapsed time meter for each pump motor enables the operator to maintain around 2000 to 4000 operating hours between pumps so that they do not all fail at the same time. More on this subject is included in Chap. 26.

10.7.4 Pump speed control

All the preceding discussion has been on programming the pumps efficiently. The second part of operating variable-speed pumps is control of the pump speed itself. Development of the variable-frequency drive has greatly changed the design, selection, and operation of pumps. The resulting variable-speed pump offers many benefits such as reduced power consumption and lower radial thrusts on the pump shafts. The variable-speed drives themselves have been reviewed in Chap. 7. Contemporary speed control of a pumping system has evolved over the past 25 years.

The basic control arrangement for these variable-speed pumps consists of a water system sensor, a pump controller, and a variable-speed drive (Fig. 10.10a). In most cases, the water system sensor can be for (1) temperature, (2) differential temperature, (3) pressure, (4) differential pressure, (5) water level, (6) flow, and (7) kilowatts. Special systems may use other physical or calculated values for varying the speed of the pumps.

The pump controller can be of many different configurations depending on the response time required to adequately control the pump speed without continual speed changes or hunting. With initial use of variable-speed drives on HVAC systems, it was assumed that since the loads did not change rapidly on these water systems, there was no need for rapid speed regulation of the pumps. It was learned

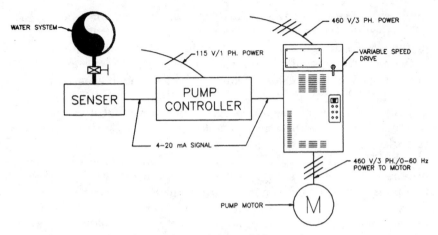

a. Basic speed control for variable speed pump.

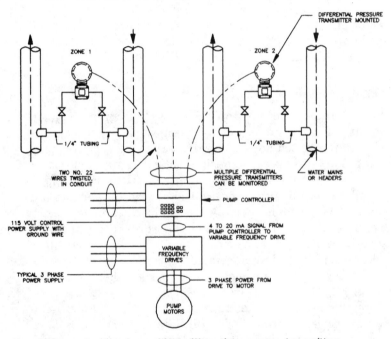

b. Piping and wiring for multiple differential pressure transmitters.

Figure 10.10   Basic speed control for variable-speed pumps.

during actual operation of these pumps on HVAC systems that rapid speed regulation was required, particularly on systems where pressure or differential pressure sensors were used to regulate the speed of the pumps. Although the load on these systems was not changing rapidly, pressure waves within the water systems created changes at the sensors that caused speed changes. Also, with digital control, after computation of the PID error, an output signal is given the variable-speed drive to increase or decrease the speed of the pump. PID error for proportional-integral-derivative control is the difference between the setpoint and the process value. This increase or decrease continues until the next signal change. Attempts to reduce the sensitivity of the pump control only caused increased variation in the actual pressure or differential pressure at the point where the sensor was installed. Currently, 500 ms has proved to be an adequate response time for most HVAC variable-speed pumping systems. *If HVAC variable-speed pumps continuously change their speed, the response time is inadequate, or another control problem exists!* Properly controlled variable-speed pumps in this industry should not change their speed so that the changes are either visible or audible.

Most pump controllers are now based on digital electronics and are of the proportional-integral-derivative type, which sense a signal, compute the error, and output the speed signal to the variable-speed drive. Since the speed of the pump is held at a discrete point by such a controller until it is updated, there can be continuous speed fluctuation when rapid control response is not provided. Standard commercial PID computer chips are available with a rate of response as fast as 10 to 500 ms. Rate of response should not be a problem on variable-speed HVAC pumps due to the digital technology that is now available for the control of such pumps. Any variable-speed pump that has continuous speed change is not being controlled properly. It may have too slow a speed response from the transmitters that are controlling the pump speed.

Although standard commercial computer chips are designed for proportional-integral-derivative control, most HVAC applications do not require any derivative value inserted in the actual control algorithm. Also, proportional-integral controllers of the analog type are still used in this industry.

10.7.5 Sensors for pump speed control

The preceding description of basic variable-speed control listed the types of sensors used for pump operation. Some HVAC systems require more than one sensor to maintain the proper speed of the pumps. This is due to shifting heating or cooling loads on the water system. The standard signal-selection technique is utilized to accommodate more than one transmitter (Fig. 10.10b). The controller selects the transmitter at which the signal has deviated the farthest from its set point. Following is information on the selection and location of sensors or transmitters.

The location of the sensors is important to ensure that the pumps can be operated at the lowest possible speed from minimum to maximum system load. See Chap. 10 for proper sensor location for specific types of water systems.

When pressure or differential pressure is used to control pumps, these transmitters must be located so that the distribution friction pressure loss of the system is eliminated from the signal. For example, assume that a distribution system for a hot or chilled water system consists of differential pressure with direct return piping and has a 20-ft loss across the largest coil, its control valve, and piping. Also assume that the distribution friction loss out to and back from the farthest coil is 80 ft. The total pump head is therefore 100 ft. If the pressure transmitter is located at the pump, it must be set at 100 ft to accommodate the system when it is operating at 100 percent load. If the transmitter is located at the far end of the loop, the distribution loss of 80 ft is eliminated from the control signal. The transmitter can be set at just 20 ft, the loss across the largest coil and its piping. The pump speed can be varied to match the load on the system, thus saving appreciable energy by operating the pump at lower speeds. The pump will operate with less radial thrust and have less repair.

The water system just described is of the direct return type; if it were of the reverse return type, differential pressure transmitters would be required at both ends of the distribution loop. Both the direct and reverse return systems and their transmitter locations are shown in Fig. 10.11a. With variable loads in a reverse return loop, if all the loads are located at one end of the loop, the differential pressure transmitter located at the other end will not provide adequate control. One transmitter can be located at the center of the loop, but one-half the distribution loss must be set into the control signal, thus forcing the pump to run at a higher and less efficient speed. This may be acceptable on small systems without much distribution friction, but it would be totally unacceptable on the preceding system with 80 ft of distribution friction. In this case, the differential pressure transmitter would be required to be set at 60 ft, causing the pumps to run at a much higher speed than when set at 20 ft with the differential pressure transmitters at each end of the loop.

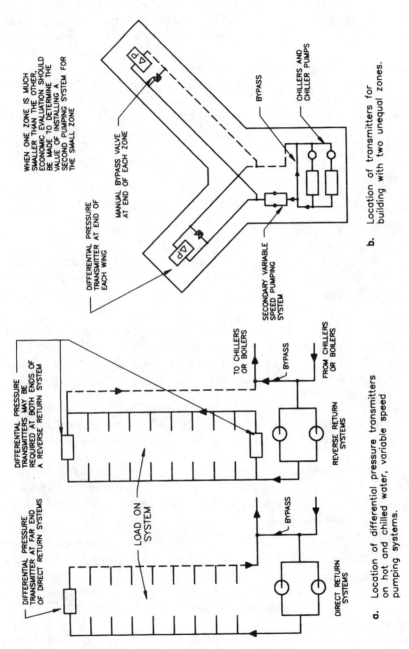

WHEN ONE ZONE IS MUCH SMALLER THAN THE OTHER, ECONOMIC EVALUATION SHOULD BE MADE TO DETERMINE THE VALUE OF INSTALLING A SECOND PUMPING SYSTEM FOR THE SMALL ZONE

DIFFERENTIAL PRESSURE TRANSMITTER AT END OF EACH WING

MANUAL BYPASS VALVE AT END OF EACH ZONE

BYPASS

CHILLERS AND CHILLER PUMPS

SECONDARY VARIABLE SPEED PUMPING SYSTEM

b. Location of transmitters for building with two unequal zones.

DIFFERENTIAL PRESSURE TRANSMITTERS MAY BE REQUIRED AT BOTH ENDS OF A REVERSE RETURN SYSTEM

DIFFERENTIAL PRESSURE TRANSMITTER AT FAR END OF DIRECT RETURN SYSTEMS

LOAD ON SYSTEM

TO CHILLERS OR BOILERS

BYPASS

FROM CHILLERS OR BOILERS

REVERSE RETURN SYSTEMS

BYPASS

DIRECT RETURN SYSTEMS

a. Location of differential pressure transmitters on hot and chilled water, variable speed pumping systems.

Figure 10.11 Connection of differential pressure transmitters.

233

Along with the proper location of the differential pressure transmitters, the transmitters must be interrogated properly by the control system. As indicated earlier, the response time of the controller must be fast enough to prevent pump hunting, but it must hold the differential pressure to an error signal of ± 1 foot or less. Any variation greater than this for the controlling transmitter is unacceptable; if a greater variation occurs, it indicates that the pumps are hunting, and energy is being lost.

Following are several special applications of differential pressure transmitters on hot or chilled water systems. All these applications with 2 to 10 transmitters will utilize a low signal selection procedure that will select the controller that has deviated the farthest from it set point.

Figure 10.11b: A two-wing building usually requires a differential pressure transmitter at the end of each wing.

Figure 10.12a: When the chillers, boilers, and pumps are located on top of a high-rise building, the differential pressure transmitters should be located at the bottom of the building to eliminate the distribution friction.

Figure 10.12b: This building is typical of reverse return systems requiring a differential pressure transmitter at each end of the loop. One differential pressure transmitter can be located at the center point of the reverse return loop if the distribution friction of the supply and return loops is less than 20 ft.

Buildings that have a great amount of glass and a pronounced sun load may require a differential pressure transmitter on both the east and west sides of the building, regardless of the configuration of the piping.

A number of control procedures are being developed to control variable-speed pumps; some have been successful, and some have not. Following are control techniques for loop-type variable-speed pumping systems that have *not* been successful:

1. *Locating the differential pressure transmitter across the pumps instead of at the end of the loops.* There is little reason for using variable-speed pumps with this control procedure. The design pump head must be the set point for the differential pressure controller; therefore, there is very little reduction in pump speed, whatever the load on the building.

2. *Temperature control by valve stem position.* This method of controlling chilled and hot water systems can be expensive and difficult to

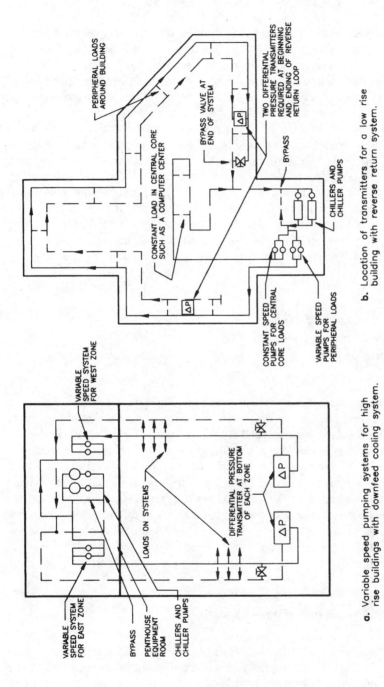

Figure 10.12 Particular applications of differential pressure transmitters to HVAC hot and chilled water systems.

a. Variable speed pumping systems for high rise buildings with downfeed cooling system.

b. Location of transmitters for a low rise building with reverse return system.

VARIABLE SPEED SYSTEM FOR WEST ZONE

VARIABLE SPEED SYSTEM FOR EAST ZONE

BYPASS

PENTHOUSE EQUIPMENT ROOM

CHILLERS AND CHILLER PUMPS

LOADS ON SYSTEMS

DIFFERENTIAL PRESSURE TRANSMITTER AT BOTTOM OF EACH ZONE

ΔP

ΔP

PERIPHERAL LOADS AROUND BUILDING

CONSTANT LOAD IN CENTRAL CORE SUCH AS A COMPUTER CENTER

BYPASS VALVE AT END OF SYSTEM

ΔP

TWO DIFFERENTIAL PRESSURE TRANSMITTERS REQUIRED AT BEGINNING AND ENDING OF REVERSE RETURN LOOP

BYPASS

CHILLERS AND CHILLER PUMPS

CONSTANT SPEED PUMPS FOR CENTRAL CORE LOADS

VARIABLE SPEED PUMPS FOR PERIPHERAL LOADS

ΔP

235

calibrate. The position of a particular valve stem has no relationship to the flow of water passing through the valve. Figure 10.13*a* describes why this is so. Valve *A* close to the pumps will have a greater flow at a specific valve position than valve *B* at the end of the loop. It is difficult to correlate an analog representing valve position to pump speed without excessive hunting. Pumps and their controls are designed to respond to pressure changes; this is why the differential pressure control method has been so successful over the past 25 years.

A combination of differential pressure control and valve position reset is possible, as shown in Fig. 10.13*b*. The differential pressure transmitter is still the principal means of controlling the pump speed, but the differential pressure set point is established by the energy management system, which changes the set point as the valves recede from full-open position. This is uneconomical unless the valve position of all the control valves is being sent to the building management system as part of the standard control system.

A typical differential pressure–valve position schedule would be

Any valve fully open:	25 ft
All valves 90 percent open:	22 ft
All valves 80 percent open:	19 ft
All valves 70 percent open:	16 ft
All valves 60 percent open:	14 ft
All valves 50 percent open:	12 ft

3. *Return water temperature control.*   Fortunately, this method of control is not seen very often. It absolutely will not work! Return water temperature has no relationship to system load on variable-volume systems; it should be maintained as closely as possible to a constant value. For example, a chilled water system operating with 44°F supply temperature and 56°F return temperature should hold these temperatures whether the load is heavy or light on the system. Therefore, return temperature cannot possibly be used as a control parameter for variable-speed pumping. Often there has been confusion of variable return water temperature that results from the use of three-way valves; these systems were, of course, constant-volume systems and not used with variable-speed pumps for most applications.

10.7.6   Communication from remote transmitters

The method of communication with remote transmitters is particularly important to ensure that a noninterrupted signal is received at the pump controller at the rate of response specified by the pump control

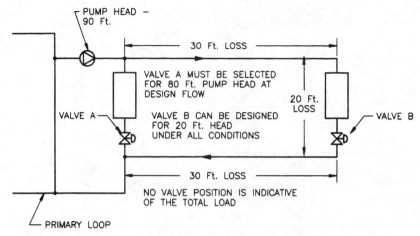

a. Problem of using valve position for pump speed control.

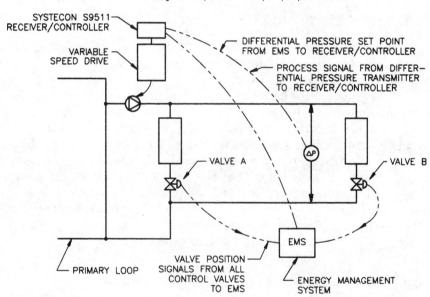

b. Differential pressure reset with control valve position.

Figure 10.13 Use of control-valve position for differential pressure reset.

manufacturer. The route that the signal takes is of importance to the water system designer.

Most signals from remote transmitters are direct current and 4 to 20 mA in strength. These signals must be protected from electromagnetic and radiofrequency interference. Therefore, the interconnecting cabling must be shielded and grounded at the pump controller end.

**TABLE 10.7   Maximum Linear
Distances for Transmitters from Pump
Controllers**

Wire size, AWG	Linear distance, ft
16	20,000
18	11,000
20	8,000
22	5,000
24	3,500
26	2,000

Table 10.7 provides typical distances between the transmitter and the pump controller that can be used with various sizes of wire.

These cables can be installed in conduit or cable trays that carry other instrument, data, or telephone signals. They must be shielded for electromagnetic and radiofrequency interference, and the shield must be grounded at the end of the cable near the pumping system controller.

On some large installations, the cable installation for pump control can become expensive; an alternate procedure is to install modems at the remote differential pressure transmitter and at the pumping system. This procedure allows the use of ordinary telephone cables for transmission of the control signals.

Care should be taken to avoid routing the control signals through building management systems that reduce the rate of response from the transmitter; the signal and data load on these central management systems may be such that the transmitter signal for pump speed control cannot be processed faster than 2 to 4 seconds. This is too slow for accurate pump speed control. This must not occur routinely or when the central management system is responding to emergency conditions such as fire or building power failure.

10.8   Effects of Water Systems on Pump Performance

The pump is the heart of every water system; when it fails, the system fails. Unfortunately, many times pump failure is due to the water system's effects on the pump's performance and physical condition. The problems with cavitation and entrained air on pump performance and damage have already been discussed in Chap. 6.

Other deleterious effects on pumps are control systems that force pumps to run in the high-thrust areas, namely, at very low or high flow rates. The pump wears rapidly, and it is not the pump's fault.

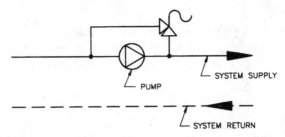

a. Unacceptable installation.

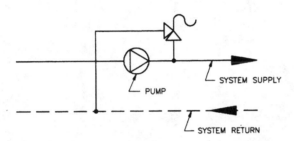

b. Acceptable installation.
(Not recommended excepting as last resort if relief valve must be used).

Figure 10.14   Pump relief valve connections.

One of the most disastrous practices in pump application is the installation of a relief valve on a pump discharge that returns the water to the pump suction, as shown in Fig. 10.14a. It should be remembered that the thermal equivalent of a brake horsepower is 2544 Btu/h. All the energy destroyed by the relief valve is returned to the pump suction. If the system is operating at low loads, where the flow through the pump is low, it is very apparent that heat will build until the pump can become very hot. Also, hot water will surge through the system when the flow does increase.

Relief valves should be avoided wherever possible on chilled water systems because every brake horsepower destroyed by the relief valve adds one-fifth of a ton load on the chillers. If there is no other way to control the water through the system than by a relief valve, it should be connected as shown in Fig. 10.14b. The heat is returned to the chiller, and the pump continues to receive and deliver chilled water.

Heating and cooling coils that are dirty on the air or water side or are operating with laminar flow increase the system flow beyond the design flow rates. Often pumps are blamed for their inability to provide these greater flows; good system control is the answer for many system problems that cause unnecessary pump wear and maintenance.

HVAC pumps should be long-lasting with very little repair. Proper installation and operation will provide many years of operation with a minimum of service and repair.

10.9 Bibliography

James B. Rishel, *The Water Management Manual,* SYSTECON, Inc., West Chester, Ohio, 1992.

James B. Rishel, *Variable Volume Pumping Manual,* Systecon Inc., West Chester, Ohio, 1982.

Pumps for Open HVAC Cooling Systems

11

Cooling Tower Pumps

11.1 Introduction

Virtually all air-conditioning systems in the HVAC industry or process cooling systems having a net cooling capacity in excess of approximately 4,000,000 Btu/hr (333 ton) transmit the heat load to the atmosphere evaporatively through a cooling tower. Evaporating a pound of water removes around 1000 Btu from the remaining water. This equates to about 8 percent of the water flow being evaporated to cool the balance of the water over a temperature difference of 10°F at design conditions. This is in contrast to the "once-through" water cooling systems (100 percent makeup), in which only 1 Btu is liberated when 1 lb of water is cooled 1°F.

Cooling towers were developed to take the place of lakes and spray ponds, which required considerable space to accomplish the same amount of evaporative heat transfer. The cooling tower industry prospered as system owners and operators were told that water was a valuable commodity that should not be wasted. Once-through water cooling systems have been outlawed in many parts of the world.

Since the evaporative process depends on wet-bulb temperature, a water-cooled refrigerant condenser is significantly more efficient than an air-cooled condenser. The air-cooled condenser must be designed for the highest dry-bulb temperature that will be encountered on each specific application.

Cooling towers have been the subject of much research to achieve the most efficient operation and the maximum amount of capacity per cubic foot of tower volume. Cooling towers are normally rated in tons of cooling, 12,000 Btu/ton. Alternate ratings are in gallons per minute with a specific temperature difference between inlet and outlet temperatures.

A detailed description of cooling tower performance is beyond the scope of this book. This chapter will be devoted to the piping and pumping of cooling towers and chiller condensers and how their performance affects cooling tower pumps and vice versa. The next chapter will deal with process cooling using cooling towers and other equipment.

For a detailed description of the thermodynamics of cooling towers and a comprehensive discussion of the various types of cooling towers, the reader is encouraged to study Cooling Tower (Chap. 37 of the *Systems and Equipment* volume of the *ASHRAE Handbook for Heating, Ventilating and Air-Conditioning, 1992.*

Obviously, there are a number of different arrangements of cooling tower fill and air patterns to achieve the most efficient and economically feasible cooling tower for the many different sizes and types of HVAC and industrial installations. Table 11.1 is a general graphic representation of the type of towers currently being used in these industries. Basically, a cooling tower is simply a container in which water and air both flow together. Most often the airflow is either forced (pushed) or drawn (pulled) through falling water. Usually, the falling water cascades over various types of fill or media that are specifically designed to slow the water fall rate and thus increase the time that the water is in contact with the air; this increases the evaporative process. There are special types of cooling towers that use neither media (fill) or air-moving systems (fans). The largest cooling towers without fans are the hyperbolic towers most often associated with electrical generating plants. The most basic categorization of towers is by the method used for air movement through the tower, whether natural or mechanical draft.

TABLE 11.1   Classification of Cooling Towers

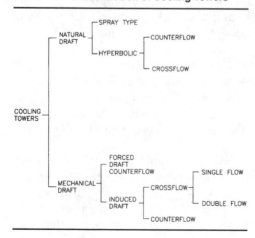

Cooling towers also can be characterized as to whether they are factory assembled or field erected and can be divided into various shapes such as rectilinear, round, or octagonal. Some cooling towers are described by their basic material of construction, such as ceramic towers.

11.2 Water Flow Conditions for Cooling Towers

A cooling tower has a very specific range of flow that provides acceptable performance. This flow range can be only 80 to 100 percent of design flow on most cooling towers. Some towers can have their flow ranges extended by the manufacturer making modifications in the cooling tower's water distribution system. If the actual flow rate is less or greater than the specified range, operating problems will develop; the water flow through the tower can become nonuniform with uneven airflow. It is the responsibility of the designer to ensure that the flow rate of the water system meets the needs of the cooling tower.

Matching the pump to the cooling tower may compromise other design requirements such as the flow rates required by chiller condensers. It must be recognized that this dichotomy exists, and the water system design must satisfy both the flow rate of the using equipment and that of the cooling towers. Revelation of this fact has led to the development of a number different chiller sumps and pumping arrangements that satisfy both the tower and the equipment using the tower water. Some of these solutions are provided later in this chapter.

The first information that must be obtained in pumping a tower is to determine the minimum and maximum flow rates required by the using equipment such as chiller condensers. In the past, there were simple rules such as 3 gal/min per ton of cooling for an electric chiller. This assumed that the heat of rejection included the 12,000 Btu/ton plus 3000 Btu/ton for the chiller compressor and auxiliaries for a total of 15,000 Btu/ton of cooling. Today, with improved chiller performance, the water rates are calculated more carefully. The flow rates should be checked for the actual chiller or other equipment under consideration at specific design water temperatures such as 95°F return and 85°F sump water temperatures. If the heat of rejection from the chiller is expressed in British thermal units per hour at a specific temperature difference, the following equation can be used to compute the water rate.

$$\text{Cooling tower flow (gal/min)} = \frac{\text{total heat of rejection (Btu/h)}}{500 \cdot \Delta T \, (°F)} \quad (11.1)$$

If the cooling load is expressed in tons of cooling, this formula becomes

$$\text{Cooling tower flow (gal/min)} = \frac{\text{heat of rejection/ton} \cdot \text{load in tons}}{500 \cdot \Delta T \; (°F)} \quad (11.2)$$

These formulas assume a specific weight for water of 62.33 lb/ft^3. The specific weight of 85°F water is 62.15 lb/ft^3, so these standard formula can be used except on special applications that may require exact calculations in gallons per minute or pounds of water per hour. In those cases, Eq. 2.4 should be used, taking in consideration the actual specific gravity and viscosity of the water at operating conditions.

Many chiller manufacturers specify an actual flow rate in gallons per minute for their machines. They do not provide a heat rejection rate, so these figures can be used in lieu of the preceding formulas.

Absorption-type steam- or hot-water–heated chillers have a higher heat of rejection per ton than electric chillers. The cooling tower flow rate is usually specified by the manufacturers of these machines at a specific temperature difference.

The use of a direct-fired chiller, where natural gas is the source of heat, requires special sizing of cooling towers. The heat of rejection also must include the combustion efficiency, so the total heat rejection per ton of cooling may be in excess of 20,000 Btu/ton. The allowable temperature difference may be much greater than the traditional 10°F, so the flow rate in gallons per minute through the tower will be different than that for the more traditional electric chiller.

11.3 Cooling Tower Piping

The most basic and conventional cooling tower pumping system is shown in Fig. 11.1a for one tower cell and one chiller condenser. This configuration is for a constant-speed pump providing constant flow through the chiller condenser. A three-way valve is included in the return piping to the tower for cold weather operation; this is a two-position nonthrottling valve. In very cold weather, the pipe to the cooling tower may freeze when the water flow is in the bypass position for a long period of time. An alternate way to pipe cooling tower bypasses is shown in Figure 11.1b. The piping is arranged so that the velocity head of the water is directed toward the bypass valve. Also, this valve should be sized so that the pressure loss across the bypass valve and the piping to the chiller sump is relatively low. This results in very little water being backed up in the tower pipe. For example, if the return pipe is 8-in steel carrying 1200 gal/min, the velocity head of the water is 0.920 ft of head. If a 10-in bypass valve is installed, the valve and exit loss to the tower sump should not exceed the velocity head of the 8-in pipe. An advantage of this two-way valve is the fact that the cooling tower piping drains when the condenser pump is stopped.

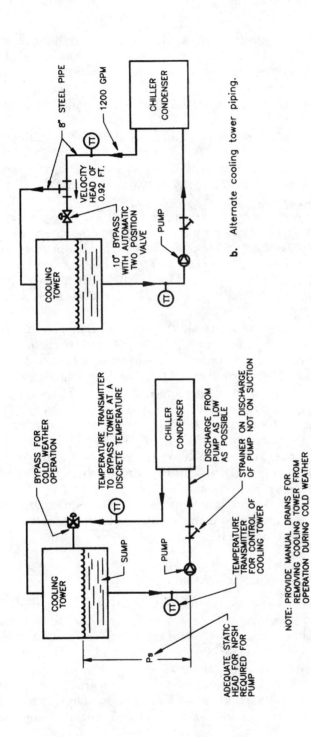

b. Alternate cooling tower piping.

a. Standard cooling tower piping.

NOTE: PROVIDE MANUAL DRAINS FOR
REMOVING COOLING TOWER FROM
OPERATION DURING COLD WEATHER

Figure 11.1 Cooling tower piping.

Figure 11.1 describes relatively simple pumping installations, since the flow through the tower is constant and equal to the needs of the condenser on the chiller. The tower can be picked directly from a cooling tower catalog or through computer software. The tower is sized to the approach temperature and actual outdoor wet bulb temperature at design conditions.

The cooling tower should be piped in such a way as to enable it to achieve optimal performance. The piping also must aid the flow of tower water to the using equipment. The water should leave a cooling tower basin so that air entrainment or water vortexing is avoided. Vortexing is discussed in Chap. 6.

Piping for cooling towers has been mostly steel. Recognizing the great amount of oxygen carried by this pipe, many designers are now using thermoplastic piping. This material is available in sizes up to 12 in in diameter. If steel pipe is selected for a cooling tower installation, the pipe should either be coated internally or the friction should be calculated using a 1.2 multiplier on the Darcy-Weisbach data included in Table 3.5.

It has been a tradition to install strainers on the suction side of cooling tower pumps. As shown in Fig. 11.1, the strainer should be installed on the discharge of the pump. The strainer is installed to protect the tubes in a chiller condenser or process equipment such as a heat exchanger; it is not there to protect the pump. The strainer in the cooling tower sump is normally adequate to keep rocks and other large debris out of the pump. A suction strainer on a cooling tower pump hazards a chance that the strainer could be obliterated with algae; the pump could then overheat and be destroyed before it could be stopped. Installing the strainer on the pump discharge protects the pump. If the strainer becomes clogged, the pump will be protected with only a small flow through the strainer. The equipment being supplied by the cooling tower pump will indicate flow problems long before the strainer is totally clogged.

11.4   Location of Cooling Tower Pumps

Cooling tower pumping is not severe duty for centrifugal pumps, but applying pumps to cooling towers is much more detailed than most HVAC pump installations. This is caused by the cooling tower loops being open-type water systems with the possibility of oxygen, airborne dirt, and chemicals existing in the water.

Every cooling tower installation should be checked to ensure that adequate net positive suction head (NPSH) available exists for the pumps selected for that cooling tower. Chapter 6 provides information on net positive suction head and how to calculate the available NPSH. Pumps should not be selected for cooling tower duty without knowing the NPSH required by them.

As the NPSH available equation (Eq. 6.8) indicates, the friction in the suction pipe to end-suction and double-suction pumps should be kept low. The pumps should be located as near as possible to the cooling towers; the suction pipe should contain a minimum of pipe fittings such as elbows (Fig. 11.2a). If there are doorways between the cooling tower and a chiller, the pump should be located near the cooling tower to avoid a loop in the suction piping (Fig. 11.2b). If the cooling tower location is such that it is difficult to provide adequate suction conditions for end-suction or double-suction pumps, an alternate is the use of turbine-type pumps located in the cooling tower sump. The cooling tower should not be planned for installation at a specific point with the hope that there is a way to pump the water from it. Pumping the cooling tower should be settled before final location is made for the tower.

Cooling tower pumps should not necessarily be located on the supply side of chiller condensers. On high-rise buildings where there is adequate NPSH available under all load conditions, the condenser pumps can be located on the discharge of the condenser, as shown in Fig. 11.3a. This reduces the pressure on the condenser water boxes and may eliminate the cost of higher-pressure construction. The maximum static pressure must be calculated to determine the working pressure for these water boxes.

11.4.1 Remote cooling tower sumps

Many of the problems involved in operating cooling towers in cold weather and with chiller condensers are solved through the use of special cooling tower sumps. These sumps can be located indoors to prevent freezing, as shown in Fig. 11.3b. When the sump is located remotely from the tower, as in this figure, the added height Z must be included in the pump total dynamic head. This indoor sump must have a drain-down capacity to handle the water of the system when the pump is stopped. Most tower manufacturers provide the drain-down volume in gallons for their towers; to this must be added any pipe capacity that is drained at the same time.

A sump that solves many application problems is the hot-well, cold-well type of Fig. 11.4a. This sump offers freezing protection and also separates the cooling tower loop from the process or chiller loop. This provides maximum flexibility for modulating the heat load that is sent to the cooling tower. The cooling tower receives the correct water flow regardless of the load on the equipment being cooled. A hot-well pump

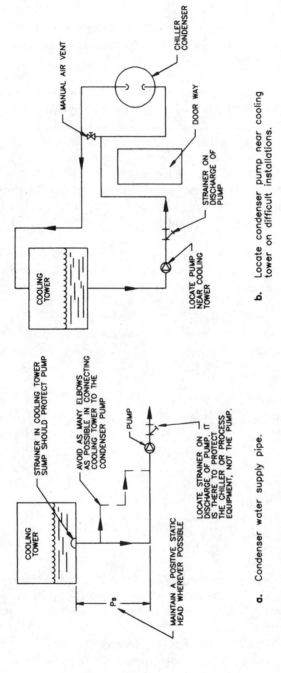

Figure 11.2 Cooling tower supply piping arrangement.

a. Condenser water supply pipe.

b. Locate condenser pump near cooling tower on difficult installations.

COOLING TOWER

CHILLER CONDENSER

MANUAL AIR VENT

DOOR WAY

STRAINER ON DISCHARGE OF PUMP

LOCATE PUMP NEAR COOLING TOWER

STRAINER IN COOLING TOWER SUMP SHOULD PROTECT PUMP

AVOID AS MANY ELBOWS AS POSSIBLE IN CONNECTING COOLING TOWER TO THE CONDENSER PUMP

PUMP

LOCATE STRAINER ON DISCHARGE OF PUMP. IT IS THERE TO PROTECT THE CHILLER OR PROCESS EQUIPMENT, NOT THE PUMP.

MAINTAIN A POSITIVE STATIC HEAD WHEREVER POSSIBLE

P_s

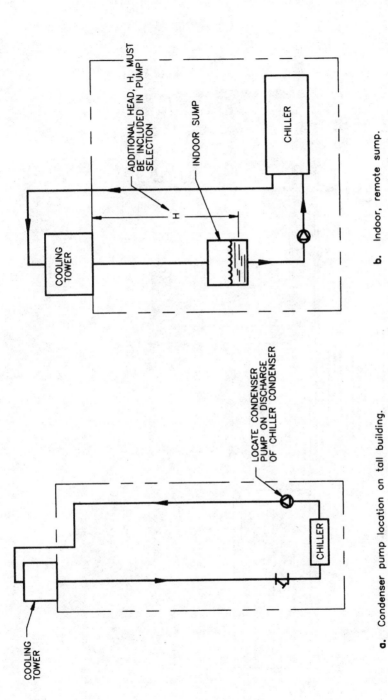

a. Condenser pump location on tall building.

b. Indoor, remote sump.

ADDITIONAL HEAD, H, MUST BE INCLUDED IN PUMP SELECTION

INDOOR SUMP

CHILLER

COOLING TOWER

LOCATE CONDENSER PUMP ON DISCHARGE OF CHILLER CONDENSER

CHILLER

COOLING TOWER

Figure 11.3 Cooling tower piping for high-rise buildings.

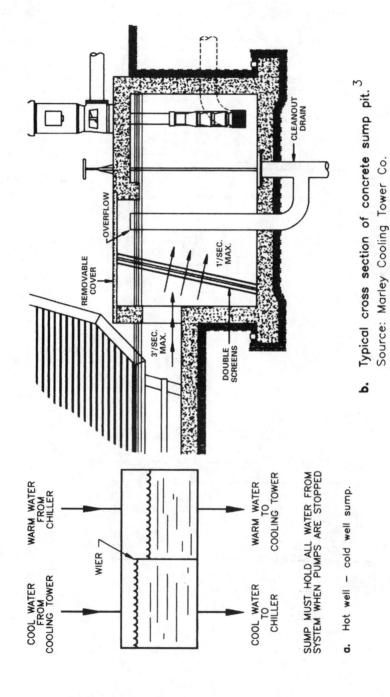

REMOVABLE COVER

OVERFLOW

3'/SEC. MAX.

1'/SEC. MAX.

DOUBLE SCREENS

CLEANOUT DRAIN

b. Typical cross section of concrete sump pit. [3]

Source: Marley Cooling Tower Co. used with permission.

WARM WATER FROM CHILLER

COOL WATER FROM COOLING TOWER

WIER

WARM WATER TO COOLING TOWER

COOL WATER TO CHILLER

SUMP MUST HOLD ALL WATER FROM SYSTEM WHEN PUMPS ARE STOPPED

a. Hot well – cold well sump.

Figure 11.4 Special sumps for cooling towers.

circulates the cooling tower, while the process or condenser pump circulates the process equipment or the condenser on the chiller. In warm climates where there is no concern about freezing, the primary-secondary circuiting that is described later in this chapter eliminates the need for this sump.

The concrete sump and turbine pump of Fig. 11.4*b* is the typical installation for large cooling towers. A battery of turbine pumps operating in parallel can develop an efficient flow program with a substantial load range.

The exit of water from cooling tower sumps should be designed carefully to avoid vortexing. Figure 6.8 should be reviewed to ensure that the possibility of vortexing is eliminated from the cooling tower water system.

11.5   Selection and Operation of Condenser Pumps

The selection and operation of condenser pumps for water-cooled chillers embraces a broad range of pump sizes and arrangements. The smaller chillers need only constant-speed end-suction pumps with a pump for each chiller (Fig. 11.5*a*). If a standby pump is needed, it can be added as shown in this figure. Normally, with this installation, a cooling tower cell is furnished for each chiller. The programming for the condenser pumps may consist of just starting and stopping each pump with its chiller. Headering the condenser pumps as shown in Fig. 11.5*b* is more expensive than the standby pump of Fig. 11.5*a*. It requires another pipe header, and each chiller must be equipped with a two-way automatic shutoff valve. On both these multiple-tower installations in cold weather areas, the piping must be arranged to prevent freezing when a tower or tower cell is taken out of service. A bypass valve may be needed for each of the towers or tower cells to prevent freezing in the tower piping.

As chiller installations become larger and consist of different types of chillers, the pumping procedures become complex and require careful and detailed analyses to achieve the most efficient and cost-effective installation. The decision to use base-mounted double-suction pumps or vertical turbine pumps in the cooling tower sump is a major one, requiring wire-to-water efficiency analysis and first-cost calculations. The height of the larger cooling towers alone may dictate the use of below-grade sumps with turbine pumps. The wire-to-water efficiency analysis of Chap. 9 should be applied to the larger condenser pumps to achieve the pumping system overall efficiency and the energy input to the condenser pump motors throughout the load range of the chillers.

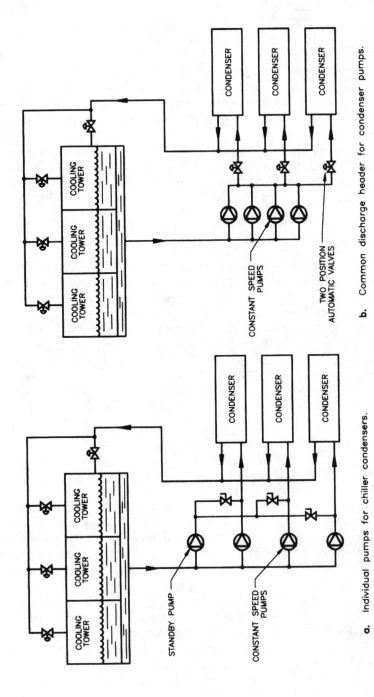

b. Common discharge header for condenser pumps.

a. Individual pumps for chiller condensers.

Figure 11.5 Constant-speed condenser pumps.

11.6 Special Condenser Water Circuits

The advent of the variable-speed pump has introduced new and innovative methods of pumping chiller condensers. The age-old problem of operating the chiller and cooling tower at their desired flow rates has been eliminated by the use of so-called primary-secondary pumping, as shown in Fig. 11.6b. The same procedure as primary-secondary pumping for hot and chilled water systems is utilized with constant-speed cooling tower pumps and variable speed condenser pumps. The constant-speed cooling tower pumps pressurize the supply to the variable-speed condenser pumps so that there are no NPSH problems with the condenser pumps. This pumping arrangement also puts all the static head of the cooling tower on the tower pumps, letting the variable-speed pumps operate on a total friction head. This may provide a higher overall wire-to-water efficiency for the condenser pumping installation. This procedure also allows the cooling towers to operate at their optimal flow rates as the water is varied through the condensers.

The number and size of the variable-speed pumps for the condenser flow can be selected for the most efficient installation. On individual chiller applications where a variable-speed pump is supplied for each chiller, the condenser pump can have its speed varied by the lift pressure or the refrigerant pressure difference between the high and low sides of the chiller (Fig. 11.6a). On an installation that consists of a number of chillers (Fig. 11.6b), the variable-speed pumps also can be controlled by the pressure drop across the condensers. As a chiller is started or stopped, the control valve on the chiller condenser water circuit opens and closes. The optimal number and size of pumps appear to be three 50 percent system capacity pumps on most of the multiple-chiller applications. Table 11.2 is a tabulated analysis of the pumps for the chillers in Fig. 11.6b.

This wire-to-water efficiency analysis indicates the number of pumps that should be run with various chillers in operation. Even though two pumps are required to handle the maximum load of 9000 gal/min, this evaluation indicates that it is more efficient to operate all three pumps at full load.

11.6.1 Operating absorption and centrifugal chillers together

A problem that the variable-speed pump is solving is the operation of an absorption and an electric chiller from the same cooling tower installation. In the past, this was unacceptable because of the difference in condenser supply water temperatures. The absorption machine required fairly constant supply water temperature, while the electric

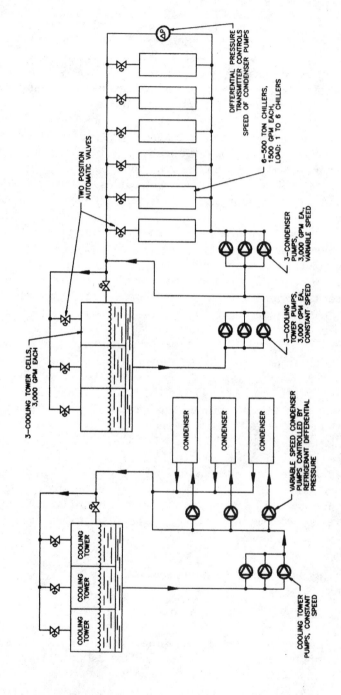

The following labels appear within the figure:

DIFFERENTIAL PRESSURE TRANSMITTER CONTROLS SPEED OF CONDENSER PUMPS

6–500 TON CHILLERS, 1500 GPM EACH, LOAD: 1 TO 6 CHILLERS

TWO POSITION AUTOMATIC VALVES

3–CONDENSER PUMPS, 3,000 GPM EA., VARIABLE SPEED

3–COOLING TOWER CELLS, 3,000 GPM EACH

3–COOLING TOWER PUMPS, 3,000 GPM EA., CONSTANT SPEED

b. Multiple chiller installation.

CONDENSER

CONDENSER

CONDENSER

VARIABLE SPEED CONDENSER PUMPS CONTROLLED BY REFRIGERANT DIFFERENTIAL PRESSURE

COOLING TOWER

COOLING TOWER

COOLING TOWER

COOLING TOWER PUMPS, CONSTANT SPEED

a. Individual variable speed condenser pumps.

Figure 11.6   Variable-speed condenser pumps.

TABLE 11.2   Pumping System Analysis for Chillers in Fig. 11.6*b*

Condenser water flow, gal/min	No. of pumps running	System bhp	Input kW	Wire-to-wire efficiency
1500	1	20.0	17.8	56.6*
3000	1	36.6	31.0	68.6*
4500	1	66.1	54.1	63.5
4500	2	60.8	52.1	66.0*
6000	2	86.3	72.0	69.8*
7500	2	121.3	99.4	70.0*
9000	2	167.8	135.7	68.7
7500	3	122.9	102.5	67.9
9000	3	161.3	132.7	70.3*

*Optimal number of pumps.

chiller preferred colder condenser water that improved its performance. With variable-speed condenser pumps, the inlet water temperature to the condenser no longer needs be a constant. With variable flow through the condenser, the control procedure now is to maintain the heat balance in the condenser like any other heat exchanger. The variable-speed pump for the electric machine is controlled by the lift pressure of the chiller, while the absorption machine is controlled by leaving water temperature and minimum flow. If the cooling tower water drops as low as 60°F, it may be necessary to reset the cooling tower sump to a higher temperature for the absorption machine. Use of any of these procedures should be approved by the chiller manufacturer under consideration.

Chiller manufacturers are evaluating the many different effects that variable flow through the condensers have on their equipment. It is obvious from the preceding examples that the optimal pumping arrangement for a specific installation requires analysis of several different pumping arrangements.

11.6.2   Avoiding ice in cooling towers

Ice in certain forms and quantities can be disastrous to cooling tower performance and to its structural integrity. Some ice is acceptable in a cooling tower that is designed for operation in freezing climates. Heavier amounts of ice that impair the cooling tower's performance or jeopardize the tower structure must be avoided.

Ice formation varies directly with the amount of air flowing and indirectly with the amount of water flowing through the tower. Therefore, pumps and pumping rates can be used to help control ice formation in a tower. Constant flow in a tower is obviously important in freezing weather. The primary-secondary pumping of Fig. 11.6 maintains a steady flow of water over the tower, regardless of the flow through the equipment using cooling tower water. The bypass valve that is used to eliminate flow from the tower should be a two-position valve for most towers and not be throttling, where the flow rate over the tower is reduced from the design rate.

Ice control techniques vary between cooling tower manufacturers. The recommendations of the manufacturer whose tower is under consideration should be the basis of design for the cooling tower pumping system.

11.6.3   Cooling tower plume abatement with pumps

Cooling towers have exit air plumes that can become quite visible, particularly in the winter. They can be reduced or their elevation above the ground can be increased by several techniques. Decreasing the temperature of the entering water can have some effect on the size and location of the plume. This can be accomplished by recirculated pumping that lowers the total return water temperature. Often, there is excess pumping capacity when the plumes occur, and some of the water can be bypassed around the chiller condensers or process by special valving back to the cooling tower.

The use of pumps to control these plumes should be as recommended by the cooling tower manufacturer.

11.6.4   Free cooling with cooling tower water

Free cooling is the name applied to processes that use the tower water for cooling chilled water systems instead of the chiller. Free cooling can be developed through a number of piping and equipment arrangements. Basically, these systems break down into two classes, direct and indirect.

Direct free cooling.   Direct free cooling defines the fact that the cooling tower water circulates directly in the chilled water system instead of through the chiller condenser. Two-way valves are provided for this change. The water passes into the chilled water system at the bypass, as shown in Fig. 11.7a. It circulates water through the chiller bypass and acts like the primary chiller pump. The secondary or distribution pumps pick up the cooling tower water and move it through the chilled water system as with the chillers in operation. The condenser pump head may be less in the free cooling mode than during normal operation with the chiller condenser. Both modes of operation must be evaluated to ensure that the condenser pump performance is adequate.

The principal drawback of direct free cooling is mixing of the cooling tower water and its chemistry with the relatively clean chilled water.

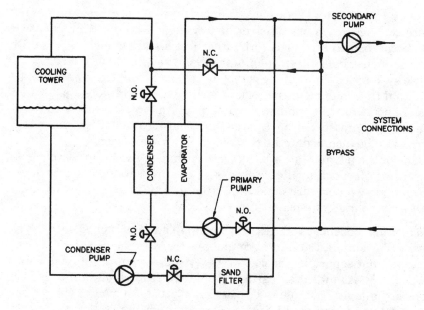

a. Direct free cooling.

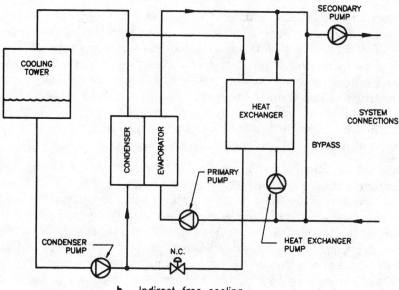

b. Indirect free cooling.

Figure 11.7 Free cooling.

Cooling tower water has to have a number of airborne contaminants, since cooling towers are efficient air washers. Cleaning cooling tower water can be accomplished with either full filtration such as sand filters or sidestream filters mounted at the cooling tower. Neither of these systems can guarantee complete removal of all the objectionable material that may be in the cooling tower water. The sidestream filter should not affect the condenser pump performance; the full sand filter may have a pressure drop greater than that for the chiller condenser. This must be included in the condenser pump evaluation.

One of the advantages of direct free cooling is the lower water temperatures that are available from the cooling tower compared with the higher temperature emanating from the heat exchanger on indirect free cooling systems.

Indirect free cooling. Indirect free cooling can be supplied in several forms. The simplest and most common is through the use of a heat exchanger that separates the cooling tower water from the chilled water (Fig. 11.7b). As indicated earlier, some temperature difference is lost across the heat exchanger. The designer must determine the difference in head between the friction loss through the chiller condenser and the loss through the heat exchange and size the condenser pump to accommodate both the friction losses.

An alternative method of indirect free cooling is described in Fig. 11.8 where the heat exchanger is located in tandem with the chiller evaporator. This installation is used where chilled water is required with outdoor temperatures of 55°F or less. Valves V_1 and V_2 modulate the water through the heat exchanger when the condenser water supply temperature is less than the temperature of the return chilled water from the chilled water system. If the condenser water is too cold for the chiller, it can be modulated by valve V_3 which is controlled by the refrigerant differential pressure in the chiller.

Refrigerant migration. Another method of free cooling is by refrigerant migration, which is used on larger chillers when they are shut down. Valves are opened on these machines that allow refrigerant vapor to pass from the evaporator to the condenser and liquid refrigerant to flow from the condenser to the evaporator. This process requires fairly cold water from the cooling tower, below 50°F. With this type of free cooling, there is no effect on the condenser pump performance.

Partial indirect cooling can be accomplished by placing the free cooling heat exchanger in series with a cooling tower. This reduces the return water temperature to the chiller and enables it to operate at a lower load. Like the heat exchanger above, the friction loss in both loops must be evaluated.

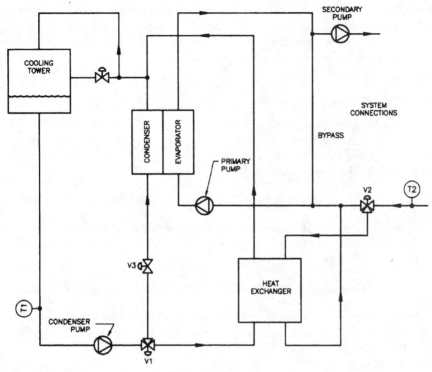

Figure 11.8 Tanden arrangement for free cooling.

All these free cooling methods require careful energy and cost evaluations by the system designer to determine the optimal procedure for a particular system.

11.7   Understanding Legionnaires' Disease

Although cooling tower water pumping in itself has little effect on the prevention of Legionnaires' disease, it does not hurt to have some specific information included in this book on this disease. *Legionnella,* the bacteria that causes it, thrives in water of temperatures from 70 to 120°F. This temperature range is the same as that for operation of most HVAC cooling towers. Therefore, it is urgent that all cooling towers be equipped with a comprehensive water treatment system that not only controls corrosion and scale but also limits biologic contamination. The system operator should become familiar with the recommendations of various authorities as contained in publications such as those of the Occupational Safety and Health Administration. Another source for this information is the 1995 ASHRAE Handbook,

Heating, Ventilating, and Air-Conditioning Applications, Chap. 44, Corrosion Control and Water Treatment. Spills of water around cooling tower pumps and leaks from cooling tower pumps should be eliminated to aid in the control of this disease.

11.8 Summary

Cooling towers remain the best method of rejecting heat in HVAC and industrial process water systems. It should be emphasized how much heat is passed out through a cooling tower. Equation 11.3 develops this as total heat per year:

$$\text{Btu/year} = \text{heat/ton} \cdot \text{tons} \cdot \text{annual hours} \tag{11.3}$$

For example, if a 1000-ton chiller has an average heat rejection of 15,000 Btu/ton and operates an equivalent of 2000 hours per year at full load, the annual heat rejection is

$$\text{Btu/year} = 15,000 \cdot 1000 \cdot 2000 = 30,000,000,000$$

This is 30 billion British thermal units per year! With an overall boiler efficiency of 80 percent and a cost of $3 per 1000 ft^3 of natural gas at 1000 Btu/ft^3, this is equivalent to 37,500 mcf of natural gas, or $112,500 per year.

Unfortunately, the average return water temperature to cooling towers is only 95°F. As HVAC water processes reduce their temperature, more of this heat will become economically retrievable. Programs are now being evaluated for raising the condenser leaving water temperature and balancing the cost of eliminating boiler operation against the higher kilowatt per ton cost incurred by running the chillers at higher leaving condenser temperatures.

This is an excellent place to again evaluate kilowatt per ton instrumentations and calculations for heat recovery systems. By measuring the kilowatt consumption of the chillers, pumps, and cooling towers, as well as the tonnage on the chilled water system, the data are achieved for measuring kilowatts per ton for the heat recovery system. Also, by determining the heat recovered from closed condenser heat recovery circuits, a total energy statement can be derived that will provide a guide for the central plant operator to program the equipment at the maximum possible efficiency. There is no way an operator can read all these system values simultaneously and develop a manual algorithm that will give the continuous and accurate data that are achievable with these computer programs.

The great amount of heat rejected by cooling towers should be remembered as we go about the business of making chilled and hot water systems more efficient. Some of the current efforts will be found in Chap. 16 on closed condenser systems. Also, systems are prime candidates for evaluation for heat recovery from cooling towers if they have chillers and boilers that operate at the same time. Chiller operation in the winter and boiler operation in the summer may be signals to evaluate the economics of heat recovery.

11.9 Bibliography

James B. Rishel, *The Water Management Manual,* SYSTECON, Inc., West Chester, Ohio, 1992.
Cooling Tower Fundamentals, The Marley Cooling Tower Company, Mission, Kansas, 1985.

12

Pumps for Process Cooling

12.1 Introduction

Although, in a strict sense, process cooling is not part of the HVAC industry, many designers of HVAC systems are also involved with this cooling due to their experience with cooling towers and chillers. This chapter reviews some of the basic cooling systems that are used in industry.

Process cooling systems, like HVAC systems, can receive their cooling from cooling towers or chillers. The actual temperature of the industrial process determines whether cooling towers or chillers are used for the cooling. A great many industrial processes require water no colder than 85°F, so cooling towers can be used for these applications.

Process cooling can consist of cooling liquid or gas streams. Many liquid cooling processes include heat exchangers with the process liquid on one side and cooling water on the other side. Such heat exchangers are used when the process liquid is dirty, toxic, difficult to handle, or too valuable to pass directly through a cooling tower. Gas streams include many special chemical processes that will not be reviewed here. Some gas streams will be discussed that are nontoxic and can be cooled by water injection or water-cooled heat exchangers.

12.2 Liquid Cooling

Liquid cooling in industrial processes can utilize cooling towers directly, but owing to the conditions listed above, many of these processes use heat exchangers of the shell and tube or plate and frame types that are cooled by cooling tower water. The piping of such heat exchangers is similar to that for heat exchangers in the HVAC field; however, there are some conditions that may be found in these industrial processes that require special circuiting of the heat exchangers. There are some industrial processes that use heating or cooling coils for gas streams. The pumping and piping connections of these coils are much like those for HVAC heating and cooling coils.

The pressure drops across industrial heat exchangers may be greater than those of HVAC applications because of the cost of the heat exchangers. It is imperative that the designer determine exactly the true pressure drop across the heat exchanger and its control valve. The manufacturer of such process equipment may provide a higher overall pressure to ensure that adequate water is available to the equipment.

Some industrial processes may contaminate the cooling tower water, and it may be necessary to provide cleaning of the water before it is returned to the cooling tower. Figure 12.1a describes a cooling system where the return water is collected in a trough and returned to a hot well where it is cleaned before being pumped through the cooling tower. The primary-secondary pumping described in this figure allows for constant flow through the cooling tower and variable flow to the process loop. This provides excellent cooling tower operation throughout the year. A bypass valve is shown for bypassing the water around the tower under very low process loads in freezing weather. One method of regulating this valve is through the installation of a temperature transmitter on the return water line to the tower. A pressure transmitter at the end of the loop controls the system pump speed and maintains the needed pressure in the supply header. The set point can be varied with the load or type of cooling being carried on in the loop.

Other industrial processes may be such that a typical closed loop can be used on the cooling tower water (Fig. 12.1b). If there is still some contamination of the cooling tower water, cleaning equipment can be provided at the cooling tower pumps. With the closed loop, a differential pressure transmitter should be used instead of a pressure transmitter for controlling pump speed to maintain the desired differential pressure for the heat exchangers. This is due to the variable pressure drop in the return header. On some applications, the pressure-gradient diagram (Fig. 12.2) may be helpful in evaluating the total installation.

The hot-well, cold-well sump of Fig. 12.1a may be required in cold areas where the heat from the process load is a variable. The bypass valve of Fig. 12.1b may be acceptable as long as there is no possibility

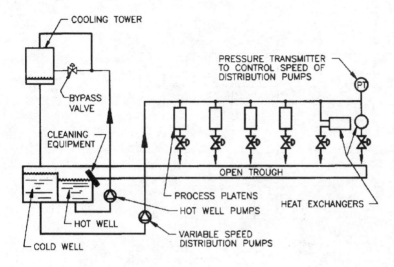

a. Industrial cooling with open return.

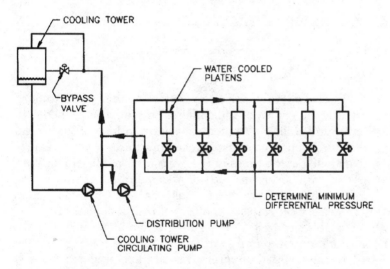

b. Industrial cooling system using cooling tower water.

Figure 12.1 General arrangement of process cooling systems.

of continuous shifting of the water from the cooling tower to the bypass connection and back again.

Cooling industrial processes with chillers is much the same as for HVAC applications. Chillers are used when water in the range of 40 to 60°F is required. Piping of the chillers should be the same as described in Chap. 14.

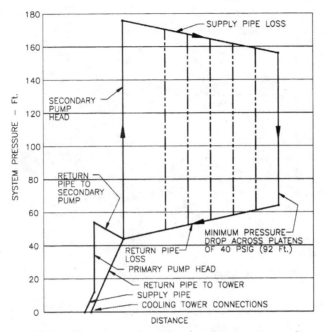

Figure 12.2 Pressure gradient diagram for a process cooling system.

There is an intermediate process temperature range of 60 to 80°F that requires both cooling towers and chillers. This is accomplished with the cooling tower and chiller in tandem, as shown in Fig. 12.3a. Special attention must be given to keeping the cooling tower water clean to avoid excessive fouling of the tubes in the evaporator of the chiller. With this arrangement, whenever the outdoor wet-bulb temperature is reduced, the chiller can adjust its load to make up only the difference in temperature from the cooling tower and the set point temperature.

It may be advisable to develop a pressure-gradient diagram as shown in Fig. 12.3b to determine pump locations and heads. Also, it may be advisable to determine maximum pressures on the piping, particularly if thermoplastic pipe is used in the cooling system.

12.2.1 Water flow for process liquid cooling

The water flow in gallons per minute for process cooling is somewhat the same as for HVAC cooling. Since the cooling water temperatures can vary over a broader range, the following formula can be used:

$$\text{Flow (gal/min)} = \frac{\text{heat transferred} - \text{Btu/h}}{(\gamma\,/7.48) \cdot 60 \cdot \Delta T\,(\degree F)}$$

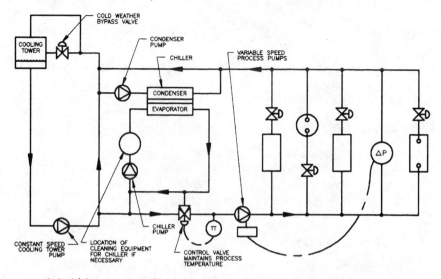

a. Industrial process cooling using both a cooling tower and chiller.

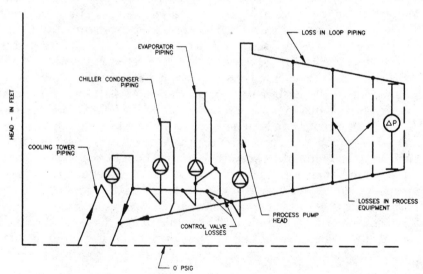

b. Pressure gradient for industrial process system of A.

Figure 12.3   Cooling tower and chiller for industrial cooling.

$$= \frac{\text{heat transferred} - \text{Btu/h}}{8.02 \cdot \gamma \cdot \Delta T \, (°\text{F})} \qquad (12.1)$$

where γ = Specific Weight of water at cooling tower return water temperature

12.3 Energy Recovery

Some industrial processes operate with higher return water temperatures to the cooling towers. This increases the possibilities of energy recovery, as discussed in Chap. 11 on cooling towers and as will be reviewed in Chap. 16 on closed condenser water systems.

12.4 Gas Stream Cooling

Gas stream cooling with cooling towers or special spray towers is limited mostly to those gases which are exhausted to the atmosphere. Typical of these are gas turbine, boiler, and incinerator exhausts. Many of these processes also scrub the gases, so the water from them will contain chemicals that may be abrasive or corrosive to pumps. This is particularly so if scrubbers are used to clean the gases and to absorb specific chemicals that are known to exist in the gases.

Specific attention should be paid to the metallurgy of the pumps for scrubbers. Erosion also may be a problem, necessitating the use of rubber-lined pumps to provide a useful life of the volute. Most pumps for such services are end-suction pumps to simplify the construction and repair capabilities. Special mechanical seals with flushing equipment may be required to keep the pumpage out of the seal cavities. Ease of maintenance, not efficiency, is often the prerequisite for these pumps, which is the opposite of that for most HVAC pumps.

The water flow rates should be determined by the user or manufacturer of the cooling equipment. The actual water flow is determined not only by cooling rates but also by flushing or filtering requirements.

12.5 Bibliography

James B. Rishel, *The Water Management Manual,* SYSTECON, Inc., West Chester, Ohio, 1992.

13

Pumping Open Thermal
Storage Tanks

13.1 Introduction

Thermal storage in the HVAC industry consists of both hot and chilled water storage. Hot water storage has not been prevalent, since it is not necessary for most hot water systems. As energy conservation grows with large chilled water plants, there may be some economic reasons to store hot water that is generated by the heat rejection from chillers.

13.2 Cool Thermal Storage

Cool thermal storage in HVAC systems has become popular in attempts to reduce overall energy costs. The driving force has been the demand charges put in place by the electric utilities as they attempted to reduce their electrical demand due to cooling load during the afternoon hours of the summer days. This demand charge necessitated the shifting of this load from the daytime to nighttime hours. Thermal storage provided this shift, using either ice, chiller water, or eutectic salt to provide cooling during the peak hours instead of electric motor–driven chillers.

The disappointing fact of this effort has been that the peak load was reduced in most cases, but the overall energy consumption was increased due to the higher kilowatts per ton of cooling required for chillers making ice or chilled water at night. The mission of many of these efforts was only peak shaving by the electric utility and demand charge reduction for the customer, not overall energy conservation.

A number of studies of the energy consumed by chillers have been conducted by various companies to determine the value of energy

TABLE 13.1   Chiller Performance
(Mean kw/ton per 5% Increment of Chiller Load)

Percent chiller load	Number of seasonal hours	Kilowatts per ton
10–15	108	1.53
15–20	542	1.32
20–25	389	1.15
25–30	303	1.03
30–35	311	0.94
35–40	209	0.87
40–45	225	0.83
45–50	262	0.79
50–55	230	0.76
55–60	137	0.74
60–65	211	0.72
65–70	128	0.70
70–75	120	0.70
75–80	122	0.70
80–85	104	0.70
85–90	32	0.67
90–95	11	0.61
95–100	4	0.59

storage in reducing on-peak demand charges and overall efficient operation of chillers. The data shown in Table 13.1 are typical of the findings of some of these studies. This information is for chillers providing cooling of the chilled water without energy storage.

The data in Table 13.1 demonstrate that the programming of chillers needs to eliminate the operation of chillers at light loads. One of the most important advantages of these cool storage systems is the "flywheel effect" they can have on chiller operation. The proper design of energy storage systems therefore should provide for chiller operation at optimal efficiency as well as to ensure adequate cooling during the peak demand charge period. The actual programming of the charging of the tanks should not be just to recharge the tanks as fast as possible but to use the storage and the chillers during light system loads to achieve optimal overall energy consumption.

For example, in Table 13.1, most of the hours of operation were from 15 to 50 percent design load on the chiller, where the kilowatts per ton varied from 1.32 to 0.79. With chilled water storage, the chiller could operate at a higher loading with the combined system and storage load. Obviously, the total economics of the installation must be evaluated before committing to chilled water storage, but the possibility of reduced chiller energy is another factor in the evaluation of chilled water storage.

Most studies concentrated on the energy consumption of the chillers alone and did not include evaluations of pumping energy. Also, many of the pumping and piping installations for these systems have not been conducive to energy-efficient pumping.

A number of pumping and piping arrangements for energy storage have resulted in too many control valves and piping circuits and inefficient pumping systems. One of the first objectives for the designer of these systems should be to evaluate the physical configuration of the installation and find ways to eliminate control valves and branch circuits. There are some very efficient and simple pumping and piping systems for energy storage. The pressure-gradient diagram provides the designer with a working tool to achieve maximum possible efficiency in design and operation.

13.3 Types of Cool Water Storage Systems

There are five basic open-type cool storage systems. They are chilled water, ice harvester, external ice melt, encapsulated ice, and eutectic salt. Each has specific advantages and disadvantages; no attempt will be made here to recommend a particular type of energy storage. Emphasis will be placed on the pumping requirements for each system. Following is a description of these systems that use an open storage tank; particular attention will be paid to the optimal methods of pumping cold water to and from these tanks.

No matter which cool storage system is selected, the pumping costs can be appreciable, particularly since the pumps must operate at periods of high electrical demand charges. Great care must be taken in the evaluation of variable-speed versus constant-speed pumps to verify if the power savings of variable-speed pumps are worthwhile.

13.4 Chilled Water Storage

Chilled water storage has the following advantages and disadvantages when compared with other types of cool storage:

1. Standard chillers can be used because the storage temperature is near that of most HVAC chilled water systems. On existing installations, the chillers can be reused in many cases.

2. The electrical power cost is lower than for all other types of cool storage except eutectic salt. This is due particularly to the higher leaving water temperature for the chillers, around 40°F instead of 26°F for the ice systems.

3. Special liquids such as the glycols are not needed.

4. The storage cost is much higher than for the smaller ice systems because of the greater volume. On large systems, the storage cost may be equal to or less that for ice systems. The storage cost is almost always less than for eutectic salt installations.

13.4.1 Chilled water system configuration

There are many different chilled water storage systems; the actual design for a specific application is often contingent on installation factors such as size and location of the chilled water storage. One of the most important design factors in any storage system is to provide the "flywheel effect" mentioned above that can be used to operate the chillers at the optimal kilowatts per ton. Figure 13.1a provides a piping circuit for a typical water storage system that uses the stored water as a chiller bypass and enables the chillers to operate at the optimal kilowatts per ton.

Table 13.2 provides typical chiller kilowatts per ton data for contemporary chillers that include the energy consumption of the condenser and primary chilled water pumps with the chiller that is delivering water at 40°F. Also, these data are for 85°F condenser water supplied to the chiller. Chillers running at lower condenser temperatures will demonstrate much lower kilowatt per ton figures.

It is obvious from the data in Table 13.2 that the chilled water storage system must be designed so that this chiller never runs below 50 percent of full load. The backpressure valves shown in Fig. 13.1a are unnecessary if the storage tank is higher than the rest of the chilled water system, as could be the case for low-rise or single-floor buildings.

In most chilled water systems, it is undesirable to reverse the flow in the chiller bypass. With the chilled water storage tank serving as the chiller bypass, the flow reverses automatically through the tank to provide the flow needed for the chilled water system when the chillers are stopped, as would be the case during high-demand charge periods. With the kilowatt per ton data of Table 13.2, the chillers should run at around 70 percent loading to achieve minimum energy consumption. In actual operation, it has been found most economical to operate most chillers as closely to full load as possible. This can be verified during actual operation if total kilowatts per ton are recorded for all percentages of system load. Operating chillers in off-peak seasons with lower condenser water temperatures also dictates the need to observe actual kilowatt per ton readings to secure optimal chiller operation. Obviously, poor operation of chillers at very low loads and high kilowatts per ton should be avoided.

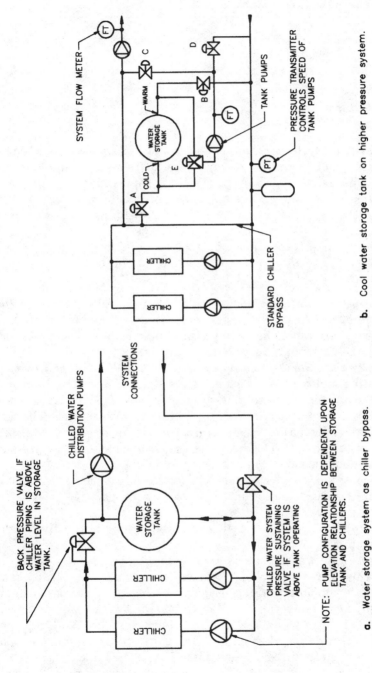

Figure 13.1 Water storage systems.

a. Water storage system as chiller bypass.

b. Cool water storage tank on higher pressure system.

SYSTEM FLOW METER

WARM

WATER STORAGE TANK

COLD

PRESSURE TRANSMITTER CONTROLS SPEED OF TANK PUMPS

TANK PUMPS

CHILLER

CHILLER

STANDARD CHILLER BYPASS

BACK PRESSURE VALVE IF CHILLER PIPING IS ABOVE WATER LEVEL IN STORAGE TANK.

CHILLED WATER DISTRIBUTION PUMPS

SYSTEM CONNECTIONS

WATER STORAGE TANK

CHILLER

CHILLER

CHILLED WATER SYSTEM PRESSURE SUSTAINING VALVE IF SYSTEM IS ABOVE TANK OPERATING

NOTE: PUMP CONFIGURATION IS DEPENDENT UPON ELEVATION RELATIONSHIP BETWEEN STORAGE TANK AND CHILLERS.

275

**TABLE 13.2   Chiller Kilowatt per Ton Data with Condenser and
Primary Pump Kilowatts Included**

Percent load	Total kW/ton	Percent load	Total kW/ton
10	1.69	60	0.69
20	1.05	70	0.68
30	0.84	80	0.69
40	0.74	90	0.70
50	0.71	100	0.72

For existing systems where chilled water storage is to be added to buildings with high static pressures, Fig. 13.1b describes another system in which the water storage system is separated from the principal chilled water system. The actual system design conditions will determine the need for this separation of the storage from the water system and the extra valving and tank pumps.

To aid in understanding the operation of the chilled water storage system in Fig. 13.1b, the five different phases of its operation are as follows:

1. *Standard chilled water system operation without the chilled water storage system.* The storage tank valves A through D are closed, and the system functions as any other chilled water system without storage; the tank pumps are stopped.

2. *Serve the chilled water system and store chilled water.* Valve A is modulating and valve D is open; valves B and C are closed. Valve E on the tank is open to the tank discharge position. Valve A is modulated to provide the flow rate desired into the storage tank. The tank pump varies its speed to maintain return water pressure in the chilled water system. This prevents the development of low pressure in the system caused by water flowing into the tank.

This control procedure allows the operator to adjust the rate of storage, and it enables operation of the chillers at optimal kilowatts per ton for the entire chilled water plant.

This operation can occur at any off-peak loading on the chilled water system.

3. *Serve the chilled water system with chillers and stored chilled water.* This operation provides peak shaving by allowing the operator to select the amount of water needed from the storage tank to achieve the desired chiller operation and reduction in power consumption.

Valve C is open while valves A and D are closed. Valve B is modulating to provide the desired flow from the storage tank. The tank valve E is open to the cold water connection of the tank.

4. *Store chilled water only.* This procedure is very similar to phase 2

except that there is no system cooling load. Again, the operator should program the chillers and operate them at a rate of cooling that will ensure minimum energy consumption in the overall kilowatts per ton in the central energy plant. Recognition should be given to the lower kilowatts per ton that can be achieved with reduced condenser water temperatures from the cooling tower during lower outdoor temperatures.

5. *Serve the chilled water system only from the storage tank.* This phase is similar to 3. excepting that the chillers are stopped, and all cooling is derived from the storage tank. Valve *B* is modulating, valve *C* is open, and valves *A* and *D* are closed. The tank valve *E* is in the cold water discharge connection on the tank. Valve *B* is modulated by the flowmeter to achieve the desired flow through the storage tank, and the tank pump's speed is controlled to maintain the desired pressure on the chilled water system.

The supply temperature to the system can be varied by blending return water with the tank water. Also, the storage tank flow rate can be paced by the system flowmeter. The type of control desired for a specific application will depend on the overall parameters that affect system operation.

Most chilled water systems do not incorporate the glycols to prevent freezing, since the water is seldom stored below 40°F. The advantage of chilled water over the other systems is lower overall energy consumption in kilowatts per ton of cooling; the disadvantage is the lower British thermal units stored per cubic foot of tank volume. If there is space for a chilled water tank, it should be used in lieu of the ice-formation types of energy storage.

13.5 Open Ice Storage Systems

For systems that do not have space for chilled water storage, open ice storage systems can be used with smaller tanks. When ice storage is under consideration, the designer must evaluate both open- and closed-type ice systems. This chapter reviews open-type ice systems, while closed-type ice systems are part of Chap. 17. Open ice systems consist of the ice harvester, external melt ice, and encapsulated ice types.

13.5.1 Ice harvester

This system offers compact installations including prepackaged ice-making machines. Its storage cost is lower than that of any other cool storage systems, but it has the most expensive chiller costs. Its power costs of generating ice are as high as for any of the cool energy storage systems. Another disadvantage pump-wise is the common use of underground tanks for storage, since the ice-making equipment must be located above the tanks. This enables the ice to fall by gravity away from the ice machine into the tank. Harvester-type ice storage systems can be provided with above-ground storage tanks and are the preferable arrangement if vertical clearance is available.

With the underground tank, self-priming pumps may be required for ease of operation. If the system is seldom shut down, standard centrifugal pumps with priming systems can be used. The disadvantage of self-priming pumps is their efficiency, which can be as much as 20 percent lower than that for standard centrifugal pumps. As mentioned above, this is not advantageous, since these pumps must run during periods of high electrical demand charges. Several very important points must be made in the use of self-priming pumps:

1. The pump suction piping must be designed with low friction losses.

2. Care must be taken that the NPSH (Net Positive Suction Head) available is adequate even when the tank water is at ambient conditions. The system must start with no ice in the tank, so the water temperature must be assumed to be equal to the surrounding air temperature.

3. If more than one self-priming pump is installed, each should have its own suction line sized to the recommendations of the pump manufacturer. Headered suction lines to pumps offer possibilities of air leaks and loss of priming water.

4. All pumps taking a suction lift, particularly self-priming pumps, always should be installed strictly in accordance with the pump manufacturer's recommendations.

There are a number of different pumping arrangements for ice harvesters. Some may have one pump for circulating the chiller and delivering water to a heat exchanger for the chilled water system. Others may have dual pumps with a constant-speed pump for the chiller and a separate variable-speed pump for the chilled water system heat exchanger. The actual pumping arrangement is contingent on the size of the cool storage system and on the load configuration.

Ice harvesters can be used in conjunction with standard chillers as peak shaving systems to reduce the overall cost of the installation. It is obvious that the economics of these systems play a great part in the final selection of equipment.

13.5.2 Encapsulated ice systems

Encapsulate ice systems of cool storage use various types and shapes of bottles or containers filled with water. A low-temperature liquid such as one of the glycols flows through the tank to either freeze the water in the containers during the charging cycle or to melt the ice during the energy-using cycle. The advantages of these systems are very low storage volumes with relatively low chiller costs. The storage cost with respect to the other cool energy systems varies with the size of the installation.

There are no specific pumping problems with these systems because they are usually installed above grade with the pumping equipment beside the tanks. The circuiting of the piping depends on the manufacturer of the encapsulation containers. Otherwise, the pumping arrangement can be like many other open tank chilled water systems.

13.5.3 External ice melt systems

Generally, these systems use an open tank that has refrigerant coils immersed in it. Some systems are provided in closed pressure tanks. Cold refrigerant passes through the coils to develop ice around the coils. When the ice has built up to a specified thickness, the refrigerant flow is stopped, and the tank is ready for use as an energy source. Water flows through the tank much like the encapsulated ice systems delivering cool water to the chilled water system heat exchanger. Also, like the encapsulated ice systems, there are no particular problems in pumping these systems. The pumping equipment is again located beside the tanks, so there usually are no NPSH problems.

13.6 Eutectic Salt Systems

Eutectic salt systems are much like encapsulated water systems except that the liquid in the containers is not water but a material that has a higher melting temperature than the 32°F of water. Some of these salts freeze at 47°F; current developments indicate that other salts with freezing temperatures of around 40°F may be available for the cool storage industry.

Eutectic salts have the advantage of using standard chillers, so their power costs are similar to those of chilled water and much less than those of ice storage systems. They have a high storage cost due to the special materials that are involved with the salts and their containers. Their tank volume is around one-half that for chilled water.

These systems again are much like encapsulated ice and external ice melt systems as far as pumping equipment is concerned. The pumping systems are located beside the tanks, so there are no problems with NPSH or priming.

13.7 Decision as to the Type of Energy Storage System

The decision as to whether to use chilled water, ice storage, or eutectic salts is beyond the scope of this book. The designer for energy storage must make this decision based on the parameters of each installation, such as first cost, space for energy storage, electrical charges, and efficiencies of various types of chilled water and ice-generation equipment. This decision requires a number of evaluations and calculations. The development of the gas-fired engine–driven chiller and absorption chillers for electric power reduction during high-demand periods has added to all the evaluations that must be made for a particular installation.

Most of these cool storage systems require some form of pumping of the water from open tanks; the location and type of pumping vary with each system, as described above. However, with proper application of the pumps, their required power should not have a great bearing on the type of energy storage system selected for a specific application.

13.8 Basics of Pump Application to Open Energy Storage Tanks

Regardless of the type of open energy storage system, there are some fundamentals on pumping that must be recognized. The following principles should be adhered to on these applications:

1. The viscosity and specific gravity of the liquids involved should be checked and used in calculations for pipe and fitting friction and pumping horsepower. Water at 32°F has a higher viscosity than water at 50°F, namely, 1.93×10^{-5} ft^2/s at 32°F compared with 1.41×10^{-5} ft^2/s at 50°F.

2. When using glycol systems, ensure that the pumps will never operate near the freeze curve or slush line of the glycol (see Figs. 2.1, 2.2, and 2.3).

3. On chilled water and all other installations except ice harvester systems, always locate the working level of the energy storage tank higher than the chillers if at all possible. This eliminates the need for backpressure valves on the chiller circuit. It is wise to run a pressure-gradient diagram on energy storage systems, since this diagram will reveal problems with water circuiting and will expose needless energy waste in control valves.

4. Check the NPSH available for pumps taking water from the open storage tank. NPSH is not a problem on most of these systems

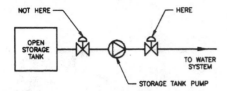

a. Location of control valve on storage
    tank pump.

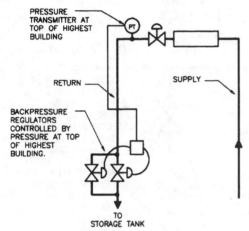

Figure 13.2   Location of valves
on open storage systems.

b. Use of backpressure regulators to maintain
    pressure in building.

unless there is a great amount of pipe or valve friction between the
open tank and the tank pumps.

5. Do not locate any control valves between the tank and the tank
   pumps (Fig. 13.2a). Control valves reduce the NPSH available for
   the pumps, and they can cause cavitation in the pumps. Three-way
   two-position valves can be located on the suction of the pumps,
   since one of the valves is always open (see Fig. 13.3a). This does
   not apply to self-priming pumps or other pumps taking a suction
   lift from the tanks, since no valves of any configuration should be
   installed in these lines.

6. When pumping closed chilled water systems from open tanks, the
   elevation of the working water level in the open tank must be compared with the minimum static pressure of the building. The difference between the two elevations must be recognized, since this

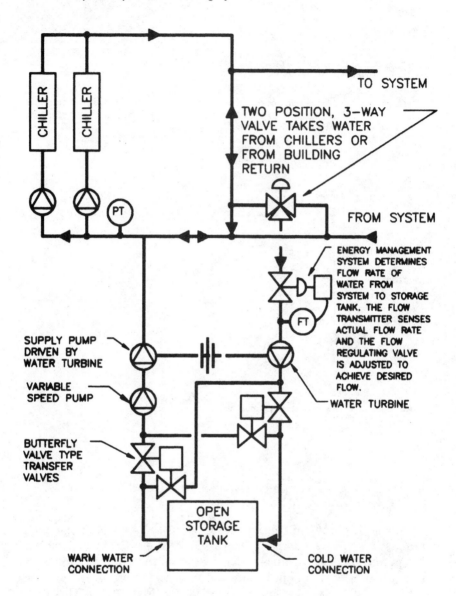

a. Energy recovery turbine on chilled water storage system.

Figure 13.3 System pressure energy recovery.

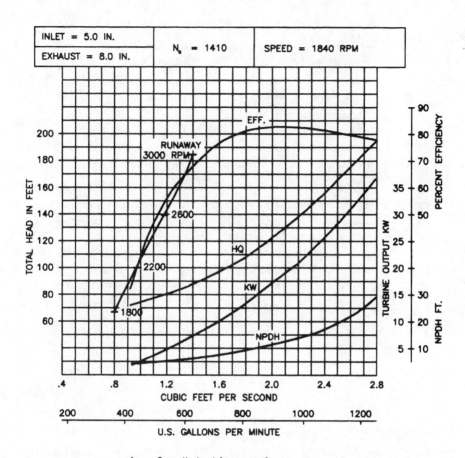

b. Small turbine performance.

Figure 13.3 (*Continued*)

static head must be accounted for in any energy evaluation of the system. The energy lost is computed as follows:

$$\text{kW lost} = \frac{\text{gal/min} \cdot \text{head} \cdot 0.746}{3960 \cdot P_\eta \cdot E_\eta}$$

$$= \frac{\text{gal/min} \cdot \text{head}}{5308 \cdot P_\eta \cdot E_\eta} \tag{13.1}$$

where gal/min = flow of water from pressurized system

head = pump head required to return the water to the system

P_η = efficiency of pump

E_η = efficiency of pump motor for constant-speed pumps or the wire-to-shaft efficiency of the motor and variable-speed drive for variable-speed pumps

The energy calculation for a system is as follows: Assume that the top of the building is at elevation 730.00 ft, and the working level in the energy storage tank is 520.00 ft. These elevations are in actual levels above sea level. Also, assume that there is an expansion tank located at the top of the building and that 10 psig is maintained in this tank by the makeup water system.

The pressure gradient required to maintain the 10 psig at the top of the building is 730 + (10 × 2.31), or 763 ft. The pressure gradient for the working water level of the storage tank is 520 ft. The pressure difference between the pressure at the top of the building, 763 ft, and the open tank, 520 ft, is 243 ft. With 15 ft of pipe friction in the piping between the storage tank and the water system, the total head required for the pump is 258 ft. If the system flow is 1000 gal/min, the pump efficiency 83 percent and the wire-to-shaft efficiency of a motor and variable-speed drive is 89 percent,

$$\text{kW lost} = \frac{1000 \cdot 258}{5308 \cdot 0.83 \cdot 0.89}$$

$$= 65.8 \text{ kW}$$

7. If the static pressure of a closed system is greater than the elevation of the open tank, some means must be used to maintain the minimum pressure at the top of the building. There are several methods of achieving this:

a. Backpressure valves (Fig. 13.2b) can be used, but they should maintain the pressure on top of the building, not the inlet pressure to these valves. Self-operating valves should not be used in many cases because the friction of the return water main between the valves and the top of the building is a variable. The valves should be equipped with a control system that measures the pressure at the top of the building and adjusts the valve position to maintain that pressure. Normally, a pressure transmitter located at the top of the building along with a proportional-integral controller can adjust the valve position to maintain the desired pressure on top of the building. Usually, this control procedure is adequate for small buildings where the energy loss from the pressurized system to the open tank is minimal.

b. On larger buildings with high pressures, it may be desirable to save this energy lost from the depressurization of the water. Figure 13.3*a* describes a procedure that can be used, that of a pressure recovery pump. This pump is driven by a similar pump equipped with a turbine wheel. A typical performance curve for such a turbine is shown in Fig. 13.3*b*. The energy recovered by such a pumping system can vary from 10 to around 67 percent of the pressure difference between the chilled water system and the open tank. A variable-speed pump responding to the chilled water system return pressure is installed in series with the recovery pump.

The variable-speed pump in series with the energy-recovery pump provides a more efficient arrangement for most applications than combining the variable-speed motor with the turbine and energy-recovery pump on one base. The difficulty of this one-base arrangement is the fact that the turbine must run at the same speed as the variable-speed pump. The turbine should operate at the speed caused by the flow of water from the system and the force on the turbine shaft created by the driven pump. The variable-speed pump runs at a speed necessary to maintain the system pressure. They are seldom the same speed.

An excellent article on the savings of the turbine-driven pump in lieu of a heat exchanger and where to use it is available in *ASHRAE Transactions* DA-88-27-2 by R. M. Tackett. The evaluations included in this article will assist the designer in the evaluation the cost of these pumps versus heat exchangers to determine the amortization periods.

c. The use of a heat exchanger to overcome the pressure difference between the open tank and the pressurized chilled water system usually is not an efficient procedure for chilled water storage systems because of the temperature drop across the heat exchanger. If it is desired to store chilled water at 40°F with a temperature approach of 3°F for the heat exchanger, chilled water must be generated at 37°F, thus increasing the kilowatts per ton of the chiller. Also, in like manner, only 43°F water is available for use by the chilled water system. The heat exchanger therefore causes a total temperature swing of 6°F. On open-type ice systems, the heat exchanger is the normal procedure because of the great difference between the water temperature in the tank and that in the chilled water system. The higher kilowatt per ton energy requirement for the chiller has already been expended to generate the ice.

13.9   Types of Pumps for Thermal Storage

When the thermal storage tanks are installed above grade or when the working water level of the tank is above the pump suction, the best type of pumps are volute, single or double suction, as would be the case for cooling towers. These pumping systems should be selected and programmed to achieve the maximum possible pumping efficiency.

When the open tank is below grade, the pumps can be vertical turbine installed in the tank with the motors above grade, or they can be self-priming pumps. Installation of the vertical turbine pumps would be much like that for large cooling towers. Usually, the self-priming pumps have a lower efficiency than the turbine pumps, so they should be used only on smaller installations. As indicated above, self-priming pumps should be installed with individual suction pipes, not with a suction header. The size of these suction pipes must be the same as that of the suction connection on the pumps themselves. This can add to the pump head and reduce further the efficiency of the self-priming pumping system. The NPSH required and dry priming capabilities of these pumps should be checked carefully with the pump manufacturer.

13.10   Operating Pumps on Thermal Storage Systems

The pumps for thermal storage systems deserve the same evaluation for efficient operation as other HVAC systems. It is wise on larger systems to run a pressure-gradient diagram at minimum and maximum loading to determine the value of variable-speed pumping.

Where there is a significant friction loss in the thermal storage and distribution system, the use of variable-speed pumps may be justified. On ice systems using glycol solutions, a careful evaluation of the viscosity range of the glycol should be made to determine the variation in friction head due to changes in viscosity of the glycol solution.

It is not within the scope of this book to evaluate the overall use of energy in the various types of open energy storage systems. Any audit of energy consumption for a particular system should include the pumping energy; the procedures outlined in this book for evaluating the efficiency of pump operation should be included in such a study. The overall kilowatts per ton or coefficient of performance should be run for the cool energy storage system under all phases of operation and at various loads on the chilled water system.

13.11   Hot Water Storage

With the development of the condensing boiler, greater opportunities are appearing for the use of rejected heat from chillers and industrial processes. Some installations may merit the storage of hot water. Whether this storage would be in open or pressurized tanks will depend on the physical characteristics of the actual installation. This will be discussed in Chap. 16 on heat recovery from chiller condenser systems.

13.12   Summary

The use of energy storage and the rejected heat of chillers offers new opportunities for achieving higher coefficients of performance for both chilled and hot water generating plants. Again, the variable-speed pump and digital control are the tools that can be applied to these more efficient systems.

Comparisons between the various types of energy storage are at best generalizations; each application must be evaluated for first cost and energy consumption and not depend on the estimates provided herein.

13.13   Bibliography

Hydro Turbines, The Cornell Pump Company, Portland, Oregon, 1986.
Design Guide for Cool Thermal Storage, American Society of Heating, Refrigerating, and Air-Conditioning Engineers, Atlanta, Georgia, 1993.

Pumps for Closed
HVAC Cooling Systems

14

Chillers and
Their Pumps

14.1 Introduction

The proper installation and operation of chillers on HVAC chilled water systems are among the most important factors in the development of efficient and cost-effective chilled water generating plants. This is obvious given the great amount of energy consumed by chillers.

Much work has been done to improve the efficiency of chillers using the latest state-of-the art-technology. For example, chiller efficiency has improved and energy consumption has dropped from around 0.86 kW/ton in 1970 to the current level of approximately 0.52 kW/ton for an efficient machine operating at its point of design peak performance. It is imperative that the chillers be installed to permit them to take advantage of this level of efficiency.

The pumps for both the chillers and the chilled water system must be installed and operated so that they aid the chillers' performance. *Pumping equipment should never be installed in a manner that will adversely affect the chiller energy consumption and efficiency.* The designer must understand how pumps or pumping can positively or negatively affect chiller performance; this is the reason for this extensive chapter on chillers and their pumps.

This review of water chillers will be limited to the installation of connecting piping, the pumping to and from them, and operation and control of the pumps and chillers as a system. If there are specific pumping requirements for one type of chiller, that chiller will be discussed with respect to its particular operation.

Briefly, an HVAC chiller has an evaporator that cools the system water and a condenser or other means that rejects the heat from the system to the cooling tower or closed-circuit cooler. All chillers need an energy source for their operation.

Chillers secure their energy for operation from electricity, natural gas, or steam. By far the most popular driver for chillers is the electric motor. Other methods of chilled water generation are becoming popular in areas where there are high electrical demand charges or shortages during the periods of maximum chilled water use.

Electric motor–driven chillers are available in four principal types: (1) reciprocating, (2) scroll, (3) helirotor (screw), and (4) centrifugal. Following are approximate ranges in tons for the various types of chillers. Reciprocating units are for smaller tonnages in the range from 10 to 200 ton, scroll units from 10 to 60 ton, screw types from 25 to 1200 ton, and centrifugal machines from 100 to 10,000 ton.

Absorption water-chilling machines are available in both steam- or hot water–heated (indirect-fired) units or fuel-fired (direct-fired) units. Fuel-fired chillers are currently available in the range of 100 to 1100 ton, while steam- and hot water–heated units are generally available in the range of 100 to 1600 ton.

Centrifugal machines are also available with engine or turbine drives in the medium to large models; they may offer energy cost savings on some installations. Although uncommon today, gas-fired engine–driven chillers or total energy plants may increase in use as cooling loads increase on electric power distribution systems. Chilled water systems under consideration for energy storage such as cold water or ice should evaluate gas engine–driven chillers as a possible life-cycle cost alternative. Currently, the consumer cost of natural gas on an equivalent basis can be one-fourth to one-fifth of that for electric power.

The condenser water operation of water chillers is described in Chap. 11 on cooling tower pumps and in Chap. 16 on closed condenser water systems.

14.2   Rating of Chillers

Water chillers for HVAC systems are rated in tons of cooling; 1 ton of cooling is equal to 12,000 Btu/h. For pumping and control procedures, it is convenient sometimes to rate chillers by the amount of chilled water that is produced by them. As described in Chap. 9, the chilled water capacity, in gallons per minute, is determined by the following formula:

$$\text{Chiller capacity, gal/min} = \frac{\text{rated tons} \cdot 24}{\text{temperature difference, °F}} \qquad (9.2)$$

where temperature difference is the water temperature drop across the chiller evaporator. For example, a 1000-ton chiller operating at a chilled water temperature difference of 12°F will have a chilled water capacity of 2000 gal/min.

14.3 Chiller Energy Consumption

The efficiency of chillers is a function of their energy consumption; for example, electric motor–driven chillers are rated in kilowatts per ton. The efficiency of absorption chillers is stated in either fuel per ton, pounds of steam per ton, or a coefficient of performance. Coefficient of performance *COP* for any type of equipment is

$$COP = \frac{\text{useful energy acquired}}{\text{energy applied}} \tag{14.1}$$

Coefficient of performance for chillers can be based on either British thermal units or kilowatts. Another ratio that may be of value is the energy efficiency ratio *EER,* which is defined by the American Society of Heating, Refrigerating and Air-Conditioning Engineers (ASHRAE) as the net cooling capacity in British thermal units per hour divided by the watt-hours applied, or

$$EER = \frac{\text{Btu/h net cooling effect}}{\text{Wh applied}} \tag{14.2}$$

The relationship between the coefficient of performance and the energy efficiency ratio is

$$EER = 3.412 \cdot COP \tag{14.3}$$

To demonstrate the calculations between kilowatts per ton, *COP,* and *EER,* assume that a chiller has an energy consumption rate of 0.52 kW/ton. This energy is equal to 0.52 · 3412, or 1774 Btu/h input. Therefore, from Eq. 14.1,

$$COP = \frac{12,000}{1774} = 6.76$$

From Eq. 14.2,

$$EER = \frac{12,000}{520} = 23.1$$

From Eq. 14.3,

$$EER = 3.412 \cdot 6.76 = 23.1$$

The energy rate for electric motor–driven chillers varies from 0.50 to 1.00 kW/ton at full load, while the steam rate for steam-heated absorption machines varies from 12 to 20 lb/ton.

14.3.1 Electric motor–driven chiller consumption

The cooling load on an electric motor–driven chiller determines to a large extent the energy consumption, although the temperatures of both the chiller water leaving the evaporator and the condenser water coming from the cooling tower or closed-circuit cooler also will affect energy usage. Following are some simple rules:

1. The kilowatts per ton for an electric motor–driven chiller are usually lowest in load ranges from 50 to 100 percent of full load and highest in load ranges below 40 percent of full load.

2. The kilowatts per ton are lowest at high leaving chilled water temperatures and highest for low leaving chilled water temperatures.

3. The kilowatts per ton are lowest for low condenser water temperatures and highest for high condenser water temperatures

These rules offer some simple guidelines for the application of chillers to standard installations as well as energy storage systems. From these rules it is obvious that the energy consumption for a specific chiller must be stated at certain chilled water and condenser water temperatures. For example, the kilowatt per ton energy consumptions for the 1000-ton chiller in Table 14.1 are for entering and leaving evaporator (chilled water) temperatures of 55 and 45°F and entering and leaving condenser water temperatures of 85 and 95°F. These figures reflect the energy consumptions of the chiller itself and do not include the energy of required auxiliaries such as cooling towers, condenser water pumps, or primary chilled water pumps.

TABLE 14.1 Typical kilowatt per ton Energy Consumptions for a 1000-Ton Centrifugal Chiller (Chiller Only)

Percent load	kW/ton	Percent load	kW/ton
10	1.140	60	0.570
20	0.805	70	0.560
30	0.683	80	0.559
40	0.633	90	0.561
50	0.594	100	0.579

14.3.2 Total energy consumption for a chilled water plant

The great emphasis has been on the efficiency of the chillers themselves; the chiller manufacturers have worked diligently to improve the efficiency of their machines. As this work was being carried out, very little was done to evaluate the energy consumption of the chiller's auxiliaries, namely, the primary chilled water pumps, the condenser water pumps, and the cooling tower fans. Often, at low-load conditions on the water chiller, the total energy consumption of these auxiliaries exceeded that of the chillers themselves. Recognizing this situation, work has been initiated in the HVAC industry to develop a total kilowatt per ton or coefficient of performance for the total chiller plant. The equation for this calculation is

$$\text{kW/ton} = \frac{\sum \text{plant consumptions (kW)}}{\text{tons of cooling on system}} \qquad (14.4)$$

where plant consumptions in kilowatts equals the energy consumption in kilowatts of (1) the water chillers themselves, (2) the condenser water pump motors, (3) the primary water pump motors, and (4) the cooling tower fan motors or the air-cooled condenser motors. The secondary or distribution pumps are not included because they are considered to be part of the chilled water system and figure in the overall chilled water system efficiency.

The load in tons of cooling on the system is computed by the formula

$$\text{Tons of cooling} = \frac{500 \cdot Q \cdot (T_2 - T_1)}{12,000}$$

$$= \frac{Q \cdot (T_2 - T_1)}{24} \qquad (14.5)$$

where T_2 = system return water temperature, °F
   T_1 = system supply water temperature, °F
   Q = system flow, gal/min

All the data required to compute the kilowatts per ton for a central chilled water plant are easily available in analog signals for digital computers to do the calculations shown above.

The significant fact about central plant kilowatts per ton figures is their cost-effectiveness for the medium- to large-sized chiller plants. The instrumentation cost is usually less than $10,000 at current pricing unless the flowmeter or a Btu meter is a required addition. An accurate flowmeter is recommended, with an accuracy of ±2 percent of flow rate through a range of 1 to 15 ft/s water velocity.

Kilowatts per ton should be displayed prominently in the equipment room to permit operators to observe the effect on the overall energy consumption when another chiller, pump, or cooling tower is put into service. Reiterating, with overall kilowatts per ton, the effect of pumps and cooling towers on central plant efficiency should be included in the plant operating displays.

The advent of metric system use in the United States has eliminated the calculation of cooling in tons. Therefore, the kilowatts per ton for plants operating with SI units become a coefficient of performance, kilowatts of useful energy divided by kilowatts expended (Eq. 14.1). Since 1 ton of cooling is approximately 3.52 kW, the coefficient of performance of these plants is usually in the range of 3 to 5.

14.4   Circuiting Chilled Water to Chillers

There are fundamentals for the circuiting of chillers that should not be violated in order to achieve maximum efficiency. Some of these are

1. Design the piping so that energy consumption of chillers is not increased.
2. Arrange the piping so that all chillers receive the same return water temperature.
3. Ensure that the required water flow through the evaporators is maintained.
4. Always install the chillers in accordance with the recommendation of their manufacturers.

Much energy is wasted because of improper circuiting of chillers. Chillers seem to be more vulnerable to incorrect installation than other equipment owing to the fact that their energy consumption is greatly affected by the actual piping arrangement. When chillers are being considered for installation, the designer must ensure that the piping arrangement does not adversely affect the energy consumption of this equipment.

The general conclusion as of this date is to operate constant- or variable-speed chillers with constant water flow through the evaporators. This has several advantages for the chiller:

1. Chances of spot freezing in the evaporator are eliminated.
2. The evaporator can operate at the design velocity in the tubes for the chilled water.

3. Control of the loading of refrigerant into the evaporator may be easier with constant water flow.

Even as this book is being written, new electronic controls for chillers are being introduced into the industry that may permit variation of flow in the chiller evaporator down to a specific velocity in the evaporator tubes. Variable flow in the evaporator is a subject that must be approved by each chiller manufacturer and depends on the chiller controls and refrigerant involved. More on this subject will be presented in Chap. 15.

The following discussion will be based on constant flow through the chiller evaporator, since that is the preference for existing chillers and some variable-speed chillers, whether driven by engines or turbines.

Equal return water temperature is probably the most important of the preceding piping fundamentals for chillers, regardless of the type of chiller or flow through their evaporators. This critical fact can be demonstrated by the information in Table 14.1. It is obvious that the chiller cannot be operated below 30 percent load without a sizable increase in energy consumption for every ton of cooling produced. If one chiller on an installation of several chillers is forced to operate at 5 to 20 percent of design capacity, the kilowatts per ton for that chiller will be excessive. There is considerable argument in the industry about preferential loading of chillers by water flow. Preferential loading will be discussed later, and it will be demonstrated how optimal loading of efficient chillers can be achieved without using special water circuits.

How can we be sure that the return water is the same for all chillers? By connecting the piping so that all return water and any water from a bypass are thoroughly mixed before any of the water enters a chiller. Figure 14.1a demonstrates the proper method of connecting chillers for variable-volume systems that include a bypass around them. This piping design uses the velocity head of the chilled water to ensure that the water flows in the correct direction in the bypass.

Figure 14.1b–d describes improper methods of connecting chillers. These connections create uneven water return to them, which results in excessive energy consumption. The bypasses are installed incorrectly with the chillers. The system return connections are not connected to the bypass before returning the water to the chillers. Table 14.2 provides typical energy consumption, in kilowatts per ton, for two chillers connected correctly in accordance with Fig. 14.1a and equally loaded and then incorrectly connected as shown in Fig. 14.1b and unequally loaded. The energy consumption is definitely increased when the chillers are sequentially loaded and when the total load on the chilled water system is 100 to 150 percent capacity of one chiller.

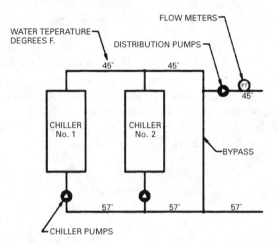

RETURN TEMPERATURE IS UNIFORM TO EACH CHILLER REGARDLESS OF
LOAD ON SYSTEM OR NUMBER OF CHILLERS RUNNING.

a. Correct chiller piping.

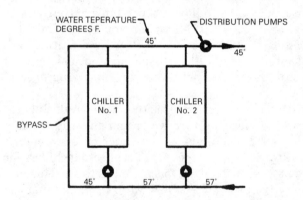

RETURN CHILLED WATER TEMPERATURES FOR CHILLER No. 1 VARIES FROM
46.2 DEGREES. AT SYSTEM LOAD OF 110% CHILLER CAPACITY TO 57
DEGREES F. AT 200% WHILE RETURN TEMPERATURE OF CHILLER No. 2
IS ALWAYS 57 DEGREES F.

b. Questionable chiller piping.

Figure 14.1 Correct and incorrect chiller piping. (*From The
Water Management Manual, SYSTECON, Inc., West Chester,
Ohio, 1992, Fig. 6.2.*)

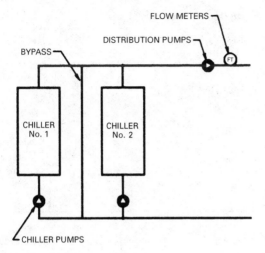

c. Incorrect bypass location.

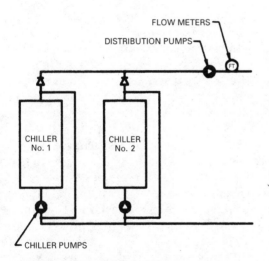

INDIVIDUAL BYPASSES DO NOT GUARANTEE EQUAL
LOADING OF ALL CHILLERS.

d. Individual chiller bypasses.

Figure 14.1 *(Continued)*

TABLE 14.2   Energy Effect of Piping Connections on Chillers

System load, percent of one chiller	kW/ton, two chillers equally loaded	kW/ton, two chillers sequentially loaded
100	0.57	Indeterminant
110	0.58	1.24
120	0.58	0.74
130	0.59	0.67
140	0.60	0.64
150	0.60	0.63
160	0.61	0.63
170	0.62	0.64
180	0.64	0.65
190	0.66	0.66
200	0.68	0.68

A detailed energy analysis will be made later in this chapter on trying to preferentially load an efficient chiller over an inefficient chiller by hydraulic procedures. Reiterating, the bypass shown in Fig. 14.1a is the proper method of installation of the chiller piping.

14.4.1   Preventing water migration in chiller bypasses

It is imperative that the system be piped and controlled so that water never flows in the reverse direction in the bypass under normal operation. The reasons for this fact are that reverse flow can start chillers inadvertently, and if warm return water gets into the supply pipe, it may take some time to get the supply water back to design temperature. This process of allowing warm system return water to get into the colder supply water is called *migration.*

The bypass must be connected to prevent migration or induction of water out of the bypass into the supply pipe. This is achieved by recognizing two facts: (1) the velocity head of water is the inertial energy that causes water to continue in one direction unless it is forced to turn, and (2) any tee is a crude eductor. Water flowing through the run of a tee can cause water to be pulled out of the branch into the run.

Both these facts must be used when connecting piping around chillers and boilers. We must go back to Bernoulli's theorem and remember that all water flowing in a pipe has velocity head. Velocity heads are shown in Table 3.4 for pipe friction. For example, if a 10-in-diameter pipe is carrying 1200 gal/min, the velocity head of the water is 0.37 ft. This does not appear to be much, but it is equal to 3700 (ft · lb)/min. Therefore, the tee connecting the supply pipe of the chilled water system to the chiller loop should be arranged as shown in Fig. 14.2; this directs the energy into the bypass pipe, not into the supply pipe, of the chilled water system. The water must be forced out

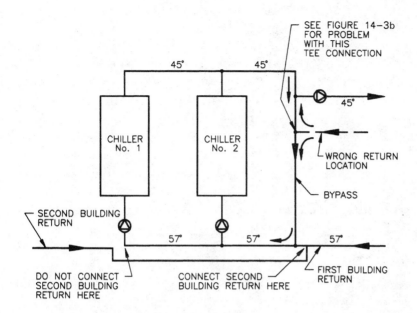

RETURN TEMPERATURE MUST BE UNIFORM TO EACH CHILLER
REGARDLESS OF LOAD ON SYSTEM OR NUMBER OF CHILLERS RUNNING.

Figure 14.2 Connection of return piping to a chiller loop. (*From The Water Management Manual, SYSTECON, Inc., West Chester, Ohio, 1992, Fig. 6.6A.*)

the bypass into the system supply pipe when this pipe is connected to the branch of the tee. This forced turn does not aid eduction of return water into the supply connection.

Likewise, the connection of the system return pipe should encourage eduction out of the bypass, and therefore, it should be connected with the system return pipe on the run of the tee and the bypass connected to the branch of the tee in accordance with Fig. 14.2. The velocity head is running through the tee and not into the bypass, as shown in Fig. 14.3*a*. The return must not be connected closely to the supply pipe where the velocity head rams into the bypass, as shown in Fig. 14.3*b*. This connection is notorious for encouraging migration of warm return water into the supply pipe.

Figure 14.1*b–d* describes some other piping arrangements that should be avoided. All return piping should be collected together with the bypass piping before any water approaches the return connection of a chiller, as shown in Fig. 14.2. Also, the return pipe should not be installed next to the supply connection, as designated in Fig. 14.2 as the wrong connection. This incorrect connection creates an unsatisfactory backflow as described above and in this figure where return water flows back into the supply water. The preceding velocity head

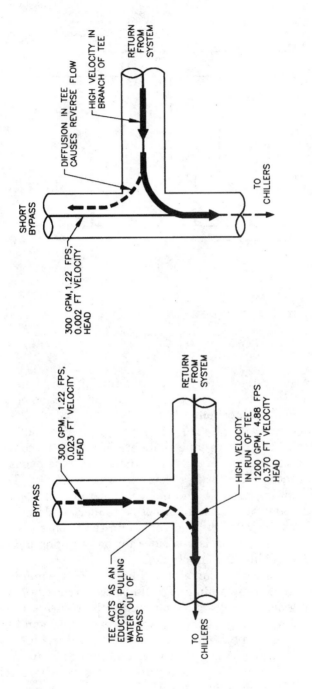

a. Water flow through run of a tee.

b. Water flow out of branch of a tee.

Figure 14.3 Velocity head in pipe tees.

rams into the bypass piping, causing this backflow or migration of return water into the supply water. This increases the supply temperature on chilled water systems and decreases the supply temperature on hot water systems.

The pipe size of the bypass has created controversy; some designers make the bypass the same size as the header, while others design the bypass to the maximum flow of water through one chiller. One important fact about sizing the headers on multiple chillers or boilers is the need to reduce the overall friction loss of the headers and bypass at full design load. If the bypass is downsized to the flow of one chiller or boiler, this may cause much greater total friction in the headers and bypass. This may result in fluctuations in flow through the chillers to where a flow-regulating device of some type is needed to maintain constant flow. On the other hand, with minimal friction in the headers and bypass, there will be very little flow fluctuation without the regulating device.

14.5 Operating and Sequencing Chillers

The efficient operation of a central energy plant consisting of more than one chiller is dependent on how this equipment is added and subtracted as the load on the system increases and decreases. There are many different methods for sequencing chillers, but the most efficient procedure has been the use of total water flow of the system as the parameter for sequencing chillers.

This procedure of using system flow is very accurate and can be applied even when temperature reset is involved. This is so because chillers can be rated in gallons per minute. For example, assume that a system consists of three 1000-ton chillers operating at a temperature difference of 12°F or 2 gal/min per ton. Therefore, each chiller has a capacity of 2000 gal/min. One chiller should operate when the system flow is below 2000 gal/min, two chillers from 2000 to 4000 gal/min, and three chillers when the system flow is in excess of 4000 gal/min. The sequence schedule would be as follows:

Chillers Running	On	Off
First chiller	Up to 1950 gal/min	—
Second chiller	1950 gal/min	3850 gal/min
Third chiller	3950 gal/min	3850 gal/min

In most cases, it is undesirable to add a chiller before the chillers on line approach 95 percent of design load. Although Fig. 14.4a and b indicates an increase in kilowatts per ton as centrifugal chillers approach design load, addition of the energy consumption of the chiller

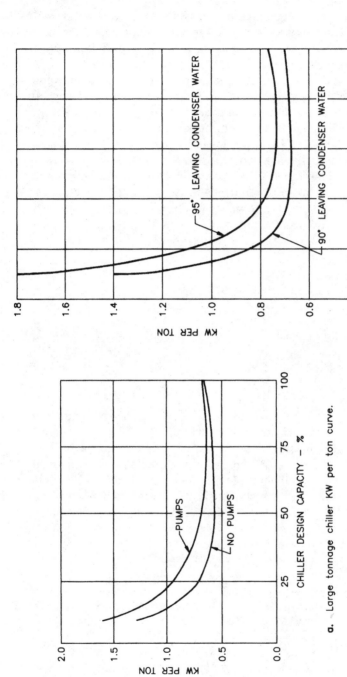

Figure 14.4 Typical kW/ton curves for a centrifugal chiller. *(From The Water Management Manual, SYSTECON, Inc., West Chester, Ohio, 1992, Figs. 6.6B and 6.8)*

a. Large tonnage chiller KW per ton curve.

b. Effect of condenser water temperature on chiller energy consumption.

auxiliaries will change the shape of this curve. If the kilowatt consumption of the condenser pump, primary chiller pump, and possibly additional energy consumption by the cooling towers are added to that for the chiller, the total kilowatts per ton curve will appear as shown in Fig. 14.4a.

If leaving temperature reset is employed to achieve high efficiency in chiller operation, flow can still be used as a means of sequencing chillers. The software of the controller involved can be set to include an algorithm that changes the gallons per minute per ton with changes in the temperature settings for chiller leaving water temperature.

14.5.1 Unsatisfactory methods of sequencing chillers

There are a number of unsatisfactory methods of sequencing chillers:

1. *By means of chiller power input in amperes.* Amperes cannot be used owing to the variation in power input with leaving condenser water temperature or leaving chilled water temperature. Figure 14.4b describes the variation in energy consumption due to changes in leaving condenser water temperature. Also, electric utilities are allowed a variation of ±5 percent in voltage, which will, of course, change the amperes consumed by the machine.

2. *By chiller leaving water temperature.* Leaving water temperature can be used to add chillers, but it cannot be used to stop a machine because the running chillers will adjust the leaving temperature to the load on the system. Stopping the chillers by return temperature can be approximate due to the fluctuation in return temperature from the chilled water system. Chillers can be added by leaving temperature and subtracted by system flow.

3. *By reverse flow in the bypass.* When reverse flow occurs in the bypass, a sizable amount of warm return water will have flowed into the supply piping before the next chiller can be started and begin to produce cool water.

Sequencing chillers by system flow has proved to be the most precise method of sequencing chillers. Also, many of the problems just described with bypasses and hydraulically loading of chillers can be eliminated by chilled water storage if there is space for its inclusion. Chapter 13 has described in detail the advantages of chilled water storage in evening out the production of chilled water by the chillers.

14.6 Preferentially Loading Chillers

This subject was deferred until the discussion on sequencing chillers was completed. Preferential loading of chillers is desirable to ensure that the most efficient chillers handle most of the cooling load. Trying to do this through water circuiting has proven to be futile because of the poor efficiency of most chillers at very low percentages of design load. The connecting of an efficient chiller with an older, less efficient chiller requires special operating procedures to ensure that the most efficient chiller is first loaded as high as possible before starting the less efficient chiller. The following example will describe this problem.

Assume that two 800-ton chillers, one old and one new, are to be operated together, and their kilowatt per ton consumptions are as shown in Table 14.3. In each case, these data include the energy requirements of both their condenser and primary pumps, which will be assumed as constant speed. The pump motors consume 52 kW for both the efficient and inefficient chillers. It will be assumed also that the cooling tower cells will be sequenced in such a manner that they are not related to one particular chiller but to the total load on the system.

Table 14.4 describes the energy consumption with preferential piping, as shown in Fig. 14.1b. The standby chiller must be started at 100 percent load on the high-efficiency machine. This may cause the standby, low-efficiency chiller to start and stop frequently because of the very low load on that chiller.

The total energy expended by preferential piping should be compared with the energy expended by equalized piping (Table 14.5).

Tables 14.4 and 14.5 demonstrate that preferential piping uses more energy than equalized piping for the two chillers in parallel. The proper method of achieving maximum tonnage from the efficient machine in a two-chiller installation is (1) to provide equal flow to the chillers and secure equal return temperatures to them and (2) to ad-

TABLE 14.3   Chiller Kilowatts per Ton with Two Unequal Chillers

Chiller load, %	Efficient chiller	Less efficient chiller
10	1.88	2.20
20	1.13	1.30
30	0.87	1.03
40	0.75	0.89
50	0.70	0.84
60	0.68	0.82
70	0.67	0.81
80	0.69	0.81
90	0.71	0.82
100	0.75	0.83

NOTE: Pump kW/ton is included in the chiller kW/ton.

TABLE 14.4   Chiller Energy Consumption with Preferential Piping

% Load of one chiller	% Load and kW on efficient chiller	% Load and kW on inefficient chiller	Total kW of two chillers
80	80–442	—	442
90	90–513	—	513
100	100–596	—	596
110	100–596	10–175	771
120	100–596	20–209	805
130	100–596	30–246	842
140	100–596	40–286	882
150	100–596	50–332	928
160	100–596	60–393	989
170	100–596	70–455	1051
180	100–596	80–519	1115
190	100–596	90–592	1188
200	100–596	100–660	1256

TABLE 14.5   Chiller Energy Consumption with Equalized Piping

% Load of one chiller	% Load and kW on efficient chiller	% Load and kW on inefficient chiller	Total kW of two chillers
80	80–442	—	442
90	90–513	—	513
100	100–596	—	596
110	55–303	55–362	665
120	60–326	60–393	719
130	65–360	65–442	802
140	70–377	70–455	832
150	75–405	75–484	889
160	80–442	80–519	961
170	85–476	85–552	1028
180	90–513	90–592	1105
190	95–553	95–626	1179
200	100–596	100–660	1256

just the chiller leaving temperatures to place more load on the more efficient chiller.

By raising the leaving water temperature 2°F on the inefficient chiller and lowering the temperature 2°F on the efficient machine, more of the load is transferred to the efficient machine. The correct leaving water temperature is still achieved because the flow is equal in each machine.

Another method to minimize energy use is to hold the percentage of load on the inefficient machine constant at the point of lowest kilowatts per ton and to transfer the remaining load to the efficient machine. There is no exact method to determine these optimized points of operation and temperature differences except by actual operation of the chillers and to record the total kilowatts with various transition points. These are the proper methods of operating efficient chillers in the same system with inefficient chillers. Utilization of the total kilowatt per ton efficiency for a chiller plant enables the operator to make such adjustments to achieve the lowest possible kilowatts per ton for the entire chilled water plant.

Note. *The data included in Tables 14.4 and 14.5 are typical of electric motor–driven centrifugal chillers of approximately 800 tons capacity. These energy consumptions are representations and should not be used for any actual installation of chillers of this size. The energy consumption of any chiller, in kilowatts or kilowatts per ton, should be secured from its manufacturer and verified at the design load point conditions.*

14.7   Fuel-Fired Absorption Chillers

Chilled water pumping for fuel-fired absorption chillers should be treated substantially like that for other chillers. The principal difference in the two types of chillers is the minimum firing rate for the burners of fuel-fired chillers. Most burners for this type of chiller have a minimum firing rate of around 30 percent of maximum fuel consumption. If the load on the chilled water system drops below 30 percent of the capacity of the smallest chiller, unsatisfactory cycling may occur. There are several methods of solving this problem. The most common method with multiple chillers is (1) to select the chillers to handle the total load on the chilled water system and (2) at minimum load on the system, to operate the smallest chiller within 30 to 100 percent of its design load. Another method that may offer a satisfactory solution is to employ chilled water storage as shown in Fig. 13.1.

With this arrangement, the normal chiller bypass becomes a chilled water storage tank that allows the chiller to run at a higher firing rate and cycle on and off in a reasonable manner by storing the excess chiller capacity in the storage tank. On small systems, a pressure-type tank can be used, which will virtually eliminate the problems caused by open tanks. On larger installations, the installed cost and space requirements of a pressurized tank may preclude its use. When used, the installation of open storage tanks should be designed in accordance with the provisions of Chap. 13.

14.8   Connecting Primary Pumps to Chillers

There are several methods of connecting primary pumps to multiple parallel chillers. Figure 14.5a describes the simplest installation with an individual pump for each chiller that can be sized to the require-

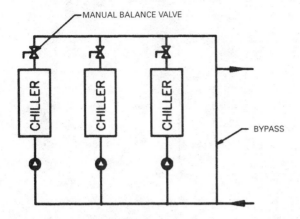

a.   Basic installation for primary pumps.

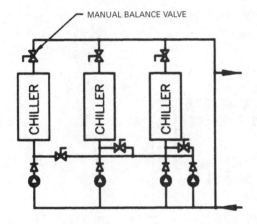

b.   Standby primary pump.

Figure 14.5   Primary pump connections for chillers. (*From The Water Management Manual, SYSTECON, Inc., West Chester, Ohio, 1992, Figs. 8.17 to 8.20.*)

ments of the individual chiller. A simple balancing valve should be installed on each chiller to set the actual flow through each unit.

Many chiller manufacturers prefer this method of connection, because each pump can be started and stopped individually by the chiller control. This enables the chiller manufacturer to program the pump to operate for a short time after the chiller has been stopped;

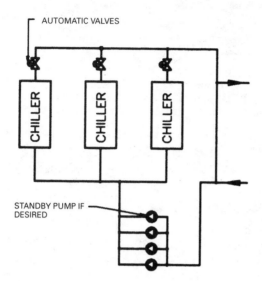

c.   Centralized primary pumps.

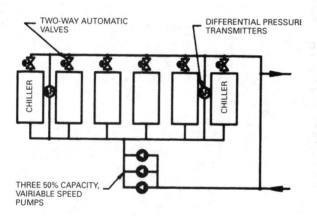

d.   Variable speed primary pumps.

Figure 14.5   *(Continued)*

this procedure helps to ensure that residual refrigerant will pass out of the evaporator. If the headers and bypass are sized for minimum friction, expensive motorized valves are not needed to control the flow through the chillers. The disadvantage to this pumping arrangement is that there is no standby pump. If one chiller plus a different pump are put out of service, only one chiller would be available for operation.

This deficiency can be remedied by installing a standby pump, as shown in Fig. 14.5b. The standby pump can be connected to any chiller automatically or manually. This is the most popular and economical method of providing standby pump capacity.

Some designers prefer to offer primary pump standby by installing the primary pumps in parallel, as described in Fig. 14.5c. This design is more expensive, since it requires three headers instead of two; it also requires a motorized valve on each chiller. If the designer is not careful, enough friction will exist in the three headers to require installation of modulating control on the individual motorized valves. There are few installations that justify this method of connecting primary pumps to chillers.

Another method of handling the primary pumping where there is a number of parallel chillers is to install variable-speed primary pumps, as shown in Fig. 14.5d. This installation with six chillers will be better served by installing three variable-speed primary pumps, each with 50 percent of the total capacity of all six chillers. With this procedure, the primary pump's speed is controlled by differential pressure transmitters that maintain the desired constant differential pressure across the chiller evaporators. The standby pumps are added and subtracted by the best efficiency or wire-to-water efficiency methods described in Chap. 10 of this book. The actual control selected will depend on the size of the primary pumps. The flow through each evaporator is constant because the differential pressure transmitters maintain a constant pressure drop across them.

There are several significant advantages to variable-speed pumping in large, multiple-chiller installations. First, the flow through each chiller evaporator is relatively constant, whether one or all of the chillers are in operation. This is due to the precise control by the differential pressure transmitter and the digital pump speed control. Second, since the pressure drop through the chiller loop piping is not included in the differential pressure set point, this loop can be reduced to a more economical size without affecting the flow through any of the chillers.

All the preceding diagrams show the primary pumps on the inlet side of the chiller. On high-rise buildings where the static pressure is great, it may be advisable to install the primary pumps on the discharge of the chiller evaporator to help reduce the total pressure on the evaporators (see Fig. 14.6a). High-rise buildings can create some special problems owing to the great amount of static head possible on equipment installed in these buildings. This static head has been de-

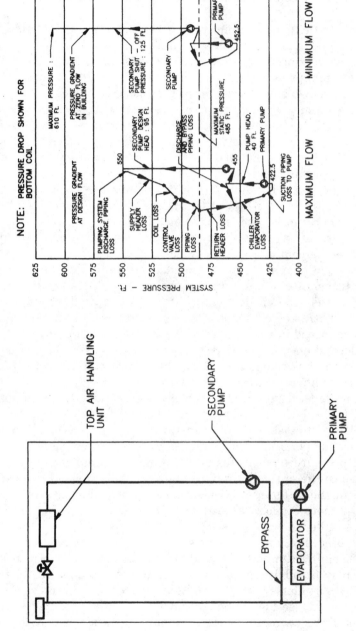

Figure 14.6 Primary pump location for high-rise buildings. (*From The Water Management Manual, SYSTECON, Inc., West Chester, Ohio, 1992, Fig. 8.21C,D.*)

scribed in a pressure-gradient diagram (Fig. 14.6b) for the building in Fig. 14.6a. For purposes of this example, this building has an overall height of 462 ft and a minimum of 10 psig (23 ft) at the top of the building for a total static head of 485 ft. The diagram shown in Fig. 14.6b provides the pressure gradients for both the maximum-flow and the zero-flow conditions. The zero-flow diagram demonstrates that the casing pressure for the pumps and the design pressure of piping and valves at the bottom of the building must be based on the shutoff head of the secondary pump. In this case, the shutoff head is 264 psig.

Locating chillers and pumps on top of a high-rise building will certainly reduce the pressure on the equipment, but the piping and coils at the bottom of the building will still be subjected to the high pressure shown in the pressure gradients of Fig. 14.6b.

14.9 Connecting Multiple Chiller Plants

A common and desirable situation encountered on campus-type installations and very large building complexes is to connect several chiller plants together to serve the chilled water requirements of all the buildings or zones. This provides significant advantages over the arrangement of individual chillers for each building; these are (1) standby capacity is provided to all buildings, (2) operating chillers on very low loads is avoided because one chiller can handle the minimum loads for a number of buildings, and (3) by connecting the chillers together, the ability to reduce the chiller plant size is often realized due to the combined large load diversity of all the buildings. This increases greatly the efficiency of chiller operation.

Figure 14.7 describes such an arrangement. This installation has four buildings with chillers and two without. The buildings without chillers are new; there is no desire or need to install chillers in these buildings because there is more than enough excess chiller capacity in the other buildings to serve these two new buildings.

To utilize this excess capacity, it is necessary to install a campus supply and return loop. The following are other design parameters that should be followed for such a facility:

1. No new central pump stations are to be used.

2. A minimum of energy-wasting devices is to be used.

3. Any building must be capable of being served by any of the chiller plants.

4. The central control center operator must have the ability to select the most efficient chillers without changing any valving in the buildings.

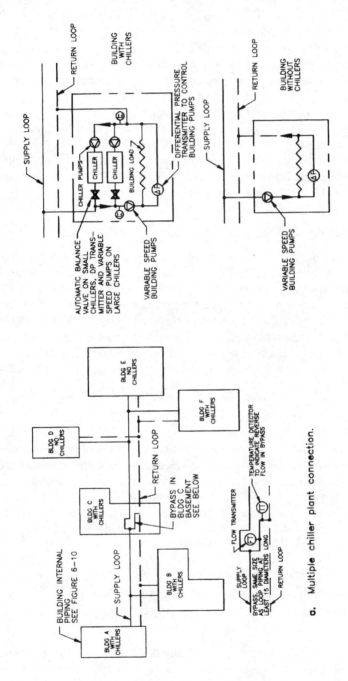

Figure 14.7 Multiple-chiller plant connections. (*From The Water Management Manual, SYSTECON, Inc., West Chester, Ohio, 1992, Fig. 6.9.*)

5. The flow in any chiller evaporator must not be restricted.

6. If a chiller fails, any building load must be picked up automatically without any changes in coil valve position.

7. Flowmeters must be installed in the loop bypass and in each building to determine individual building loads and to indicate the amount of excess capacity occurring in the bypass. A temperature transmitter also must be located in the bypass to indicate reverse flow that should not occur under normal operation.

The building connections are described in Fig. 14.7a for buildings both with and without chillers. The chiller pumps are constant-speed pumps, and the building pumps are variable-speed pumps. The building pumps are controlled by differential pressure transmitters located properly in each building. The flow transmitters also provide the central control operator with data on each building's load and the flow from the supply loop that is occurring in each building.

The central bypass, as described in Fig. 14.7b, is necessary to ensure that the flow in any chiller evaporator is not reduced. Sudden changes in campus load could drastically reduce the flow in a chiller evaporator and cause possible freezing in the tubes without this bypass.

Questions often arise on how to calculate the head for the chiller pumps and the building pumps as well as sizing the supply and return loops. Both the chiller and building pumps must have the capacity to pump to and from the loop bypass under the most diverse load conditions that can occur on the entire installation. Likewise, the design engineer must evaluate all the possible load conditions and select the one that requires the largest flows in the loops.

This will enable the system designer not only to select the head for both the chiller and building pumps but also to size the loop piping. In actual operation, the normal load on the loops is much less than the maximum possible load. This explains the need for the variable-speed building pumps.

If there is a great head variation on the chiller pumps, they also should be variable-speed pumps with differential pressure transmitters across the evaporators to maintain constant flow through each evaporator. If the variation of this flow is expected to be moderate, the chiller pumps may be constant-speed pumps with an automatic flow control valve on each chiller.

The maximum head for the chiller pumps and that for the building pumps must be checked to ensure that the chiller or loop head never approaches that of the buildings. Otherwise, the chiller pumps have the possibility of overpressuring the building pumps. If there is a possibility of this, a pressure gradient should be developed for the condition to verify the pressure relationship.

If designer determines that significant energy can be saved by pumping a building with its chiller pumps, the building pumps should be equipped with a bypass. This enables the building pumps to be shut down when adequate pressure is developed by the chiller pumps. A pumping system with a bypass is shown in Fig. 15.4c.

There are several advantages to this method of connecting multiple chiller plants over those systems which depend on computerization of flow and temperature transmitters to interconnect the chiller plants: (1) the central bypass ensures that adequate flow will be provided for all the chiller evaporators that have been placed in operation, (2) the system can operate with the central computer totally shut down, and (3) the central bypass method is much simpler, less expensive, and more reliable than the computerized control system.

There are other procedures for interconnecting multiple chiller plants. Their use depends on the actual physical arrangement of the entire facility. Some multiple-chiller systems have been designed to use energy wasters such as three-way temperature-control valves or return temperature-control valves; as a general rule, where possible, these systems should be avoided.

14.10 Bibliography

ASHRAE Terminology of Heating, Ventilation, Air Conditioning and Refrigeration, 2d ed., ASHRAE, Atlanta, Ga., 1991.
Engineers Newsletter, 23(1), 1994, The Trane Company, La Crosse, Wis.
The Water Management Manual, SYSTECON, Inc., West Chester, Ohio, 1992.

15

Chilled Water Distribution Systems

15.1 Introduction

Chapter 8 described basic uses of water in HVAC systems, while Chap. 14 reviewed the proper application of chillers and primary or chiller pumps. This chapter describes the various methods of distributing chilled water in a building or group of buildings. It pertains principally to the second zone of a chilled water system, that of transporting the chilled water from the chiller plant to the terminal units or coils.

The principal emphasis of contemporary design of chilled water systems is to move the water with the highest efficiency and lowest energy consumption possible within the economic constraints of the installation. This is of greater importance for chilled water than for hot water because of the larger volumes of water required with chilled water systems. Because of their extensive piping, chilled water systems usually have higher pump heads and larger pump motors.

Chilled water distribution systems should be configured to avoid interfering with the efficient operation of the chillers. This was discussed in Chap. 14 and must be emphasized again here. There have been efforts to produce complicated pumping systems to save pumping energy at the expense of chiller energy consumption. *Resetting chiller leaving water temperature upward almost always affords greater energy savings than resetting zone water temperatures to save pumping energy.* The added friction of zone control valves and extra piping often nullifies much of the savings in pumping energy achieved by zone temperature reset. It is the responsibility of the system designer to ascertain this on each prospective installation. It should be emphasized that the designer must always ensure that the system chilled water temperature is low enough to satisfy the latent load on the cooling coils.

Chapter 8 provides some specific information on the energy wasted in extra piping, fittings, and valves when three-way valves or zone reset with circulators is used on chilled water. Table 8.3 provides information for a specific application. Almost every detailed analysis of zone reset proves that energy is lost as a result of extra friction losses and loss of energy savings in the chillers themselves.

There are a number of different chilled water distribution systems; all the major types will be reviewed in this chapter. The principal types are (1) primary, (2) primary-secondary, (3) primary-secondary-tertiary, and (4) distributed. The discussion will review pumping energy, first cost, ease of control, and general operating considerations.

15.2 Primary Systems

The primary chilled water system is a constant-volume system that is used with three-way valves (Fig. 15.1a). When energy was relatively cheap, this system was used. It had several energy-wasting conditions that are no longer acceptable. For example,

1. Constant-speed pumps were used; any overpressure caused by selecting too high a design head for the pumps was eliminated in the balance valves.

2. The balance valves consumed energy even though the pump head was selected correctly; this is due to the differences in distribution friction between the cooling coils.

3. As the cooling loads subsided on this system, the return temperature mounted as a result of bypassing of cold water around the cooling coils. This caused the chiller or chillers to operate at very low loads and very high energy rates in kilowatts per ton.

Because of these energy problems, this system should not be used on most chilled water systems. On top of the energy losses, this is an expensive system in terms of cost because of the three-way valves and their piping. In some cases, actual bids for this system have proved it to be higher in first cost than a variable-speed primary-secondary system of the same size.

Another problem with this system involves rebalancing the flow through it whenever a major load is removed from operation. As the system aged or changes in the piping occurred, rebalancing also was required. This rebalancing was eliminated by automatic balancing

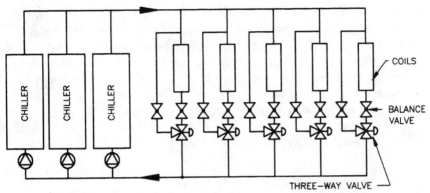

a. Constant volume, primary system.

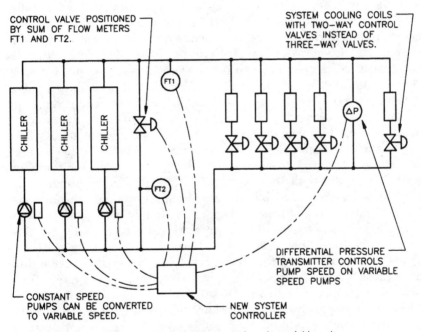

b. Conversion of constant volume system to variable volume.

Figure 15.1 Primary chilled water systems.

valves that contained springs and orifices which limited the flow through a particular coil or terminal unit to its design flow. This solved the rebalancing problem but added appreciable friction to the water system. Further, full design flow required by the chiller was not achieved under all load conditions. The basic primary system has been modified to eliminate some of its problems.

15.2.1 Small primary pumping system

An exception to the preceding statements about primary three-way valve systems is a relatively small system consisting of a chiller and air-handling units located near the chiller. In such an installation, the distribution friction is very low, so there is no great energy loss resulting from the use of three-way valves. Concern should still be given to the chiller operation, since a single chiller could cause very poor chiller efficiencies under low loads on the air handlers. The use of a multiple-compressor reciprocating chiller may solve this problem.

15.2.2 Variable-volume primary systems for existing chillers

There is a primary chilled water system for multiple-chiller installations that uses contemporary digital electronics that may be an economic answer in converting existing chilled water systems to variable volume. This is shown in Fig. 15.1b in its final design. Often, there is no physical room to change to other systems such as primary-secondary pumping, which requires the addition of secondary pumps. The changes required in the system in Fig. 15.1b include (1) replacing the three-way control valves on the coils with two-way valves and (2) installing a chiller bypass with a control valve and flowmeter.

With these changes, the system becomes a variable-volume system with the opportunity to achieve the design return water temperature. The bypass with its control valve and flowmeter provides the minimum flow required by the chiller manufacturer through the chiller evaporator. Control of evaporator flow is developed by summing the values of the two flowmeters and adjusting the bypass flow until this desired flow is achieved. If the chiller manufacturer approves a minimum evaporator flow, such as 80 percent of design flow, this can be set easily into the control algorithm. This system functions best on multiple-chiller installations. If constant flow is required on a single-chiller installation, energy is still saved because the system head is reduced appreciably when there are light cooling loads on the system.

The pumps on this primary system may be constant-speed pumps when the total system head is less than 50 ft. Higher system heads usually result in the selection of variable-speed pumps; this, of course, is a judgment decision by the designer. Variable-speed pumps are controlled by remote differential pressure transmitters that hold the desired pressure differential across the cooling coils, their control valves, and the branch piping. With constant evaporator flow, the efficiency of this system is less than that for a secondary pumping system, since the bypass water passes through a control valve that has a loss equal to that of the pump head less the pressure loss of the chiller evaporator and its piping.

15.2.3 Variable-volume primary system with contemporary chillers

As this book is being written, new computer systems for controlling chillers are being offered to the industry. These control systems utilize a proportional-integral-derivative control that provides close operation of the chillers so that the flow can be varied through the chiller evaporators. The chiller manufacturer specifies a minimum flow or velocity through the evaporator tubes in feet per second that must be maintained by the variable-speed pumps; the bypass valve arrangement, as shown in Fig. 15.1b, varies its position to maintain the desired flow in the evaporator. These systems will replace many of the primary-secondary pumping arrangements offered below when chillers are available with precise control of their refrigeration and leaving water temperature.

15.2.4 Variable-volume primary system with bypass valve

Some designers use a pressure-relief valve in the bypass on this primary system with two-way control valves, as shown in Fig. 15.2a. This is not recommended because it merely eliminates the rise to shutoff of the pump head and does not guarantee proper flow through the chillers. Also, this relief valve maintains high differential pressures in the control valves on the cooling coils when the system friction subsides at low loads. This may cause lifting of the valve stems or wire cutting of the valve seats. This is a wasteful practice because a great amount of energy is lost with this relief valve. An almost-constant-volume system results; the pumping energy remains substantially that required at full system flow and head.

Another problem with this system is improper control of the relief valve. So often the relief valve just maintains a chiller leaving pressure; with this control, any slight change in system pressure causes the relief valve to function improperly. If it is absolutely necessary to use this wasteful practice, the relief valve should maintain a differential pressure between the supply and return mains of the chilled water system shown in Fig. 15.2a, not a fixed leaving pressure from the chillers.

As an example, the 2500 gal/min water system that was used in Chap. 10 for the description of wire-to-water efficiency using both

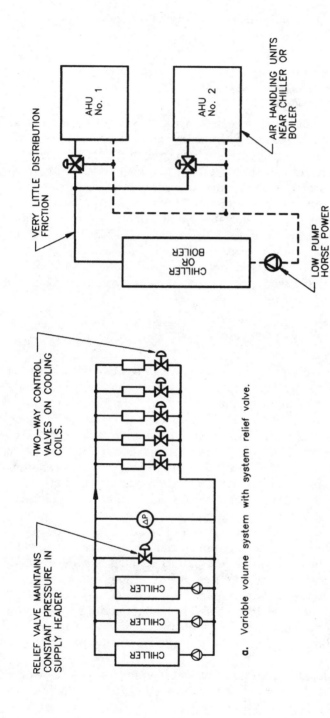

a. Variable volume system with system relief valve.

b. Small constant volume system.

Figure 15.2 Special primary pumping systems.

TABLE 15.1   Wire-to-Water Efficiency for a Pressure-Relief System

System gal/min	System head, ft	Water hp	Overpressure,* ft	Pump input kW	Wire-to-water efficiency, %
500	23.6	3.0	66.4	59.5	3.8
625	25.4	4.0	64.6	59.5	5.0
750	27.5	5.2	62.5	59.5	6.5
875	30.0	6.6	60.0	59.5	8.3
1000	32.9	8.3	57.1	59.5	10.4
1125	36.0	10.2	54.0	59.5	12.8
1250	39.4	12.4	50.6	59.5	15.5
1375	43.2	15.0	46.8	59.5	18.8
1500	47.2	17.9	42.8	59.5	22.4
1625	51.5	21.2	38.5	59.5	26.6
1750	56.2	24.8	33.8	59.5	31.1
1875	61.1	28.9	28.9	59.5	36.2
2000	66.3	33.5	23.7	59.5	42.0
2125	71.8	38.5	18.2	59.5	48.3
2250	77.6	44.1	12.4	59.5	55.3
2375	83.7	50.2	6.3	59.5	62.9
2500	90.0	56.8	0.0	59.5	71.3

*Assumes that the pressure-relief valve is set to maintain 90 ft of differential pressure on the system at the pumping system discharge.

constant- and variable-speed pumps (Tables 10.3, 10.4, and 10.5) also can demonstrate the great energy losses that occur with the relief valve in Fig. 15.2a. With the pressure-relief valve, the wire-to-water efficiency for this system with constant-speed pumps will be much lower. Table 15.1 describes this. This table demonstrates not just the possibilities of energy waste with pressure-relief valves but the need for system analysis to determine the most efficient method for pumping chilled water systems.

15.3   Primary-Secondary Systems

The most popular chilled water system today is the primary-secondary system because it separates the generation zone of a chilled water system from the transportation or distribution zone. This system really should be called a *chiller/distribution system,* but the industry has settled on the words *primary-secondary.* This system was developed to separate the chillers from the distribution system, as shown in Fig. 15.3a, and to reduce the differential pressure drops across the coil control valves. With this system, the chiller piping can be designed to provided optimal performance. The flow through the chiller can be constant or whatever flow the chiller manufacturer desires in the evaporator.

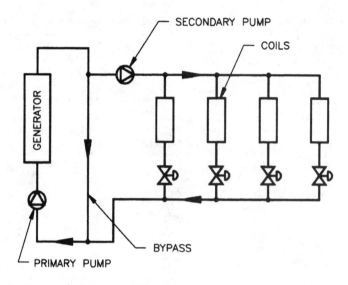

a. Primary/secondary system.

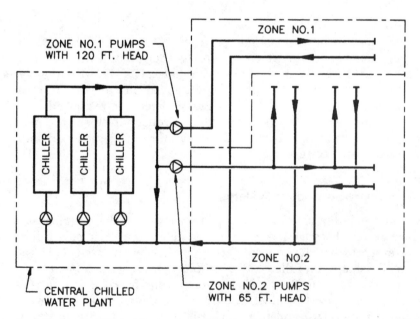

b. Two zone primary/secondary chilled water system.

Figure 15.3 Primary-secondary distribution systems.

The drawbacks of this system are that (1) two sets of pumps are required, and (2) the chiller bypass water is sent back to the chillers unused. With proper chiller sequencing, the energy lost to the bypass water is minimized; most of the head for a chilled water system is provided by the secondary pumps, not the primary or chiller pumps. The energy lost in bypassing the chilled water is minimal if the chillers are sequenced correctly. The piping of the chillers must be as described in Fig. 14.1a to avoid migration of the warmer return water back into the cooler supply water.

Primary-secondary pumping systems with system heads over 50 ft and secondary pump motor horsepowers over 10 or 15 hp are normally variable-speed systems. This is a decision again that must be made by the designer, whose evaluation must include seasonal load factors. If the decision is to use variable-speed pumps, the variable-speed drives are controlled by remote differential pressure transmitters, as discussed in Chap. 10.

Some primary-secondary systems require more than one set of secondary pumps when the head in one part of the system is much greater than that in the remainder of the system. For example, in Fig. 15.3b, one part of the system requires 120 ft of pump head because of the distance of that zone, while the second zone only requires 65 ft of pump head. The coil control valves on the first zone must be able to accept a pressure differential of 150 ft (pump shutoff head), while the second zone coil control valves must withstand only 80 ft of differential pressure against pump shutoff head. Using only one secondary pump system would require high-pressure control valves on all of the coils, and there would be a continual energy waste due to the high pressure loss in the control valves of the second zone.

Wire-to-water efficiency analyses should be run for secondary pumping systems requiring motors in excess of 20 hp. There are so many configurations possible with a primary-secondary chilled water system that it may be difficult to select the correct system without such an analysis of the pumping system. Again, the availability of computers eliminates the drudgery and time consumption of such evaluations.

15.3.1 Primary-secondary zone pumping

As indicated above, there are many versions of primary-secondary pumping to achieve optimal pumping conditions. Often, there are special design conditions to fit a specific application. One special system is zone pumping on campus-type installations (Fig. 15.4a). The secondary pumps in the central energy plant serve the core area of the

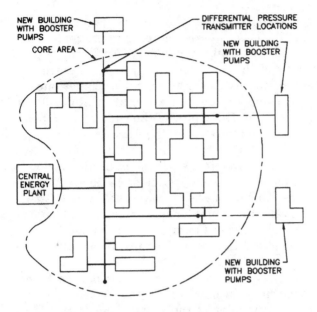

a. Secondary pumping for core area with booster pumping system.

Figure 15.4   Secondary pumping systems with booster systems.

campus, while booster pumps provide the added load required by the outlying buildings. This eliminates overpressuring of the core area to reach the far buildings.

Figure 15.4*b* is a pressure-gradient diagram for this system. It demonstrates how the booster pumping concept reduces the system pressure in the core area. Only the far building is shown for the core area for clarity. As shown, the booster pumping procedure reduces the maximum discharge pressure from 277 to 212 ft. This also reduces the total energy consumption of the system and cuts the first cost required for high-pressure control valves and accessories.

The booster pumping systems must be designed to operate in conjunction with the secondary pumps in the central plant. If the secondary pumps are constant-speed pumps, it may be necessary to equip the booster systems with bypasses and building control valves (Fig. 15.4*c*). The bypass around the booster pumps enables the building to be served by the central plant secondary pumps under light loads when the secondary pumps may produce adequate head to serve the building. This results in added pumping efficiency, since the booster pumps can be shut down during this period. The building con-

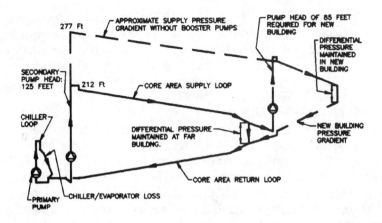

b. Pressure gradient for primary/secondary system with booster pumps.

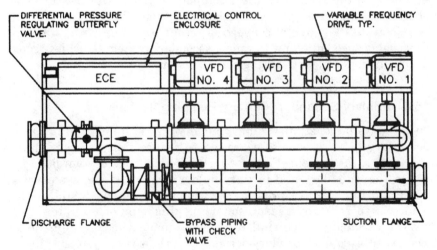

c. Addition of variable speed pumping system to existing water system.

Figure 15.4 (*Continued*)

trol valve is necessary if the central plant secondary pumps can over-pressure the building. This valve is controlled by the same remote differential pressure transmitter that controls the speed of the booster pumps when they are in operation.

If the secondary pumps are variable-speed pumps, they are controlled by differential pressure transmitters located remotely at key points in the central zone (Fig. 15.4*a*). The bypasses and control valves required with constant-speed secondary pumps are not required on most installations where all the distribution pumps are variable-speed pumps. Typical of this is the hospital installation shown in Fig. 15.5, where the secondary pumps in the central energy plant transfer the water to the hospital.

Figure 15.5*b* describes the series pumping for this installation. An age-old rule of not running pumps in series on HVAC systems appears to be violated here. If these secondary pumps were constant-speed pumps, there would be pump operating problems. Since all the secondary and booster pumps are variable-speed pumps with each pumping system being controlled by its own differential pressure transmitters and digital controllers, there have been no operating problems over a number of years of service.

This review of secondary pumps demonstrates how important a pressure-gradient analysis of an existing campus-type chilled water system is in the evaluation these complex systems. No major changes should be made in such an installation without this analysis. It will demonstrate the value of converting secondary pumps from constant to variable speed and how to zone a secondary system to achieve maximum efficiency in distributing chilled water.

15.4 Primary-Secondary-Tertiary Pumping Systems

In the days of constant-volume chilled water systems, the primary-secondary-tertiary system was used to reduce the pump pressures throughout the system (Fig. 15.6*a*). By splitting the system head between the secondary and tertiary pumps, no part of the system was subjected to excessive pressure in properly designed installations. There is seldom any reason for using this type of pumping with today's variable-volume chilled water systems. These systems must not be confused with primary-secondary with booster systems. On the primary-secondary-tertiary system, all cooling loads are equipped with tertiary pumps, as shown in Fig. 15.6*a*.

The overall efficiency of these systems is much less than that of the primary-secondary systems, particularly those with a central zone and booster pumps, as described above. The efficiency of existing primary-secondary-tertiary systems may be improved by converting the tertiary systems to variable-speed booster systems with the bypass and control valve as shown in Fig. 15.4*c*. Also, additions to existing primary-secondary-tertiary systems can be made by eliminating this costly arrangement and using a variable-speed pump with bypass, as shown in Fig. 15.6*b*.

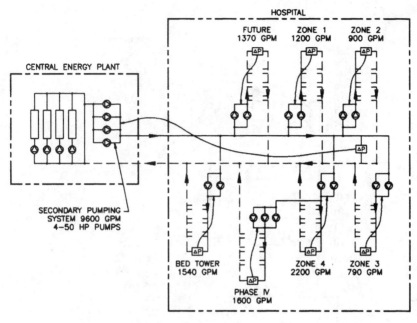

a.   General arrangement.

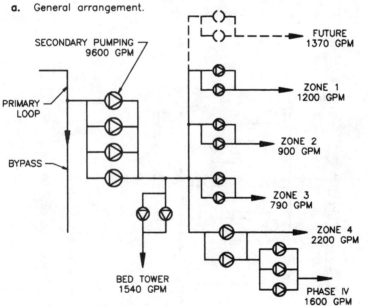

b.   Pumping schematic.

Figure 15.5   Distributed pumping in a large hospital.

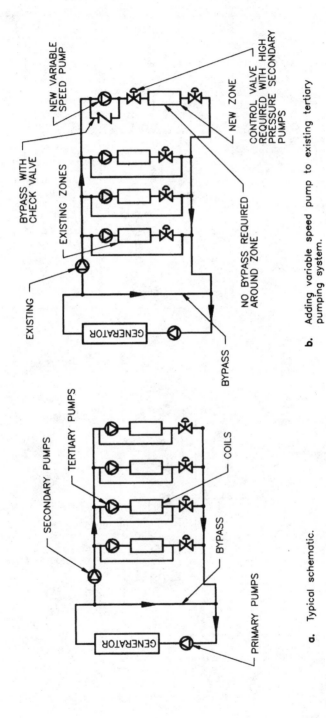

a. Typical schematic.

b. Adding variable speed pump to existing tertiary pumping system.

Figure 15.6 Primary-secondary-tertiary distribution systems.

15.5   Distributed Pumping

Distributed pumping was developed around 1967 to secure a simpler pumping system for campus-type installations where some buildings were near the central energy plant and others were far from that plant. Distributed pumping also was developed to eliminate many of the energy consumers such as crossover bridges, three-way control valves, and return temperature-control valves. It is obvious from the preceding descriptions of other types of systems that much energy is wasted in pumping circuits. Sizable reductions in pump horsepower, pumping energy, and initial cost have been achieved with this type of pumping. In most cases, the system friction and pump motors have been large enough to always justify variable-speed pumps in the buildings. There is no reason why this principle cannot be used in small systems with relatively low pump head and constant-speed pumps.

Distributed pumping is a very simple pumping arrangement as well as the most efficient pumping system for (1) large, multiple-zone buildings and (2) multiple-building systems with central energy plants. It eliminates all overpressure caused by differences in pump head requirements between zones and buildings. Figure 15.7 describes the schematic piping for a distributed pumping system; its significant feature is the lack of secondary pumps at the central energy plant. Its design is based on the Bernoulli theorem that demonstrates that any kind of pressure in a water system can be used to overcome pipe and fitting friction and therefore, cause water to move through the system against that friction. Distributed pumping utilizes the water system pressure in the central energy plant that is maintained in the expansion tank by the makeup water equipment.

Figure 15.8 is a typical three-building installation with a central energy plant that will be used to demonstrate how distributed pump-

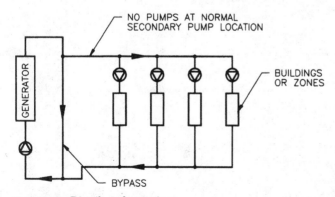

Figure 15.7   Distributed pumping.

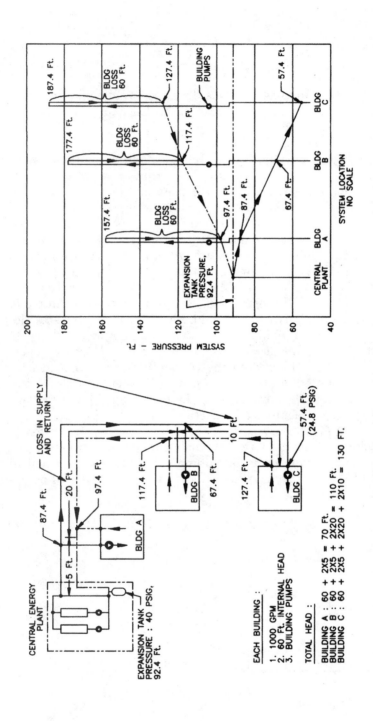

b. Pressure gradient diagram for typical, three building system.

a. Typical distributed pumping, three building system.

Figure 15.8 Typical three-building distributed pumping system.

332

ing is developed for a prospective water system. Figure 15.8*b* describes the pressure-gradient diagram for this installation. The actual system pressures are shown on this figure to illustrate utilization of Bernoulli's theorem to transmit the water from the central plant to the buildings. It is apparent from Fig. 15.8 that the supply loop to the buildings operates at a lower pressure than the return loop from the buildings. This may be confusing to some designers, but this is how distributing pumping functions and is perfectly acceptable as long as this fact is recognized when additions are made to the system.

The basic design procedure for distributed pumping requires that the ultimate buildout of a water system should be developed for loop pipe sizing. This, of course, should be done for any type of distribution system. A significant advantage to distributed pumping is the ability to install only the distribution pumps required initially, not secondary pumps with flow and pressure capacities for an ultimate system that may never be built.

One of the most important advantages of distributed pumping is the elimination of overpressuring of the zones or buildings near the central plant. This will be described by the use of the two building installations of Fig. 15.9. Figure 15.9*b* is a hydraulic gradient diagram for this system as a primary-secondary system at full design load. It is obvious that building 1 is being overpressured by the secondary pumps in their effort to provide adequate flow and head to building 2. Figure 15.10 consists of hydraulic gradient diagrams for the same system configured as distributed pumping, and it demonstrates that there is no overpressuring of building 1 with distributed pumping. Figure 15.10*b* is a gradient diagram of the same system but under the condition of light load on the near building 1 and full load on the far building 2. Again, there is no overpressuring of building 1. This demonstrates a very important fact—there is no interaction of one building with another with proper control. The load on all the buildings does change the friction loss in the supply and return loops, but this is corrected easily by the pump speed control in each building. The distributed pumping system is actually a number of small systems operating independently of each other but all of them using the same distribution piping. Contrary to the claims of some manufacturers of balancing valves, there is absolutely no reason to install balancing valves on the various buildings or zones of a distributed pumping system.

15.5.1   Variable chiller flow with distributed pumping

If the new chillers with computer control of their operation are available, the flow can be varied through the evaporators with variable-speed primary pumps. The bypass is still supplied in the central ener-

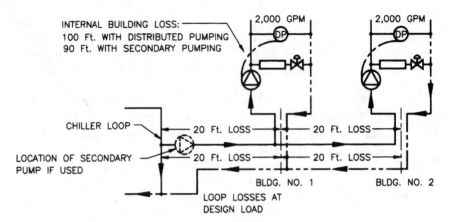

a. General arrangement.

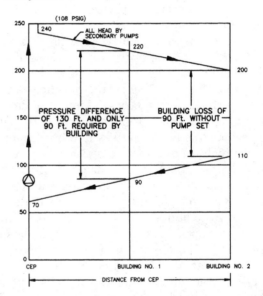

b. Pressure gradient diagram for two building system with secondary pumping.

Figure 15.9 Two-building distributed pumping system.

gy plant to ensure that minimum flow is always available to the chillers. With proper control, very little water flows in the bypass pipe under normal operation. This provides the highest possible level of pumping energy conservation. The minimum flow through the chiller evaporator should be specified in gallons per minute or minimum velocity in the tubes in feet per second by the chiller manufacturer.

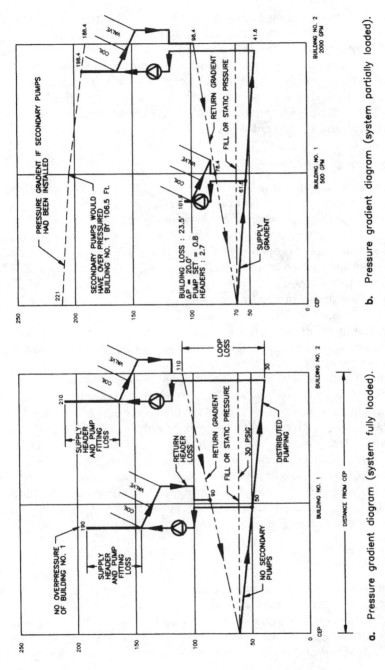

a. Pressure gradient diagram (system fully loaded).

b. Pressure gradient diagram (system partially loaded).

Figure 15.10 Pressure gradients for the two-building distributed pumping system.

335

15.5.2   Water reset in distributed pumping

On distributed pumping systems, where there is a need for reset of the water temperature over that supplied from the central energy plant, a return water control valve can be installed to change the supply water temperature in accordance with some schedule such as load or outdoor temperature. A hydraulic analysis must be made to determine whether this control valve is a two-way valve installed in the bypass or a three-way valve controlling the return water to the building or zone and to the central energy plant.

15.5.3   Use of distributed pumping

Distributed pumping has been used on both large and small installations. It offers a great advantage on systems where there is a great difference in distribution friction between zones or buildings. Typical of the large systems is the international airport shown in Fig. 15.11a; in this case, three secondary pumps with 1000-hp motors were eliminated by distributed pumping. Building motors from 50 to 150 hp were installed in lieu of the secondary pumps. A 26 percent reduction in installed pump motor horsepower was achieved by using distributed pumping.

A small installation is shown in Fig. 15.11c that is for a regional college campus. A sizable reduction in overall pumping requirements was achieved by replacing a primary-secondary-tertiary system with distributed pumping. The water flow rate in the campus was cut to half that required by the old system. The proposed addition of chiller capacity was canceled when it was found that the existing chillers could handle the campus with ease with the greatly reduced chilled water flow.

Distributed pumping offers excellent energy savings for new systems. Unfortunately, it cannot be applied to existing campus installations where there are many buildings; this would result in great cost for modifying existing pumping, piping, and control. Usually, there is an alternative method such as primary-secondary zone pumping or modifications of distributed pumping.

15.6   Consensus

It is obvious from this discussion that there is no one type of chilled water distribution system that fits all applications. The pressure-gradient evaluation of a proposed or existing system is a valuable tool to guide the designer toward the optimal system that provides maximum efficiency under the economic constraints of that particular installation.

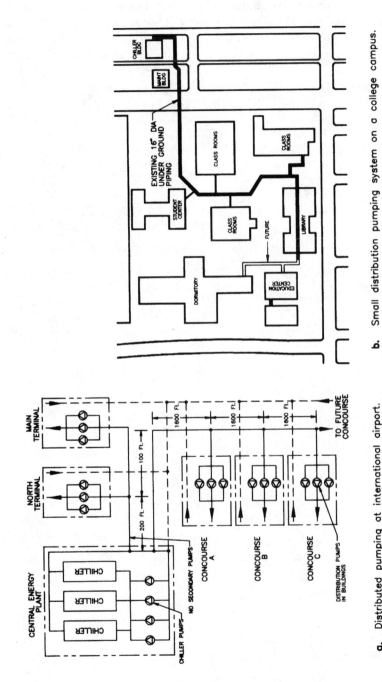

a. Distributed pumping at international airport.

b. Small distribution pumping system on a college campus.

Figure 15.11 Typical uses of distributed pumping.

Every effort should be made in the design of chilled water distribution systems to use the computer technology that is available today. This should eliminate much of the drudgery of design that can accompany the evaluation of some of these water systems. Further, all the energy-wasting devices such as balance valves, pressure-regulating valves, and most tertiary pumping should be eliminated by simplification of the distribution design. The advent of the variable-speed pump and the computer-controlled chiller offers reduced overall chilled water plant kilowatt per ton or coefficient of performance values never achieved in the past.

15.7   Bibliography

James B. Rishel, *Twenty Years Experience with Variable Speed Pumps*, transaction no. OT-88-09-2, vol. 94 (2), American Society of Heating, Refrigerating and Air-Conditioning Engineers, Atlanta, Ga., 1988.

James B. Rishel, *The Water Management Manual*, SYSTECON, Inc., West Chester, Ohio, 1992.

James B. Rishel, *Distributed Pumping for Chilled and Hot Water Systems*, Transaction No. 94-24-3, vol. 100(1), American Society of Heating, Refrigerating and Air-Conditioning Engineers, Atlanta, Ga., 1994.

16

Closed Condenser Water Systems

16.1 Introduction

Closed condenser water systems are those which reject thermal energy to heat exchanger surfaces, not cooling towers, which are open systems. Open-type condenser systems are included in Chap. 11 on cooling towers. The heat exchanger can be a plate and frame or shell and tube type. The heat exchanger can be operated with a cooling tower (Fig. 16.1a,) or be part of a closed-circuit cooler (Fig. 16.1b). The basic circuitry of a closed-circuit cooler consists of a cooling coil (the heat exchanger), fans to move air over the coil, and a circulating pump to wet the coil surfaces. These coolers can be operated wet with the circulating pump running or dry with the pump stopped. The closed-circuit cooler is also known as an *evaporative cooler* because most of the heat transfer is by evaporation. It can be an induced-draft unit, where the fan is located on top of the unit and pulls the cooling air across the coils containing the warm water, or it can be a forced-draft unit, which pushes the cooling air through the unit. This type uses blower fans that usually require more fan horsepower. Forced-draft units are used mainly for ducted systems, where it is necessary to transport the moist leaving air to a specific location.

16.2 Reasons for Closed Condenser Systems

Closed condenser water systems are used for several reasons:

1. The condenser circuit must be closed or pressurized.

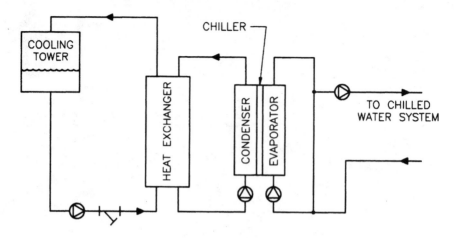

a. Basic closed condenser system with heat exchanger and cooling tower.

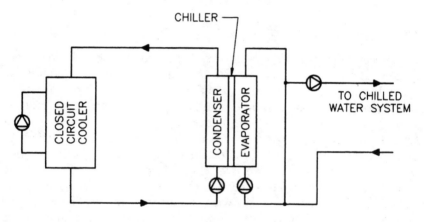

b. Basic closed condenser system with closed circuit cooler.

Figure 16.1 Closed condenser water systems.

2. The chemicals or particulate matter seen in open condenser water systems cannot be tolerated.

3. Other physical factors involved with cooling towers may make them unacceptable for the installation.

4. There is a desire to capture and use the heat of rejection from the chiller or process.

5. Chemicals such as glycol are used that cannot be passed through a cooling tower.

16.3 Types of Closed Condenser Systems

There are standard closed-circuit condenser systems, special-use systems, and heat-recovery condenser systems. The most elemental pumping arrangement for a closed condenser circuit is shown in Fig. 16.1*b* for a chiller rejecting its heat to a closed-circuit cooler. No attempt is being made in this case to use the heat of rejection. A small circulator is supplied with the closed-circuit cooler for the spray water. Figure 16.1*a* describes another simple closed system using a heat exchanger and a cooling tower. This system, of course, requires both condenser and cooling tower pumps. In both these cases, the pumping systems are very straightforward and have been described in other chapters of this book. The decision as to whether cooling towers with heat exchangers or closed-circuit coolers are used depends on the costs and installation factors of each application. The size of this equipment may be reduced through the use of heat-recovery equipment described below.

The heat exchanger and closed-circuit cooler are water-to-air heat-rejection systems. A third system using underground coils has been developed for closed condenser systems; this is a water-to-ground heat-rejection system. Such an installation is particularly suitable for variable-speed pumps owing to the greater friction of the underground coils.

16.3.1 Special closed condenser systems

There are special-use closed condenser systems that are designed to accommodate specific types of equipment that must be cooled. A special use of closed condenser circuits is the pumping of condenser water for computer air-conditioners. A typical installation is shown in Fig. 16.2*a*. Normally, this water system would be filled with a glycol solution if it must sustain a broad range of temperatures from summer to winter operation. Such a temperature variation with glycol may produce a sizable change in viscosity and therefore in pumping head and energy. A simple method of solving this problem is through the use of variable-speed pumps that maintain the required differential pressure at the condenser coils of the computer air-conditioners regardless of the temperature and viscosity of the glycol solution. An energy evaluation must be run at minimum and maximum temperatures to determine the variation in pump head due to the changes in viscosity and specific gravity of the glycol solution. Although this is substantially a constant-volume, variable-head system, the variable-speed pump allows the system to run at the true diversity of the system under all load conditions. Also, individual air-conditioners or

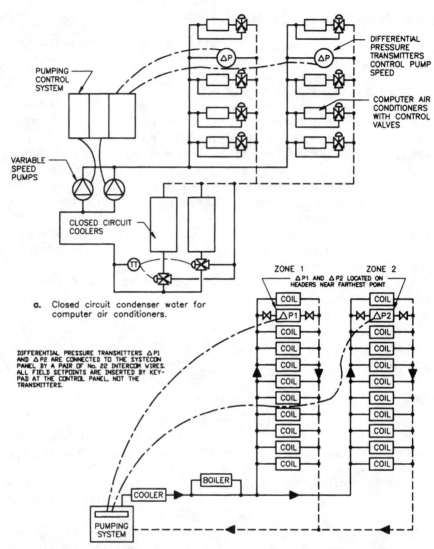

a. Closed circuit condenser water for computer air conditioners.

DIFFERENTIAL PRESSURE TRANSMITTERS ΔP1 AND ΔP2 ARE CONNECTED TO THE SYSTECON PANEL BY A PAIR OF No. 22 INTERCOM WIRES. ALL FIELD SETPOINTS ARE INSERTED BY KEYPAD AT THE CONTROL PANEL, NOT THE TRANSMITTERS.

b. Typical variable volume condenser water for water cooled heat pumps.

Figure 16.2 Special closed condenser water systems. (*From Variable Water Volume Is Hydro-Electronics, SYSTECON, Inc., West Chester, Ohio, 1986, p. 30.*)

groups of them can be shut down, and no readjustment of the cooling system is required.

16.4 Heat-Recovery Condenser Systems

There has been a desire to capture the heat of rejection from water-cooling equipment such as chillers. This has resulted in heat-recovery equipment such as water source heat pumps and heat-reclaim chillers. Water source heat pumps can be terminal-type located near the heating and cooling loads or they can be central-type located in major equipment rooms.

The amount of heat recoverable is appreciable; for example, a 1000-ton chiller operating over a cooling season of 2000 hours with an average heat rejection of 15,000 Btu/ton has a total heat rejection of 30 billion Btu! If only 50 percent of this energy is reclaimed, this amounts to 15 billion Btu, which, at an overall boiler efficiency of 80 percent, is equivalent 18,750,000 ft^3 of natural gas with a heating value of 1000 Btu/ft^3. At a cost of $3 per thousand cubic feet, the annual savings would be over $56,000. An audit should be made on cooling tower installations to determine the economic feasibility of energy recovery. The development of plate and frame heat exchangers and condensing boilers has resulted in low-temperature thermal operations that can enhance the recovery of heat from chiller condensers.

16.4.1 Terminal-type water source heat pumps

Another special use of closed condenser systems for specific equipment and the simplest heat-reclaiming system in terms of pumps is the terminal-type water source heat pump, where the heat of rejection from a unit under cooling load is transferred to a unit with a heating load. Figure 16.2b describes a typical installation of a number of small water source heat pumps utilizing a central boiler and closed-circuit cooler for any needed heat addition or rejection. The boiler should be designed for this low-temperature duty or be circuited in accordance with the boiler manufacturer's instructions. The new condensing boilers are ideal for this service.

The circulating pumps can be constant- or variable-speed pumps; variable-speed pumping is shown in this figure with control by means of differential pressure transmitters located at the top of each building. With high diversities, which can be common for installations of water source heat pumps, there can be substantial savings through the use of variable-speed pumping. Such systems that contain a number of water source heat pumps must have a water-regulating valve on each water source heat pump. If constant-speed water pumps are used, an automatic balance valve must be used on each water source heat pump. With variable-speed pumping, the water-regulating valve can be self-operating, responding to head pressure in the heat pump, as shown in Fig. 16.3a, or it can be electric of the on-off type, as in Fig. 16.3b. Usually an automatic balancing valve is furnished with the electric valve to provide the amount of water required by the heat pump. The actual coil connections should be made in accordance with the requirements of the manufacturer of the heat pump. Some manufacturers require close control of the flow rate through the condenser coil.

16.4.2 Central plant heat-reclaiming equipment

The water source heat pumps shown in Fig. 16.2b consist of units located out near the actual heating and cooling loads. Other heat-recovery equipment can be central heat-recovery machines operating in conjunction with conventional chillers and boilers, or this equipment can be heat-recovery chillers with second condensers incorporated that utilize the heat of rejection. Figure 16.4a describes the elementary circuit for the central heat-recovery water source heat pump, while Fig. 16.4b provides the circuit for a heat-recovery chiller. The decision as to which type of equipment to use for heat recovery on a specific installation depends on the economics of each application and is not within the scope of this book.

16.5 Uses of Heat Reclaimed from Chillers

Some of the uses for the recovered heat are

1. Space heating
2. Reheat coils
3. Outdoor air heating coils for air-conditioning or industrial process
4. Radiant surface heating
5. Domestic water heating
6. Snow melting
7. Process heating
8. Natatoriums
9. Dehumidification

The piping and pumping, as well as control of the various circuits for use of the heat, must be arranged so that the most efficient use of

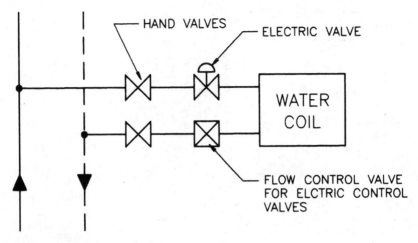

WATER SYSTEM

a. Typical coil connections for water source heat pump with electrical water regulating valve.

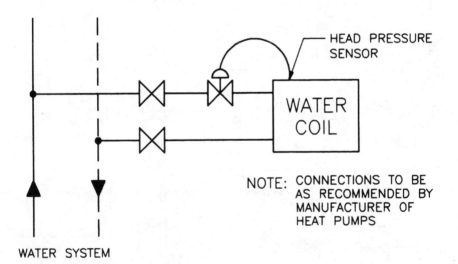

WATER SYSTEM

b. Typical coil connections for water source heat pump with self—operating water valve.

Figure 16.3   Coil connections for water source heat pumps.

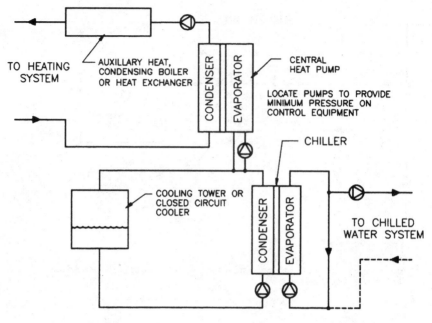

a.   Central heat pump.

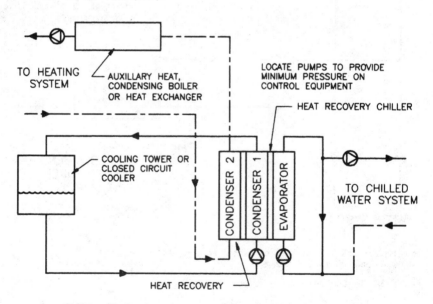

b.   Chiller with heat recovery condenser.

Figure 16.4   Central heat reclaim systems.

this heat is achieved. Also, the circuiting must be such that the operation required by the chiller is achieved. There must be positive control of (1) the performance of the chiller from a cooling standpoint and (2) achievement of the optimal heat balance for the entire installation. For example, under some load conditions it may be most economical to use the heat for reheat, whereas in other situations it may be better to use the heat for snow melting. Also, the heat rejection from the chiller may be such that an increased leaving water temperature from the chiller condenser is achieved at the expense of the chiller efficiency, but a separate boiler operation is eliminated.

Following are some basic methods for saving the 30 billion Btu per year that was discussed in this chapter and in Chap. 11 on cooling towers.

16.6 Condensing-Type Boiler Improves Heat Recovery

Another development that is increasing the use of heat-recovery equipment is the emergence of the condensing boiler, which operates most efficiently at water temperatures around 100°F or less. Conventional gas-fired boilers that are noncondensing cannot be operated below 140°F without corrosion in their flue collectors. This lower-temperature boiler operation with the condensing-type boiler offers greater savings with the heat-recovery equipment described above. The efficiency of the condensing boiler with different leaving water temperatures is described in Chap. 19.

The reduced temperature at which condensing boilers operate enables them to be installed without heat exchangers or pumped subloops on these heat-recovery systems. The entire heating efficiency of the hot water system is greatly improved with operation from 80 to 140°F. This is particularly true with radiant surface systems in floors and ceilings for heating.

16.7 Pumping Aspects of Heat Recovery

What is important from a pumping standpoint is the location of the pumps and their operation in these heat-recovery circuits, which can become complex. Pumps should be located to eliminate as much pressure as possible on the operating equipment. Since these are closed systems, adequate pressure can be provided on pump suctions that will eliminate possibilities of cavitation. Variable-speed pumping with proper control can result in efficient and often simplified circuits for these heat-recovery systems. Constant or variable flow through the chiller evaporators for heat recovery is determined by the chiller manufacturer.

The evaluation of (1) when to use heat-recovery systems and (2) what type of heat-recovery equipment is most economical is a complex subject. Proper pumping systems and their location can have a substantial effect on the success of such heat-recovery applications. Following is a list of some of the factors that must be considered when evaluating the pumping for these systems:

1. Consider both constant- and variable-speed pumps. Variable-speed pumps offer so much more flexibility on these systems, which may have complex piping arrangements.

2. Avoid, wherever possible, energy-consuming devices such as balance valves, pressure regulators, and bypasses.

3. Locate pumps to keep pump pressure off the central equipment wherever possible (see Chap. 19 on the location of pumps with boilers).

4. Consider distributed pumping on multiple-building systems for both the hot and chilled water systems.

Figures 16.5a describes a typical heat-recovery circuit using auxiliary heat from a condensing boiler or a heat exchanger heated by a conventional steam or hot water boiler. It should be noted that these systems describe pumps in series, which is acceptable where the

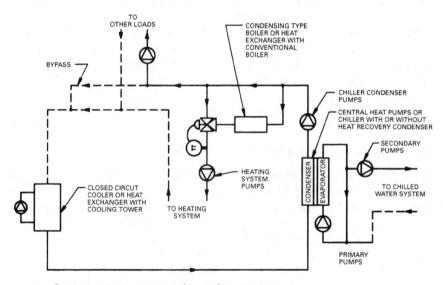

a. Condenser heat recovery with auxillary heat.

Figure 16.5 Circuits for heat reclaim from closed condenser water systems

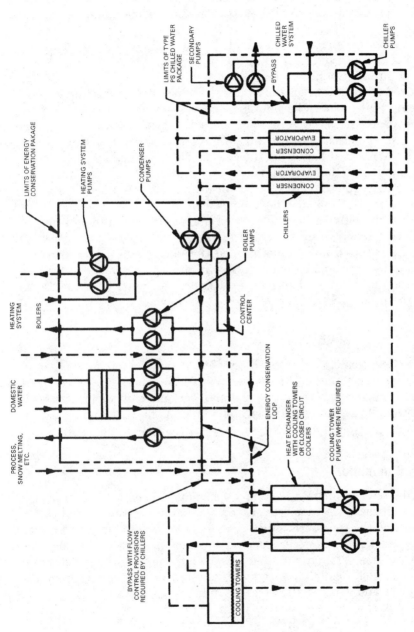

b. Energy management systems for heat recovery from chiller condensers.

Figure 16.5 (*Continued*)

349

pumps are variable speed and equipped with the proper speed control. Pressure-gradient diagrams should be run for such circuits to verify the actual pressures that exist in these systems under various load conditions.

Some of the heat-recovery installations where various uses are made of the rejected heat from the chillers can be complex in piping and pumping. Simplification of the field piping and control can be achieved through the use of factory-assembled energy-management systems that include the pumping, piping, control, and energy-management operations. Figure 16.5b describes such circuitry and equipment. All the circuits are arranged for proper control and priority of use of the rejected heat from the chiller. The total management of the water and heat flow must include any minimum flow and temperature limits required by the chillers or heat-using equipment.

All these diagrams include a bypass for the condenser pumps. If the condenser pumps are variable-speed pumps utilizing temperature or pressure control within the chillers, a bypass is still needed to ensure that the required flow is maintained through the chiller condensers. Some condenser applications utilize variable-speed condenser pumps; in these cases, a complete hydraulic statement must be made to ensure the proper distribution of the recovered heat.

16.8   Storage of Hot Water

The reclamation of heat from chillers has so many interesting possibilities that there may be cases that justify the storage of excess heat rather than waste of the heat through a closed-circuit cooler or a cooling tower. Open storage of hot water would, as far as pumping is concerned, be very similar to cold water storage, as described in Chap. 13.

16.9   Summation

Closed-circuit condenser systems offer many different possibilities for energy-efficient systems. The energy audits require substantial calculations and the consideration of a number of different circuits. Energy storage in the form of chilled or hot water also can play a part in these systems. From a pumping standpoint, the variable-speed pump offers so much flexibility without balance valves or pressure-regulating valves and without any excessive pumping pressure.

The closed condenser system will become much more popular than it is today because of the heat-recovery possibilities. What better method of plume abatement of cooling towers is there than to prevent the heat from getting to the closed-circuit cooler or cooling tower.

16.10   Bibliography

James B. Rishel, *Variable Water Volume Is Hydro-Electronics,* SYSTECON, Inc., West Chester, Ohio, 1986.

Cooling Tower Fundamentals, 2/e, The Marloy Cooling Tower Company, Mission, Kansas, 1985.

17

Pumps for Closed Energy Storage Systems

17.1 Introduction

Closed energy storage systems consist of pressure tanks for both hot and chilled water as well as ice systems with open tanks that are described as ice-on-tube systems. The growth of energy-conservation efforts has resulted in the use of a number of different procedures for storing energy. Closed energy storage systems differ from the open storage systems of Chap. 13 in the storage of the energy at system operating pressures; this results in the ability to maintain system pressure throughout the energy storage system.

17.2 Pressure Tanks for Energy Storage

Pressure tanks have not been popular because of the cost of the tanks themselves. Of course, it is the responsibility of the water system designer to determine the most economical means of storage of energy on a specific installation.

Closed tanks offer the specific advantage of operating at system pressure. All the pressure-control devices needed in open storage tanks are not required for these systems. The tank can be located in the system at the point most advantageous for the operation of the equipment and the storage of energy.

For example, on chilled water systems, the pressure tank can be installed in the chiller bypass, as shown in Fig. 17.1. With this arrangement, the chillers can be operated at peak efficiency and, therefore, at the lowest kilowatts per ton. On very light loads, the chiller shuts

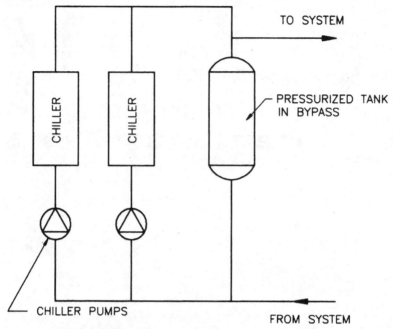

Figure 17.1   Pressurized chilled water storage.

down, and chilled water is drawn from the tank; this eliminates chiller operation at loads less than 50 percent of design capacity, where the overall kilowatts per ton are high. This arrangement provides a very simple but effective means of storing chilled water without control valves, storage pumps, and all the other appurtenances required with open tank systems.

When pressurized chilled water is used for energy storage, the volume of the chilled water piping system must be addressed, since some chilled water systems have many tons of cooling stored in the normal piping. On older systems that may have been oversized in terms of piping, a sizable tonnage of cooling could exist in the piping itself. The volume of water per foot of pipe in gallons has been included in Table 2.7 to assist in this evaluation. The tonnage stored is calculated by Eq. 17.1, where the specific heat of water is assumed to be 1 Btu per pound per degree F.

$$\text{ton} \cdot \text{h stored} = \frac{V_s \cdot 8.34 \cdot (T_2 - T_1)}{12,000}$$

$$= \frac{V_s \cdot (T_2 - T_1)}{1439} \tag{17.1}$$

where V_s = volume of the system in gallons

T_2 = system return water temperature, °F

T_1 = system supply water temperature, °F

There have been some unusual installations such as airports where the total gallons of water in the distribution system provided enough storage that there was no need for additional chilled water storage for energy curtailment.

Eutectic salts can be used to reduce the volume of the pressurized storage tanks. The eutectic salt tanks can be as much as one-third to one-half the size of those for straight water storage.

17.3 Closed Energy Storage Systems with Open Tanks

On chilled water systems that cannot justify pressure tanks for energy storage, the closed water circuit that uses open ice tanks provides an economical option for many systems. This type of system passes glycol through tubes in an open tank and either creates or melts ice on the external surfaces of the tubes as required by the system demands. A storage tank for this system is shown in Fig. 17.2. The storage tanks can be single, large tanks or modular systems comprised of a number of small tanks.

Such a system is called an *internal melt ice-on-coil system* because the glycol passes through the pressurized coils in the tank and ice forms on the outside of these coils. The design of this system offers

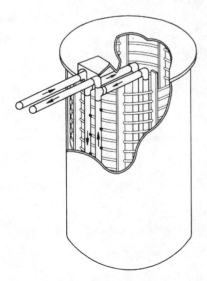

Figure 17.2 Modular internal melt ice storage tank.

some general advantages over the other types of ice storage systems that were described in Chap. 13. For example,

1. Chiller cost is much the same as for chilled water and the other forms of ice storage other than the ice harvester system, which has a much higher chiller cost.

2. Tank volume is equivalent to that of most of the other forms of ice storage except the ice harvester system, which requires 10 to 20 percent more storage volume.

3. Storage cost is similar to that for external ice melt and encapsulated ice; it is greater than for the ice harvester system and much less than for the eutectic salt system.

4. Power costs per ton-hour are comparable with those for all the other ice storage systems; they are much higher than for chilled water storage or for the eutectic salt system.

5. A great advantage of this system is the ability to run the chillers at normal chilled water system temperatures such as 44°F leaving water and 56°F return water when ice is not being used or generated. This brings the power costs down to close to that of ordinary chilled water operation.

The arrangement of the piping and pumping in this method of energy storage is very important in order to achieve efficient and trouble-free operation. All the different phases of operation for the chilled water and storage systems must be recognized and evaluated to achieve the desired results from this energy system. These phases are

1. Operate the chilled water system without the ice storage system.

2. Serve the chilled water system and generate ice.

3. Serve the chilled water system with the chillers and from the ice storage.

4. Serve the chilled water system from only the ice storage.

5. Generate ice only.

Not all systems will require all five of these phases. It is up to the designer to develop the phases or cycles required for a particular application.

17.4   Circuiting Internal Ice Melt Systems

There are a number of different methods of circuiting these energy storage systems; three of these are shown in Fig. 17.3. Figure 17.3a de-

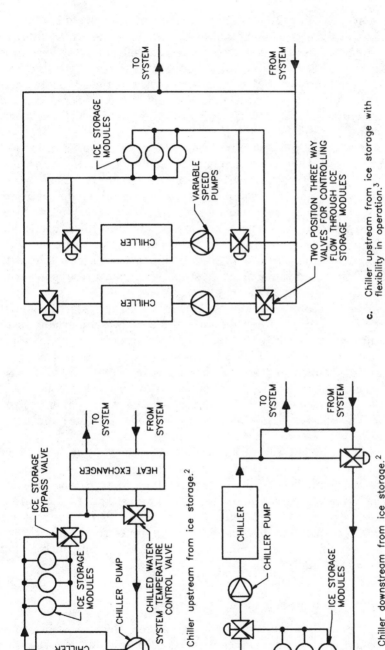

a. Chiller upstream from ice storage.[2]

b. Chiller downstream from ice storage.[2]

c. Chiller upstream from ice storage with flexibility in operation.[3]

Figure 17.3 Chiller location with respect to ice storage.

357

scribes a system with the chiller upstream from the ice storage tanks, while Fig. 17.3b provides the arrangement with the chillers downstream from the ice storage tanks. A variation of this latter hookup is shown in Fig. 17.3c with two chillers in parallel operation. In the first two designs, a three-way valve is shown for varying the supply temperature to the chilled water system. The system of Fig. 17.3c is designed to receive either a constant low-temperature glycol under all loads or a higher temperature when the stored ice is not being used.

Installing the chiller upstream from the storage reduces the power consumption in kilowatts per ton because the water in the evaporator is higher in temperature; the total storage may be reduced because of the temperatures of the system. Installing the chiller downstream from the storage may increase the overall kilowatts per ton as a result of the lower temperatures in the evaporator. The total system storage may be improved because of the colder temperatures leaving the ice storage tanks. In any of these variations, the total kilowatts per ton of the chilled water plant must be run under all loads to verify the total cost of energy for each system.

Figure 17.3c is a variation of the downstream chiller arrangement that provides flexibility in system operation. Either chiller can make ice simultaneously with the other chiller providing the system water temperature as required. The chiller making ice could be operating at leaving water temperatures of 26°F and at around 0.90 kW/ton, while the other chiller could be making chilled water for the system at 44°F and at 0.60 kW/ton. Another advantage of this arrangement is the ability of the chillers to operate at any desired load and maintain a constant leaving temperature as the ice is depleted.

Like other water systems containing glycol, these internal ice melt systems must be equipped with heat exchangers if it is desired to keep the glycol out of the chilled water system. Figure 17.3a includes such a heat exchanger, while Figs. 17.3b and c do not. This is a decision for the designer and the client. Generally, chilled water systems utilizing large air-handling systems do not require the heat exchanger, while systems with a number of small terminal units such as fan coil do require them.

17.5 Pumping Internal Ice Melt Systems

The variable-speed pump offers some appreciable advantages in pumping internal ice melt systems because of the sizable changes in viscosity of the glycol solution as it varies from 26 to 55°F. The variable-speed pump can be controlled by any desirable parameter such as flow through the chiller evaporator. The flowmeter for this circuit provides a

4- to 20-mA signal to the pump controller that varies the pump speed and maintains the desired flow as the glycol temperature changes.

The friction loss in the various loops for this ice system varies as to whether ice is being used or bypassed. The variable-speed pump automatically adjusts its speed to accommodate these variations caused by the control valves that determine the phase of operation for the energy storage system.

Further, as indicated in Fig. 17.3b, the use of variable-speed pumps on the chiller circuit eliminates other pumps that might be required with heat exchangers or the ice storage itself.

17.5.1 Pumps for internal ice melt systems

This type of ice storage system requires a minimum of special provisions and controls for the pumping system, since it is a closed system where system pressures are easily controlled. There should be no concern for cavitation in the pumps other than the usual precautions that should be taken for any system.

Most of the pumps for these systems are either single- or double-suction volute-type pumps. Standard bronze-fitted construction is very adequate for these applications.

Many of the standard pumping and distribution systems for ordinary chilled water systems are usable with internal melt ice storage systems.

17.6 Bibliography

ASHRAE Handbook Fundamentals, American Society of Heating, Refrigerating, and Air-Conditioning Engineers, Atlanta, Georgia, 1993.
Design Guide for Cool Thermal Storage, ASHRAE, Atlanta, Ga., 1993.

18

Pumps for District Cooling and Heating

18.1 Introduction

Although this chapter has been included in this part on pumps for closed cooling systems, district heating is included here because it is so similar to district cooling in layout and pumping. Much of the following discussion on pumping district cooling can be applied to district heating systems. Usually, district hot water capacities are much smaller than those for district cooling.

District cooling and heating is the name for the centralization of chilled and hot water plants that will supply water to a number of independent buildings. District cooling and heating is increasing in popularity because of the cost advantages and efficiencies that can be achieved in the generation of the chilled water.

One of the important parts of these installations is the cost of moving the chilled water from the central energy plant to the buildings. Pumping energy must be evaluated using computer programs to secure the optimal method of pumping. Most of these systems have utilized primary-secondary pumping to distribute water to the buildings. Primary-secondary pumping was explained in Chap. 15, as was distributed pumping. Distributed pumping is an alternative procedure for district hot and chilled water systems. It uses the central energy plant water system pressure to distribute the water to the buildings. As mentioned before, it is based on the Bernoulli theorem, which demonstrates that there are three different pressures in a water system, namely, static head, velocity head, and pressure head. Any of these three can be used to move water through a system of pipes.

Computer analysis may demonstrate that distributed pumping provides the most efficient system for district cooling or heating. It offers some decided advantages for the entirely new system. This will be explained using a typical model system.

18.2   Model System for District Cooling

The following model system is based on district chilled water; district hot water would be similar but smaller in flow. Figure 18.1 describes a district cooling system of 16,000 tons. The system differential temperature is assumed to be 12°F or requiring 2 gal/min per ton for a maximum flow of 32,000 gal/min. This system will be modeled using primary-secondary and distributed pumping, demonstrating the energy requirements for this system when using both these pumping procedures. The friction losses in feet of head for each segment of the system are shown in parentheses in Fig. 18.1.

The chilled water plant in the model system is using an internal bypass that separates this plant from the distribution system. The distribution system is designed on the basis that any connected building with a height over 115 ft must be equipped with a heat exchanger to prevent excessive static pressure on the district chilled water system. This height is, of course, determined by each individual district system and is contingent on the building load in tons.

This model system is assumed to be on flat terrain to simplify the pressure gradients. In an analysis of an actual system, the static elevations of the buildings above sea level, as well as the central energy plant, would be factored into these diagrams. To use the actual elevations in this model system would have made description of the hydraulics of the distributed pumping more difficult.

This evaluation of the two types of pumping for this system will study the distribution friction and horsepower only. The actual internal head and pump horsepower for each building will not be included to clarify the differences in distribution horsepower. Also, the pumping horsepower of the primary pumps or chiller circuit will not be included, since the pump horsepower is the same for the two systems. Water horsepower is used in these evaluations to remove any disparities that may be caused by actual selection of pumps and motors.

System distribution such as is described here often relies on a complete loop, as is shown in Fig. 18.1b. This is an expensive procedure in most cases, since the extra cost of the connecting loop may never be amortized in energy savings. Distributed pumping proves that complete loop-type construction is seldom justified on most district heating and cooling projects. This, of course, is an evaluation that the de-

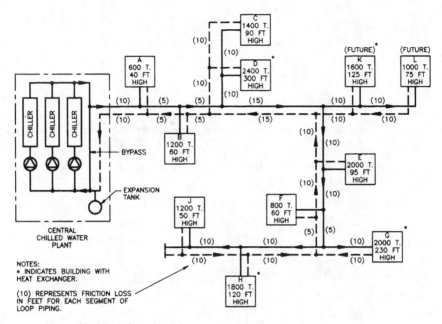

a.   Model system layout.

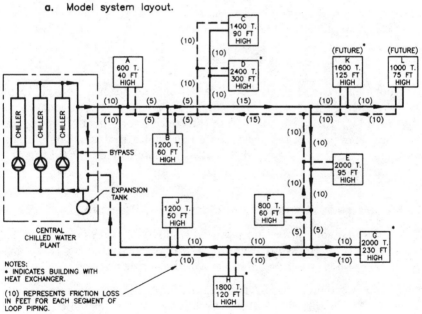

b.   Looped system.

Figure 18.1 Typical district cooling systems.

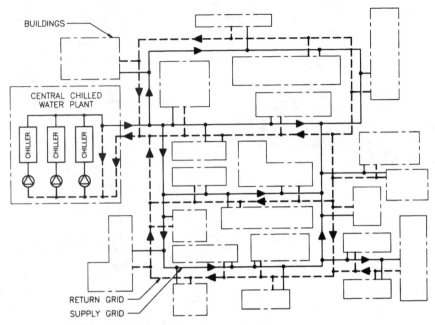

Figure 18.2 Grid-type district cooling system.

signer must make in the first stages of district or campus water system design.

Other district heating and cooling systems do not have simple loops as described in Fig. 18.1; instead, they are a grid-type system as shown in Fig. 18.1a. Calculation of the friction for a grid of piping can be made using computer programs for network analysis. The distributed pumping procedure can be used on such grids by taking the maximum grid loss that can occur to each and every building.

18.3 Calculating Central Energy Plant Static Pressure

There may be some question as to how the static pressure in the expansion tank at the central energy plant is calculated for these two systems. Following is an explanation of these calculations.

Secondary pumping system. The expansion tank pressure must be selected by the height of the tallest building plus the needed cushion on top of that building, which in this case is 20 ft. The tallest building directly connected to the system without a heat exchanger is building *E,* with a height of 95 ft plus the 20-ft cushion, which equals 115 ft, or 50 psig.

Distributed pumping system. The expansion tank pressure must be selected using the building which with its height and loop pressure loss produces the highest expansion tank pressure. Every building must be evaluated; in this case, only the highest and farthest buildings will be evaluated.

Farthest building. Building J with a height of 50 ft plus the 20-ft cushion plus 80 ft of loop friction = 150 ft total.

Highest building. Building E with a height of 95 ft plus the 20-ft cushion plus 45 ft of loop friction = 160 ft.

The highest building requires the most expansion tank pressure of 160 ft, or 69 psig.

Note: During actual design, the pressure gradient diagram using building elevations will determine the required expansion tank pressure.

18.4 Primary-Secondary Pumping System Analysis

The primary-secondary system is described in Fig. 18.3a, and its hydraulic gradient is shown in Fig. 18.3b. The pump head for this system includes 160 ft of loop friction plus 10 ft of loss in the secondary pumping system fittings plus 10 ft of differential pressure maintained at the end of the longest loop for a total of 180 ft. As arranged in Fig. 18.3a, at design system load, the total water horsepower to overcome loop friction only is (from Eq. 6.10)

$$\text{Water horsepower} = \frac{Q \cdot H \cdot s}{3960} \tag{6.11}$$

$$\text{Secondary water horsepower} = \frac{32{,}000 \text{ gal/min} \cdot 180 \text{ ft} \cdot 1.0}{3960} = 1454.5 \text{ hp}$$

One of the disadvantages of the primary-secondary system is the great overpressure that is imposed on the near buildings. As shown in Fig. 18.3a and b, building A has a supply pressure of 275 ft and a return pressure of 125 ft for a differential pressure of 150 ft. This is much more head than is needed in this building for overcoming the internal piping and cooling coil losses. A pressure-reducing valve or some other energy-consuming mechanical device must be incorporated to destroy this energy when the system is under an appreciable load.

This over-pressure may force the use of high-pressure or industrial-grade control valves on the cooling coils. For example. as indicated above, building A has 150 ft (65 lb/in^2) of differential pressure imposed on its control valves. Most commercial control valves have a

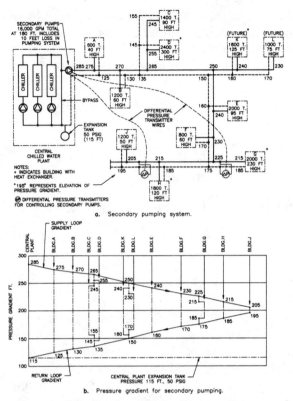

a. Secondary pumping system.

b. Pressure gradient for secondary pumping.

Figure 18.3   Secondary pumping for district cooling.

maximum allowable differential pressure of around 35 lb/in^2. Therefore, with secondary pumping, this building must be equipped with
high-pressure control valves on its cooling coils.

This overpressure also complicates the building pumping system. If
the total load on the system subsides, this high differential will recede, and pumping may be required in this building since the central,
variable speed pumps will be running at a lower speed. A pumping
system with a bypass as described in Fig. 15.4c may be an answer for
pumping this building when the other buildings are inactive.

Another disadvantage of the secondary pumping system is the requirement that the total system operate as a single chilled water distribution system. This requires the location of differential pressure
transmitters at the far ends of the distribution system, as shown in
Fig. 18.3a. Direct connection must be made from these transmitters
to the pump controllers to ensure smooth operation of the pumps
without hunting or erratic speed changes.

The overall operating pressure can be higher for the secondary pumping than that for distributed pumping, which will be discussed next. In this case, the maximum pressure for the secondary pumping system is 285 ft or 123 psig, while the distributed pumping system has a maximum pressure of 240 ft or 104 psig. Some distributed pumping systems show a pronounced reduction in system pressure over secondary pumping. The pressure gradient must be evaluated for each actual system to calculate the maximum pressure, since the location of the tallest building in the system, in this case, determines the actual set pressure for the expansion tank of the distributed pumping system.

18.5   Distributed Pumping System

The distributed pumping system is shown in Fig. 18.4. The loop pressure losses are assigned to each building and are added to the head of the building pumps. The calculation of loop horsepower for distributed pumping must be done on a building-by-building basis. Table 18.1 provides this information.

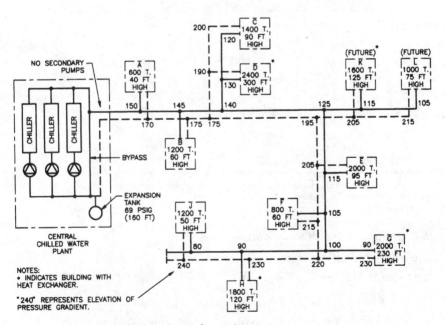

a.   Distributed pumping system.

Figure 18.4   Distributed pumping for district cooling.

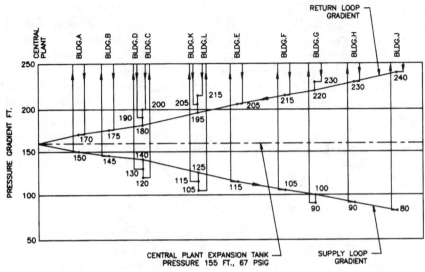

Figure 18.4 *(Continued)*

TABLE 18.1   Distribution Horsepower for Distributed Pumping

Building	Tons	Gal/min	Loop head, ft	Water hp
A	600	1,200	20	6.1
B	1,200	2,400	30	18.2
C	1,400	2,800	80	56.6
D	2,400	4,800	70*	84.8
E	2,000	4,000	90	90.9
F	800	1,600	110	44.4
G	2,000	4,000	150*	151.5
H	1,800	3,600	150*	136.4
J	1,200	2,400	160	97.0
K	1,600	3,200	100*	80.8
L	1,000	2,000	110	55.6
TOTALS 16,000 ton, 32,000 gal/min			822.3 hp	

*Ten feet of pump head has been added to these buildings for the friction loss for a pumping system that may be needed for the heat exchangers.

18.5.1   Comparison of water horsepowers

The great reduction in water horsepower for this example, 822 from 1454 hp, is typical of what may be achieved in HVAC systems based on distributed pumping. The distribution pump motor horsepower for the Denver International Airport was reduced from 2600 to 2020 hp initially and from 3600 to 2670 hp ultimately. This horsepower reduction is achieved by not overpressuring any building with distributed pumping. The energy figures acquired with this model system should not be used for an actual system. If distributed pumping is considered for an actual system, the computer calculations and pressure gradient should be run for that system as it was done here with this model system.

18.5.2   Features of distributed pumping

One of the most significant features of distributed pumping is its simplicity. There are no crossover bridges, balancing valves, or pressure regulators that consume energy. Another feature is total independent control for each building. The building pumps are usually variable-speed pumps because of the tremendous variation in total head from minimum to maximum flow. Speed control is maintained for the pumps in each building. The only effect that one building has on another is the change in the loop pressure that is caused by a load change in a building. This is picked up quickly by the pump controllers in each building without hunting or rapid changes in pressure. There is absolutely no reason to install balance valves on each building; these systems are completely self-balancing.

The principal disadvantage of a pure distributed pumping system is its inability to be applied to an existing installation, particularly if there are many buildings on the system. In some cases, the installed cost of the existing pumping and piping makes it prohibitive to try to utilize the principles of distributive pumping. Because of this, there are many versions of existing primary-secondary pumping for district cooling and heating systems operating with distributed pumping in newer buildings. For example, distributed pumping as booster systems simulates distributed pumping in new sections of a system. This can be called *zone pumping* where the existing central energy plant has distribution pumps with enough head to pump the central zone of a campus. This was demonstrated in Figs. 15.4, 15.5, and 15.6. Instead of boosting central plant pressures to accommodate new buildings or extensions, booster pumping systems each with a bypass can be installed. The booster pumping systems usually employ variable-speed pumps, and they provide the additional head required under peak load conditions when the central pumps cannot meet the head requirement of the peripheral buildings. Typical of these is the system shown in Fig. 15.4.

As mentioned elsewhere in this book, there is great fear about running variable-speed pumps in series. This has been done successfully on many installations. The important factor is proper control of both pumps; with this control, there should not be any concern over the first pump forcing the second pump to run inefficiently, and vice versa.

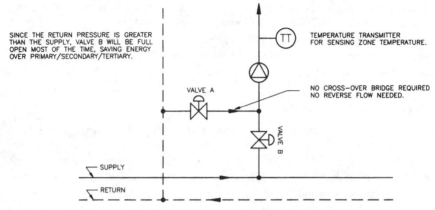

SINCE THE RETURN PRESSURE IS GREATER THAN THE SUPPLY, VALVE B WILL BE FULL OPEN MOST OF THE TIME, SAVING ENERGY OVER PRIMARY/SECONDARY/TERTIARY.

TEMPERATURE TRANSMITTER FOR SENSING ZONE TEMPERATURE.

NO CROSS-OVER BRIDGE REQUIRED NO REVERSE FLOW NEEDED.

VALVE A

VALVE B

SUPPLY

RETURN

Figure 18.5 Blending or chilled water reset with distributed pumping.

18.6 Differential Temperature

The question always arises as to what differential temperature should be used in the transportation of water from the central plant to the buildings on the system. This is a function of piping costs and chilled water generation costs. As described elsewhere in this book, reducing the chilled water temperature leaving a chiller increases the kilowatts per ton for that chiller. This is an overall economic evaluation that must be made by the designers and operators of a district cooling system. If it is in the best interests to market low-temperature cooling water such as 40°F, the distributed pumping system offers a simple procedure for blending the water at the building to a higher temperature such as 45°F. Figure 18.5 describes the blending-valve installation on a distributed pumping system.

The distributed pumping system has a particular advantage over other systems when blending water because the return pressure is always higher than the supply pressure. This reduces the energy lost in the blending procedure because the supply valve is normally open.

18.7 Building Connections

Building connections are very important in district cooling. Figure 18.6a describes suggested connections for a building without a heat exchanger, while Fig. 18.6b provides the connections with a heat exchanger. It should be noted that there is no return temperature-control valve included in Fig. 18.6, as is often suggested. Use of this valve can cause some very complex problems.

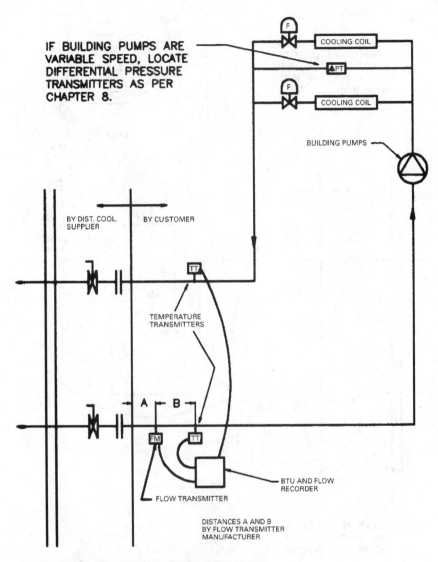

IF BUILDING PUMPS ARE
VARIABLE SPEED, LOCATE
DIFFERENTIAL PRESSURE
TRANSMITTERS AS PER
CHAPTER 8.

COOLING COIL

COOLING COIL

BUILDING PUMPS

BY DIST. COOL.
SUPPLIER

BY CUSTOMER

TEMPERATURE
TRANSMITTERS

A   B

BTU AND FLOW
RECORDER

FLOW TRANSMITTER

DISTANCES A AND B
BY FLOW TRANSMITTER
MANUFACTURER

NOTE: ISOLATION VALVES, EXPANSION TANKS
AND OTHER PIPING EQUIPTMENT NOT
SHOWN FOR CLARITY.

a.   Building connections without heat exchanger.

Figure 18.6   Building connections for district cooling.

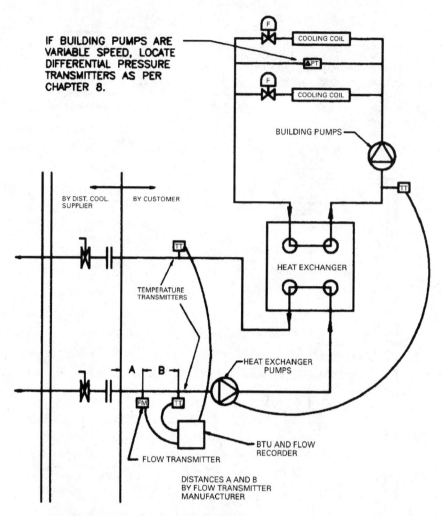

IF BUILDING PUMPS ARE VARIABLE SPEED, LOCATE DIFFERENTIAL PRESSURE TRANSMITTERS AS PER CHAPTER 8.

NOTE: ISOLATION VALVES, EXPANSION TANKS AND OTHER PIPING EQUIPTMENT NOT SHOWN FOR CLARITY.

b.   Building connections with heat exchanger.

Figure 18.6   (*Continued*)

There should be no attempt to artificially control the return water temperature in a particular building. This has been tried with disastrous results by installing a return temperature-control valve that limits the return temperature to a value such as 57°F or higher. Total loss of building humidity control can result due to the high supply water temperature that is caused by this valve that bypasses return water back to the supply.

The return temperature from each building is the result of the quality of design and maintenance in that building. Low return temperature is caused principally by operating cooling coils in the laminar flow range or allowing them to become dirty on the water or air side. It should be the responsibility of the building management to ensure a minimum return water temperature. If they are lax in their design or maintenance, they should pay for it.

18.8 District Heating

Much of the information on district cooling also applies to district heating. Similar evaluations for primary-secondary and distributed pumping can be made for hot water systems. One advantage of heating over cooling is the ability to transmit hot water at much higher differential temperatures such as 40 to 150°F. This reduces the pumping costs and may require more heat exchangers in the buildings served. For example, high-temperature water at 300°F may be delivered to a building requiring only 160°F for its heating system. The added thermal cost of generating higher-temperature water must be considered in the economic evaluation of the system.

The building connections for hot water are, again, the same as those for chilled water, as described in Fig. 18.6. Obviously, piping for high-temperature water is different from that for chilled water, but the general configuration is much the same. Thermal expansion provisions obviously are greater for medium- or high-temperature water than for chilled water.

18.9 Conclusions

Recognizing that return water temperature from a building is a variable should prompt the operators of district chilled water systems to market their energy as is done in the electric power distribution industry. Both flow and differential temperature at each building should be measured with a calculation of total British thermal units. Each building management should pay a commodity charge for their British thermal unit consumption plus a demand charge based on their maximum flow in gallons per minute. This places the responsibility on them for maintaining the highest possible return water temperature.

The development of digital electronics and the variable-speed drive has offered the district cooling industry new tools in the design and operation of their distribution systems. The most important part of this review of district pumping is the use of pressure-gradient diagrams to evaluate and achieve the most efficient method of water distribution in a particular district cooling system. The pressure-gradient diagram provides the designer with a graphic representation of the system's pressures, pressure losses, and overpressure. The equations for system energy consumption (Eqs. 8.4 through 8.7) should be used to determine the most efficient method of distributing water in district cooling and heating systems.

18.10   Bibliography

James H. Ottmer and James B. Rishel, Airport's pumping system horsepower take nose dive, *Heating, Piping and Air-Conditioning Magazine,* vol. 65, no. 10, pp. 51–55, October 1993.

James B. Rishel, Pumping district chilled water systems, Ninth Annual Cooling Conference, International District Energy Association, October 1994.

Pumps for HVAC Hot Water Systems

Steam and Hot Water Boilers

19.1 Introduction

This chapter reviews all HVAC boilers as well as the circuiting and pumping of hot water boilers. Condensate pumping and steam boiler feed will be discussed in Chap. 22. A brief synopsis will be given of the types of boilers that are used in the HVAC industry today.

19.2 Classification of Boilers

Boilers are governed by the American Society of Mechanical Engineers' (ASME) *Boiler and Pressure Vessel Code,* of which there are two sections, Sec. 4, "Rules for Construction of Heating Boilers" (low pressure), and Sec. 1, "Rules for Construction of Power Boilers" (high pressure). The limits between these two codes are by temperature as well as pressure. Heating boilers cannot exceed 15 psig steam, 160 psig water, or 250°F temperature. This temperature is the evaporation or saturation temperature for water at 29.83 psia or approximately 15 psig pressure. All boilers, water or steam, operating above 250°F must be designed to the ASME power boiler code.

Clarification should be made of the term *hot water boilers.* Normally the word *boiler* indicates a piece of equipment that boils a liquid to the gaseous or vapor state. Originally, these boilers were called *hot water generators.* This created confusion in the HVAC industry because heaters of domestic water also were called *hot water generators.* It is now the American Society of Heating, Refrigerating and Air-Conditioning Engineers' (ASHRAE) definition that a boiler is a vessel in which a liquid is heated with or without vaporization. For the HVAC industry, the liquid is water; if it is a fluid such as air, the vessel is called a *furnace.*

19.3   Types of Boilers by Materials of Construction

Boilers, both steam and hot water, are manufactured of cast iron, steel, copper, and other materials that may be used in condensing-type boilers. For many years, low-pressure heating boilers were manufactured from cast iron, while high-pressure boilers were of riveted steel design. As welding technology progressed, all steel boilers were welded and became available for low-pressure heating applications. Copper-tube boilers were developed for both low and high pressures because of their flexibility.

Today, most HVAC boilers are of steel, copper, or cast iron construction. Steel and copper boilers can be made for both low- and high-pressure service, while cast iron boilers are designed for 15-psig steam or 30-psig water working pressures. Both steel and cast iron boilers are made in a number of configurations for the firing of specific fuels or to fit the thermodynamic design of the manufacturer. Electrically heated boilers are usually of steel or alloy construction.

Steel boilers can be of *fire-tube,* with the products of combustion in the tubes, or of *water-tube* construction, where the water or steam is in the tubes and the products of combustion surround the tubes. Both types of boilers are made in small and large sizes; the largest steam boilers are of water-tube design.

Some boilers are made of a rigid construction that makes them susceptible to thermal shock. *Thermal shock* is an expression in the boiler industry that denotes uneven expansion in a boiler that causes leakage around the tubes. This can cause severe damage to the boiler in some cases. The boiler manufacturer will stipulate the differences between supply and return water temperatures and may specify a maximum rate of temperature change in the supply water. The boiler manufacturer may provide a circulating pump that will solve this problem or will specify the flow requirements of a primary or boiler pump furnished by others. This will determine whether primary or primary-secondary pumping systems are required with certain boilers.

Medium- and high-temperature water boilers must be designed to the ASME power boiler code because they operate above 250°F. They can be steam boilers, where the steam is used principally as an expansion cushion, or they can be water-tube boilers with special circuits for high-temperature water. Further description of these boilers will be made in Chap. 21 on medium- and high-temperature water.

19.4   Types of Boilers, Condensing, or Noncondensing

Hot water boilers also can be classified as to their operation, whether condensing or noncondensing. For many years, hot water boilers were operated noncondensing. This means that the water vapor in the products of combustion does not condense within any part of the boiler. To avoid condensing the water vapor, the boilers were designed so that any of the water next to the heating surfaces of the boiler never dropped below 140°F. For most natural gases, the dew point of water vapor at 20 percent excess air is around 130°F. Therefore, a 10°F temperature difference between the water and the flue gases was maintained at 140°F to avoid condensation.

Why is condensation undesirable in these boilers? Many of them are made of ferrous products (steel or cast iron). A mixture of carbon dioxide and water vapor produces carbonic acid, which attacks steel or cast iron and causes severe rusting. For many years, control of combustion and boiler design prevented the use of condensing boilers.

Why is condensing of water vapor desirable in boilers manufactured of materials that withstand attack by carbonic acid? An investigation of the steam tables will reveal that the heat of evaporation for water vapor at the partial pressures at which it exists in flue gas products is over 1025 Btu/lb. Most of this heat is transferred to the water being heated if condensation occurs. Figure 19.1 describes the efficiency of a boiler that can be operated condensing. This figure is approximate and is considered to be for a boiler operating at full firing rate. Boilers with modulating burners may provide even higher efficiencies.

The maximum possible efficiency of a gas-fired boiler without condensing is around 87 percent; this efficiency may be difficult to achieve without condensing in the boiler itself or in the flue system. Condensing boilers operating with low return water temperatures from the heating system can produce combustion and thermal efficiencies in excess of 95 percent. It is obvious from these efficiency differences that it is imperative that every effort be made to reduce operating water temperatures where possible to take advantage of these improved boiler efficiencies.

Field studies of seasonal efficiency for HVAC boilers have revealed that they are well below the certified combustion efficiencies for these boilers. One study of four commercial hot water boilers produced average apparent seasonal gas boiler efficiencies of 51.2 to 65.3 percent. This demonstrates the need for reduced hot-water temperature as a means of achieving higher seasonal boiler efficiencies.

Continued development of gas-fired condensing boilers should produce much greater efficiency in the operation of hot water systems. It

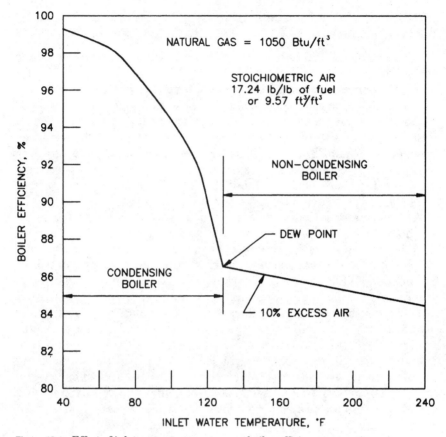

Figure 19.1 Effect of inlet water temperature on boiler efficiency.

also aids in the use of heat rejection from chillers. When condenser water leaves chillers at temperatures around 90 to 100°F, it can be used in off-peak load conditions in hot water systems as well as other heating requirements such as reheat on cooling systems, domestic water heating, and snow melting.

19.5   Rating of Boilers

Low-pressure steam boilers are rated in pounds of steam per hour from and at 212°F, British thermal units per hour, or boiler horsepower. Boiler horsepower is not used as often as in the past; it is equivalent to 34.5 lb of steam per hour from and at 212°F or 33,475 Btu/h. The expression *from and at* means from feedwater at 212°F to steam at 212°F.

Hot water boilers are rated in thousands of British thermal units transferred per hour, which is abbreviated as MBH. This rating may be specified by the manufacturer at some specific temperature such as 180°F. Boiler ratings are established by boiler manufacturers' associations and government agencies. They can be so-called gross or net ratings. For the designer of a water system, from a pumping standpoint, the only rating of concern is the maximum boiler output or heat flow to the water system in MBH.

19.6 Computing Boiler Water Flow

The standard formula for calculating water flow in gallons per minute is as defined in Chap. 9:

$$\text{Hot water system gal/min} = \frac{\text{system MBH} \cdot 1000}{500 \cdot \Delta T \, (°F)} \qquad (9.1)$$

This formula assumes that the specific weight of water is always 62.3 lb/ft^3, which is not true. If precise flow is desired, the true specific weight γ at the operating temperature must be used in the following formula:

$$\text{Hot water system gal/min} = \frac{\text{system MBH} \cdot 1000}{(\gamma/7.48) \cdot 60 \cdot \Delta T \, °F}$$

$$= \frac{125 \cdot \text{system MBH}}{\gamma \cdot \Delta T \, (°F)} \qquad (19.1)$$

where 7.48 = U.S. gallons per cubic foot
γ = specific weight of water from Table 2.3

For example, assume that system load is 5000 MBH with a 200°F supply water temperature and a differential temperature of 40°F. From Table 2.3, the specific weight of water at 200°F is 60.13 lb/ft^3. Thus

$$\text{Gal/min} = \frac{125 \cdot 5000}{60.13 \cdot 40}$$

$$= 259.9 \, \text{gal/min}$$

This formula assumes an average specific heat of water to be 1 Btu per pound per degree F. For most hot water operations below 250°F, this is acceptable; it cannot be used for medium- and high-temperature water systems, where the actual specific heat must be factored into the equations for water flow.

19.7   Connecting Pumps to Boilers

Condensate return and connecting pumps to steam boilers will be reviewed in Chap. 22. Of great concern to the hot water system designer is the location of the pumps with respect to the boiler and whether a circulating pump is required to maintain a minimum flow rate through the boiler regardless of the flow to the hot water system. Both these factors must be settled before the design of the boiler room is completed. The cost of the installation and its successful operation are very dependent on these two decisions.

19.7.1   Minimum flow through the hot water boiler

As indicated earlier, the boiler manufacturer whose boiler is under consideration must provide information as to minimum flow rate and, in some cases, the maximum change in flow rate. These two factors may or may not be important. The boiler manufacturer should guarantee the boiler against thermal shock under specific flow conditions.

Thermal shock is avoided by providing a minimum flow rate in the boiler. Minimum flow rate is maintained by providing a circulating pump on the boiler or by primary-secondary pumping. Other manufacturers depend on bypass valves rather than the boiler or primary pump. It is the boiler manufacturer's responsibility to provide detailed pumping and piping instructions for the boiler installation to ensure that thermal shock does not occur in the boiler.

19.7.2   Locating pumps at boilers

The question always arises as to the location of the pump with respect to the boiler. There are some accepted rules in the HVAC industry that have no foundation. The selection of the pump location with respect to the boiler must be determined by the physical conditions of the actual application. For example,

1. The pumps should be located, if possible, on the cooler return water to the boiler. This enables the pump internals such as the mechanical seal to have a longer useful life.

2. Locating the pumps on the cooler return water also does not expose the pumps to the air released from the water in the boiler (see Table 2.5).

3. The static pressure of the building determines the required operating pressure of the boiler. If putting the pump on the inlet to the boiler increases the design pressure of the boiler, either the boiler

construction should be changed or the pump should be located on the discharge of the boiler.

Often, boilers constructed of steel can be equipped with pressure-relief valves set at pressures required by the installation with the pump installed on the return water connection to the boiler. Location of the pump with respect to the boiler is a decision that must be made by the designer, taking in consideration all the conditions that exist on a particular installation.

19.8 Reset of Boiler Temperature versus Zone Reset

A controversy exists in the HVAC industry over the relative merits of boiler temperature reset and the reset of zone temperatures. It is important that this be addressed in this chapter, since many designers are losing sight of what achieves lowest possible energy consumption in a hot water system. Operation of the boilers at the lowest acceptable leaving water temperature provides the lowest energy consumption in a hot water system. Figure 19.1 demonstrates the great increase in efficiency secured by operating a boiler at low entering water temperatures. The condensing-type boiler increases these opportunities.

Zone reset of water temperature may appear to have some advantages, but on most installations, the economies of boiler reset and eliminating zone reset valving more than offset the energy saved by zone temperature reset. Specific installations where some zones have much more glass load than others might justify reset of the zone temperature on a particular zone. Chapter 8 on the use of water in HVAC systems has specific information on the pumping costs of zone reset.

19.9 Boiler Controls

A brief review of boiler control should be included, since boiler sequencing and interfacing with pump controls are often encountered on HVAC heating systems. There are three basic types of boiler control:

1. *On-off control.* The operating controller, such as a thermostat or pressure control, starts and stops the boiler firing equipment. The boiler operates at design firing rate. This is the least efficient control system, since there is a standby loss whenever the boiler is not operating. Also, the flue gas loss is highest at design firing rate.

2. *High-low-off control.* This control procedure is more efficient than the on-off control, but it does not eliminate all inefficient operation. Instead of turning on and off, the burner returns to a low firing

rate before stopping. With proper adjustment, the boiler can operate between low and high firing continuously on much of the load range.

3. *Modulated firing*. This control varies the firing rate and combustion airflow with the load and eliminates much of the starting and stopping of the boiler. The load range for most noncondensing boilers is around 30 through 100 percent of full-load firing. Some condensing boiler manufacturers quote operating ranges as broad as 8 to 100 percent. This great range reduces the time the boiler is shut down, and therefore, the standby loss is reduced appreciably.

Modulating firing is the most efficient boiler control because it eliminates much of the standby loss and accomplishes most of the operation at an off-peak firing rate. This type of control is easy to interface with boiler reset and other control algorithms, since the firing controls can accept a standard 4- to 20-mA signal from the other system controls.

Interfacing boilers efficiently with hot water systems has become much easier with digital electronics; piping has become simpler too. Studies have shown that existing, older boilers have seasonal efficiencies as low as 50 percent. With the newer digital controls and the contemporary boilers, particularly condensing boilers, the seasonal boiler efficiencies should rise dramatically.

19.10 Bibliography

Systems and Equipment Handbook, ASHRAE, Atlanta, Ga., 1992.

Low-Temperature Hot Water Heating Systems

20.1 Introduction

Low-temperature hot water heating systems comprise the majority of HVAC hot water heating systems. The obvious reason for this is the lack of a need for water temperatures in excess of 250°F on most of these systems. Also, the advent of the condensing boiler has driven down the design water temperature of many systems. As was discussed in Chap. 19, lower operating water temperatures increase appreciably the boiler efficiency. This fact and the development of the variable-speed pump constitute two of the greatest improvements in the efficiency of space heating in commercial, governmental, educational, institutional, and industrial buildings. Residential heating by hot water will not be included herein, since this is a specific field by itself. Also, the pumps for residential heating consist of small circulators whose horsepower is not of major consideration in terms of energy use.

20.2 Classification of Hot Water Systems

Hot water systems in HVAC can be classified as to the source of their heat as well as by use of the hot water. These systems receive their heat from (1) fuel-fired boilers, (2) low- and high-pressure steam, (3) medium- and high-temperature water, (4) district heating, and (5) heat recovery. Hot water is used for (1) space heating, (2) heating of outside air, (3) reheat in cooling systems, and (4) process heating in industry. Heating of domestic water is not considered to be an HVAC hot water system.

20.3   Sources of Heat for Hot Water Systems

20.3.1   Hot water boilers

Most of the discussion of hot water boilers was included in Chap. 19. From a hot water heating system standpoint, the boiler type should be selected that offers optimal efficiency, simplicity in the design of the hot water system, and ease of installation. First cost is, of course, a consideration on many installations.

The static pressure imposed on the hot water system will have some effect on the type of boiler selected. For example, high-rise buildings have high static pressures, and steel boilers designed to 100- or 160-psig working pressure are suitable for these applications. If the boilers are installed in equipment rooms on top of these buildings, low-pressure boilers can be used.

20.3.2   Low- and high-pressure steam

Steam systems provide an important source of heat for hot water heating systems. Both low- and high-pressure steam are used for heating hot water. There is very little difference in the configuration of heat exchangers for generating this hot water. The traditional method for connecting these heat exchangers is shown in Fig. 20.1. The principal

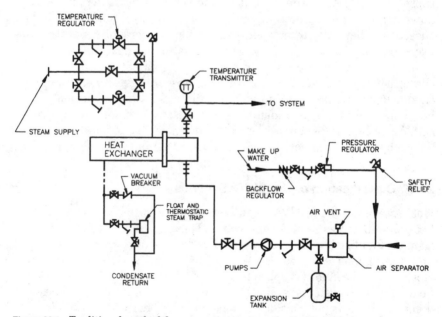

Figure 20.1 Traditional method for connecting a steam-heated heat exchanger.

difference between the use of low- and high-pressure steam is in the selection of the safety relief valves that protect the hot water heating system from the higher pressures or temperatures of the steam. Great care should be taken in sizing and configuring these safety valves. Various governing agencies have specific regulations on the selection of such valves. The designer must review local codes to ensure that the installation of the heat exchanger complies with all applicable regulations. Obviously, the heat exchanger, piping, and appurtenances must be rated to accommodate the temperatures and pressures of the installation. Also, the discharge of these relief valves must comply with these codes. The steam emitted by them must not be allowed to escape into the equipment room. It must be vented into a stack or outdoors at an elevation that will not create harm or destruction. If vented to an area that can be at temperatures below freezing, the relief valve must be equipped with a standard drip-pan elbow.

Most heat exchangers for this service have been of the shell and tube type. Some plate and frame manufacturers are offering their products for use on low-pressure steam. One important factor that must be addressed is the possibility that a vacuum may be imposed on the heat exchanger; the plate and frame heat exchanger must be designed with gasketing that will withstand this vacuum. Heat exchangers can withstand variable hot water flow through them (Fig. 20.2), so there is no need for special piping arrangements such as primary-secondary pumping on most installations.

The traditional method of connecting these heat exchangers is shown in Fig. 20.1. The pumps, expansion tank, and makeup water equipment must be installed on the cooler return water side of the heat exchanger. This reduces the operating temperature of the pumps and offers longer mechanical seal life. There is little advantage to lo-

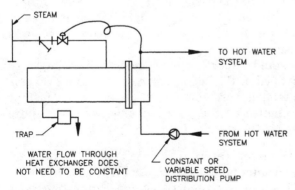

Figure 20.2 Primary pump with heat exchanger.

cating the air separator and the pumps on the discharge side of the heat exchanger; a review of Table 2.5 will reveal that only a small amount of air is removed with additional heating of the water. Most of the air coming into the system is from the makeup water. The use of chemical treatment, as discussed in Chap. 9, is the proper method of eliminating air and therefore oxygen from these systems.

The traditional arrangement of the heat exchanger and its piping as shown in Fig. 20.1 has had some operational difficulties on variable-volume hot water systems. The use of variable-speed hot water pumps and the resulting low flow through the heat exchanger have created operational problems. These problems are as follows:

1. At very low heating loads, the steam temperature regulator reduces the steam flow until a vacuum occurs in the heat exchanger. This causes the heat exchanger to fill with air or condensate through the vacuum breaker. If air is injected into the heat exchanger, this increases the possibility of corrosion in the internal parts of the heat exchanger. Control of the temperature of water required by the system is lost.

2. If the heat exchanger is filled with condensate, any increase in load in the water system will cause the steam valve to open wide in an attempt to recover the water temperature. Full steam pressure may be exerted on the condensate in the heat exchanger to drive the condensate out of the heat exchanger. This can cause hammering and pounding in the heat exchanger.

3. The steam trap on the heat exchanger is overloaded and may limit the flow of condensate out of the heat exchanger.

An attempt has been made to solve these problems by using two steam temperature regulators, one for one-third of the design load and the other for two-thirds of the design load. This has solved these problems in some installations, but a hunting problem may exist when the steam control transfers from the one-third control valve to the two-thirds control valve.

An alternative to the standard method of piping and pumping these heat exchangers is described in Fig. 20.3. A three-way mixing valve is used to control system temperature, and on most systems, a constant steam pressure is maintained on the heat exchanger. Following are solutions to the problems of the standard hookup:

1. Since a constant steam temperature is maintained on the heat exchanger, there is no vacuum produced in it, and it is not filled with air or condensate. This provides a constant-temperature source for the hot water system.

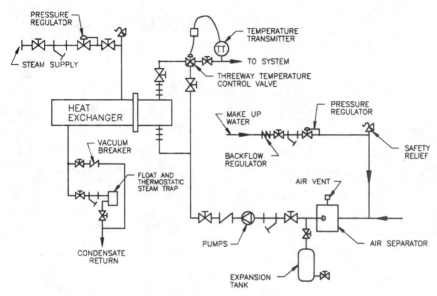

Figure 20.3 Heat exchanger installation with three-way valve control.

2. The steam valve can consist of one valve in most cases, not the more complex one-third, two-thirds valve arrangement.

3. The steam trap receives a steady flow of condensate that is based on the load on the hot water system; its overloading is eliminated.

The steam pressure required in the heat exchanger is determined by the condensate system that exists at the heat exchanger. If the condensate drops to a condensate pump, a minimum steam pressure of 1 or 2 psig gauge pressure may be adequate. If the condensate is sent to a condensate return pipe, the elevation of that pipe with respect to the elevation of the heat exchanger will determine the steam pressure. A rule of thumb that is used in the industry is to allow 1 lb of steam pressure per foot of elevation difference. This accounts for the elevation plus pipe friction. A more precise method is to calculate the pipe friction and the elevation change to arrive at the exact steam pressure in the heat exchanger.

There is a requirement for the sizing of the three-way control valve on these systems, and that is minimum flow. Manual control valves should be located out in the system at its far ends to (1) maintain water temperature in the supply mains, (2) provide minimum flow through the pumps, and (3) ensure some flow through the three-way valve to aid in the control of water temperature. Usually, this flow

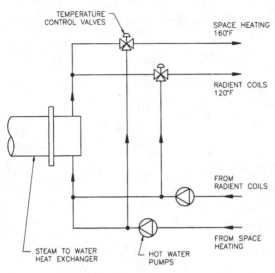

Figure 20.4   Multiple-temperature water services.

can be calculated, and normally, it is a low percentage of the full flow of the hot water system.

With the availability of digital control, it may be possible to vary the steam pressure with the load on the hot water system, which will reduce the size and cost of the heat exchanger and steam control valve. For example, the minimum steam pressure may be calculated at 2 psig. If the steam pressure is changed on a schedule with the load on the heating system so that 10 psig is on the heat exchanger at full load, the size of the heat exchanger and control valve may be reduced appreciably.

Another value of the three-way valve system is the ability to generate more than one temperature of water on one heat exchanger system. Figure 20.4 describes the use of two three-way control valves to produce two different temperatures of water.

20.3.3 Medium- and high-temperature water

Low-temperature hot water heating systems can receive their heat from medium- and high-temperature water systems that may be available on campus-type installations. Heat exchangers of the shell and tube type are usually the heat transfer medium rather than blending water. Blending water may create temperature or makeup water problems.

The heat exchangers for this service are much like those discussed above for low- and high-pressure steam. The problems with condensate and vacuums that were reviewed do not exist on medium- and high-temperature water sources. The heat exchangers must comply with any applicable codes for pressure vessels of the size and temperatures involved.

20.3.4 District heating systems

District heating systems can be a heat source for low-temperature hot water systems. Chapter 18 provides information on such low-temperature hot water systems.

20.3.5 Heat-recovery processes

Heat recovery from various sources can be applied to low-temperature hot water systems. Utilization of the heat of rejection from chillers is reviewed in detail in Chap. 16. Heat recovery from exhaust air or other gas streams can be a source of heat for these water systems. Special hot water heating coils are provided for these applications; usually, they are installed upstream from the main hot water heating coil. The metallurgy of the heating coil must be designed for the conditions of the gas stream. Dirty process gases may require cleaning equipment ahead of the heating coil.

20.4 Uses of Hot Water in HVAC Systems

Low-temperature hot water is used for (1) space heating, (2) heating of outside air, (3) reheat in cooling systems, and (4) process heating in industry. All these uses of hot water employ a heating coil when the receiving fluid is air. Some industrial heating processes may use a heat exchanger to heat other liquids.

The same problems with laminar flow and freezing that occur in cooling coils also exist in heating coils. The discussion in Chap. 9 that covers most of these problems also should be applied to hot water coils.

The distribution of hot water to all the preceding uses is much the same. Some industrial processes might require special cleaning or sensors to detect unacceptable material in the return water.

20.5 Distribution of Low-Temperature Hot Water

The distribution of low-temperature hot water is substantially the same as that for chilled water, as developed in Chap. 15. Greater concern must be given to air elimination, thermal expansion, and corrosion control.

20.5.1 Primary systems

There are more primary systems in hot water than in chilled water owing to the ability of some boilers to accept variable flow through them. Most fire-tube boilers are subject to thermal shock without primary pumps of some type.

Some boiler manufacturers provide a circulating pump with the boiler. This gives such boilers the ability to control minimum flow. Other boiler manufacturers require standard primary pumps with constant flow through the boilers similar to that of chillers. Also, control for limiting return water temperature may be recommended and furnished by some boiler manufacturers. As discussed in Chap. 19, it is imperative that the boiler manufacturer's recommendations on installation and operation be adhered to without exception.

20.5.2 Primary-secondary systems

As indicated earlier, primary-secondary pumping should be used for boilers that require constant flow. This includes some cast iron and steel fire-tube boilers. There is very little difference to this method for piping low-temperature hot water boilers from that described in Chap. 15 on chilled water distribution systems. The principal difference for hot water is the ability to use much higher differential temperatures than is possible for chilled water. Differential temperatures as high as 100°F are designed into some hot water installations.

20.5.3 Primary-secondary-tertiary systems

There is even less reason to use primary-secondary-tertiary pumping on hot water than on chilled water owing to the lower water flows and pump horsepowers. There are very few hot water systems that should utilize this complicated pumping system. Some retrofit projects may justify it, but even in these instances, variable-speed zone pumping or a modification of distributed pumping may offer more energy savings.

20.5.4 Distributed pumping

The advantages of distributed pumping as fully developed in Chap. 15 on chilled water systems also apply to campus-type hot water systems. The energy savings may not be as dramatic as those for chilled water because the pumps are usually smaller. There are no special design conditions for hot water over those already described for chilled water.

20.5.5 Special distribution systems

There are some special hot water systems that utilize manometer fittings that divert water into coils from a single pipe loop; with the development of digital electronics and other contemporary controls, the use of these systems seldom can be justified except, possibly, on small residential heating systems.

20.6 Hot Water Design and Differential Temperatures

With the development of the condensing boiler, the classic water temperature difference of 20°F for low-temperature hot water systems can be expanded to as much as 100°F. In the past, in an attempt to secure higher differential temperatures, operating temperatures varied from 240°F at design load down to 140°F at minimum load. This temperature range reduced greatly the pumping energy, but it elevated the boiler's stack temperature and resulted in a maximum combustion efficiency of around 80 percent. It should be pointed out that the reset of water temperature with percentage of load is usually accomplished through the use of outdoor temperature as the parameter for changing water temperature. With digital electronics, any range that is desired can be inserted into the control algorithm. In days of mechanical control, only two or three ranges were available.

It is now recognized that for most installations it is better to design with lower water temperatures to secure higher combustion efficiencies. If noncondensing boilers are utilized, a typical temperature range would be 200°F at design and 140°F at minimum load. At a 60°F differential temperature, efficiency in pumping has been achieved with a better combustion efficiency for the boiler, smaller pipe, and lower heat losses from the pipe surface.

The condensing boiler goes beyond this overall efficiency and offers a new level of efficiency of operation that was never attained with noncondensing boilers. For example, low-temperature hot water systems can be operated with a 60°F differential temperature with a maximum temperature of 150°F at full load and 90°F at minimum load on the hot water system. At 150°F water temperature, the boiler efficiency would be around 86 percent and at 90°F water temperature, 94 percent.

The use of these lower operating temperatures may require all heating coils to be of fan type; most types of radiation lose much of their heating effect below 140°F. It also opens the way for the use of radiant floors, ceilings, or walls where unusually high boiler efficiencies can be achieved with their low water temperatures. It is obvious from these facts that the selection of operating temperatures for low-temperature hot water systems is much more complex than it was with just noncondensing boilers.

20.6.1 Reset of boiler water temperature versus zone reset

As demonstrated in Chap. 8 on the use of water in HVAC systems, in most cases the reset of water temperature at the boiler offers greater energy savings than does resetting water temperature in various zones of a hot water system. The preceding dramatic increases in boiler efficiency with condensing boilers illustrate what can be done with resetting the boiler temperature. Also, the coil circulating pumps for hot water systems are small with low pump efficiency.

On heating coils, a substantial savings in pumping energy can be achieved by using central pumps with their higher pump efficiencies; small circulators for individual coils have much lower efficiencies. If there is a pronounced difference in zones for heating in a hot water system, such as a north zone and a south zone on a building, an alternative could be the use of two sets of central pumps, as described in Fig. 15.5 for chilled water.

20.7 Bibliography

James B. Rishel, *The Water Management Manual,* SYSTECON, Inc., West Chester, Ohio, 1992.

Pumps for Medium- and High-Temperature Hot Water Systems

21.1 Introduction

Medium- and high-temperature water systems are economical where large quantities of hot water must be transported over an appreciable distance. The ability to operate at differential temperatures in excess of 100°F has reduced greatly the volume of water that must be pumped in these systems.

The efficiency of boilers for these systems cannot be as great as that of some of the contemporary low-temperature hot water boilers; this is due to the higher stack temperatures of the boilers for these higher-temperature waters. Most high-temperature water boilers operate with stack temperatures 100 to 150°F higher than the leaving steam or water temperatures. The actual stack temperature for a specific application should be certified by the boiler manufacturer at the design load and temperature. Boilers for these systems must be designed to the American Society of Mechanical Engineers' (ASME) power boiler code.

The decision to use these systems must weigh the pumping energy saved against the energy lost due to the lower efficiency of the boilers. Likewise, the savings in the cost of reduced pipe sizes achieved with the high differential temperatures must be balanced against the higher cost of the high-pressure boilers, valves, and other appurtenances.

Medium-temperature water systems are those which have water temperatures in excess of 250°F, the maximum for low-temperature water systems. Many of the medium-temperature water systems have a maximum operating pressure of 125 psig. This enables the use of valves, piping accessories, and pump casings designed for 125 psig and the actual operating temperature. This includes some of the class 125 cast iron fittings and the class 150 steel fittings. Depending on the cushion pressure required at high points in these water systems, the maximum temperature for them is around 320 to 340°F. The tables in Chap. 2 for temperatures and allowable pressures for cast iron fittings should be consulted as well as valve manufacturers' catalogs for temperature-pressure ratings for valves. Copper and brass materials generally are not considered suitable above 300°F, so this may be a point of differentiation between medium- and high-temperature water for some designers.

High-temperature water systems are all of those hot water systems that exceed the limits defined above for medium-temperature water systems. The practical upper limit for high-temperature water systems appears to be in the range of 430 to 450°F. The vapor pressure for 450°F water is 422.6 psia, so 500-psig pump casings may provide a practical limit for high-temperature water systems. For heating processes requiring higher temperatures, there are a number of special liquids manufactured by chemical companies for heat-transfer service with special heat exchangers. Seldom are they used in HVAC heating systems.

21.2 Calculation of Water Flow

The water flow for medium- and high-temperature water is calculated with recognition of the variation in specific heat and specific gravity of water at these higher temperatures. The water flow is often expressed in pounds per hour:

$$\text{lb/h} = \frac{\text{system Btu/h}}{\Delta T\,(°\text{F}) \cdot c_p} \qquad (21.1)$$

where c_p = specific heat of water at maximum operating water temperature
ΔT = design differential temperature, °F

The basic formula for calculating required pump flow in gallons per minute is

$$\text{Pump gal/min} = \frac{\text{system Btu/h}}{(\gamma/7.48) \cdot 60 \cdot \Delta T\,(°\text{F}) \cdot c_p}$$

$$= \frac{0.125 \cdot \text{system Btu/h}}{\gamma \cdot \Delta T\,(°\text{F}) \cdot c_p} \qquad (21.2)$$

TABLE 21.1   Specific Heat of Pure Water at Higher Temperatures

Temperature, °F	Specific heat, Btu/°F	Temperature, °F	Specific heat, Btu/°F	Temperature, °F	Specific heat, Btu/°F
200	1.005	280	1.022	370	1.062
210	1.007	290	1.025	380	1.070
212	1.008	300	1.032	390	1.077
220	1.010	310	1.035	400	1.085
230	1.012	320	1.040	410	1.090
240	1.013	330	1.042	420	1.096
250	1.015	340	1.047	430	1.107
260	1.018	350	1.052	440	1.115
270	1.020	360	1.057	450	1.121

SOURCE: Keenan and Keyes, *Thermodynamic Properties of Steam,* John Wiley & Sons, New York, 1936, pp. 30–32. Used by permission.

where γ = specific weight of water at maximum operating temperature, lb/ft^3 (see Tables 2.3 and 2.4)

Equation 21.2 can be used for any water system of any water temperature where very precise water flows in gallons per minute are required, but on most low-temperature water applications, the specific heat of water is assumed to be 1.00. Table 21.1 provides the specific heat for pure water at temperatures from 100 to 450°F.

A sample calculation would be as follows: Assume:

System Btu/h = 50,000,000 Btu/h

Operating temperature = 400°F

Differential temperature = 150°F

From Tables:

γ = 53.65 lb/ft^3 at 400°F

c_p = 1.085

$$\text{Pump gal/min} = \frac{0.125 \cdot 50{,}000\,000}{53.65 \cdot 150 \cdot 1.085} = 716 \text{ gal/min}$$

21.3   Calculation of Pump Head

Rather than using pipe friction tables that are acceptable for low-temperature water, the calculation of pump head for these higher-temperature water systems requires actual calculations using the Darcy-Weisbach formula, Reynolds number, and a Moody's diagram for developing the Colebrook friction factor f. Thus the reduction in the viscosity of water at these higher temperatures can be accounted for in the piping design.

For example, assume that the preceding flow is to commence in a 6-in steel pipe with an internal diameter of 6.065 in (0.505 ft) and a velocity of 7.95 ft/s. The kinematic viscosity of water at 400°F is 0.16×10^{-5} ft²/s. Thus

$$\text{Reynolds number} = \frac{V \cdot D}{v} = \frac{7.95 \cdot 0.505}{0.16 \cdot 10^{-5}} \tag{3.6}$$

$$= 2.51 \cdot 10^{6}$$

From the Moody chart (Fig. 3.2), with a Reynolds number of 2.51×10^{6} and 6-in pipe, the friction factor f is 0.015.

Using the Darcy-Weisbach equation (Eq. 3.3),

$$H_f = f \cdot \frac{L}{D} \cdot \frac{V^2}{2g} = 0.015 \cdot \frac{100}{0.505} \cdot \frac{7.95^2}{64.4}$$

$$= 2.92 \text{ ft/100 ft}$$

Adding 15 percent for variation in pipe diameter, construction, etc.,

$$H_f = 3.36 \text{ ft/100 ft}$$

Although these calculations may be laborious when done by hand, the computer programs now available simplify the use of Reynolds numbers to determine friction losses in pipe and fittings more accurately. The entire medium- or high-temperature water system can be computed thusly to determine the required pump head.

21.4 Medium- and High-Temperature Water Generators

The generation of heat for these water systems can be by steam boilers with an internal steam cushion for expansion or hot water boilers with external expansion tanks. The technology is such that it is beyond the scope of this book to review all the methods of hot water generation. Excellent texts are available to the designer on the construction and operation of these boilers and generators. Some medium-temperature water systems receive their heat from high-temperature water systems through water-to-water heat exchangers and others from ordinary high-pressure steam boilers and steam-to-water heat exchangers.

21.5   Distribution of Medium- and High-Temperature Water

Most of the principles of low-temperature water distribution apply to these systems (see Chap. 20). Added caution must be maintained to ensure that cavitation does not occur in any part of these water systems. The operating temperatures are so much higher that adequate pressure must be maintained in all parts of a system. For example, for the preceding system with 400°F water, from Table 2.4, the evaporation pressure is 247.31 psia. At this temperature, the system pressure should not drop below 250 to 275 psig wherever the water is at 400°F.

The pressure-gradient diagram is an excellent tool for the design of these water systems. It enables the designer to verify that cavitation will not appear in any part of the water system. The pressure heads used in these diagrams must account for the lower specific gravity of the high-temperature water. This can be done by using Eq. 4.1, where the pressure is determined by $144 \div \gamma$, the specific weight of water in pounds per cubic feet.

Distributed pumping should be considered for these systems because of the distances normally encountered between buildings or individual loads.

21.6   Pumps for Medium- and High-Temperature Water Systems

Pump selection for high-temperature water systems is much like that for other HVAC water systems, namely, to achieve maximum efficiency with respect for the economic factors of the installation. There is one additional factor that is not encountered in the other water systems that must be recognized on these higher-temperature water systems. This is cooling of the mechanical seals that are normally used on these pumps.

Mechanical seal cooling must be designed to eliminate as much waste of cooling water as possible. Many pump manufacturers supplying high-temperature water pumps are experienced in providing the correct type of mechanical seal and cooling medium. The use of a heat exchanger mounted on the pump base and furnished by the pump manufacturer with interconnecting piping and control provides the optimal seal cooling system for most applications. If domestic water is used for this cooling, the heat exchanger must be on the load side of an approved backflow preventer.

Because of the need for mechanical seal cooling and the relative low pump capacities, single-suction pumps with one mechanical seal are normally selected for these services. The metallurgy of the pumps must meet the higher pressures and temperature of these water systems. Cast iron volutes are seldom used, whereas ductile iron may be acceptable for medium-temperature systems. Most high-temperature water systems require cast or welded steel volutes with steel or ductile iron brackets. Although bronze impellers may be acceptable below 300°F temperature, the pump manufacturer should specify an impeller material such as iron or stainless steel no. 316. The minimum clearances between the impeller and the case wear rings may need to be increased for stainless steel because of possible galling. This increases bypassing and may decrease pump efficiency.

Most of the pumps for these higher water temperatures should be equipped with expansion joints or means on the pump connection flanges to ensure that piping stress is not imposed on the flanges. These expansion means must have the correct pressure and temperature ratings and must be installed in a position that will ensure that no pipe stress is imposed on the pump flanges.

Variable-speed pumping offers energy savings for these systems that have a sizable variation in the heating loads in them. Again, the pressure-gradient diagram may be a valuable tool for evaluating system pressures with variable-speed pumps.

Some manufacturers of hot water generators for these systems have very specific requirements for circulating the water through their equipment. These circulating pumps may be mounted on the generator itself, or they can be field mounted. Any field-mounted boiler circulating pump must comply strictly with the boiler or generator manufacturer's requirements.

21.7 Bibliography

Following are some sources for additional information on high temperature hot water design and types of generators for these systems. These texts were written before the current technology was available on variable frequency drives and their application to HVAC water systems.

Nils R. Grimm and Robert C. Rosaler, *Handbook of HVAC Design,* McGraw-Hill, Inc., New York, N.Y., 1990.

Erwin G. Hansen, *Hydronic System Design and Operation,* McGraw-Hill, Inc., New York, N.Y., 1985.

Joseph H. Keenan and Frederick G. Keyes, *Thermodynamic Properties of Steam,* John Wiley & Sons, Inc., New York, N.Y., 1936.

22

Condensate, Boiler Feed, and Deaerator Systems

22.1 Introduction

Steam is the name given to the vapor phase of water and is produced by boiling water in steam boilers. It was the principal method of transmitting heat in HVAC systems before the development of hot water systems and hot water pumps. Although steam and process heating systems are not as prevalent as in the past, they are still used extensively in HVAC operations.

Heat is given up to steam-heated systems by the condensing of the steam back to the liquid state of water. The condensate from these systems should be returned to the boilers. Some low-pressure steam systems return the condensate directly to the boilers without pumps. Other low-pressure systems and all high-pressure steam systems utilize condensate, vacuum, or boiler feed pumps for return of the condensate.

22.2 Review of Steam Systems

Steam systems are either low- or high-pressure systems in accordance with the American Society of Mechanical Engineers' (ASME) boiler codes. Low-pressure steam systems are 15 psig or less; all systems higher than 15 psig are high-pressure steam systems. Most HVAC steam heating systems are low-pressure systems; high-pressure steam is used in hospitals and buildings where other high-pressure steam heating services are required. High-pressure steam is used in district heating systems or other campus-type installations where long distances are required for the transmission of steam. Today, medium- and high-temperature hot water systems have replaced high-pressure steam on many large systems with long piping runs to remote buildings.

22.2.1 Rating of heating systems

Steam heating systems are rated in (1) MBH (thousands of BTU per hour) like hot water systems, (2) in pounds of steam per hour from feedwater at 212°F to steam at 212°F, or (3) in square feet of radiation. A square foot of steam radiation is equal to 240 Btu/h. Since a pound of steam at 212°F has a heat of evaporation of 970.3 Btu, it has an approximate equivalent of 4 ft^2 of steam radiation. When steam was the principal heating medium, cast iron and other radiators were rated in square feet of radiation; today, most steam heating coils are rated in MBH like other heating coils.

22.3 Low-Pressure Steam Systems

Low-pressure steam systems can range in pressure from vacuum systems up to 15 psig. Before the advent of hot water heating systems, many buildings were heated with vacuum-type steam systems. Older tall buildings still are heated successfully and very efficiently by vacuum steam systems. These vacuum systems allow a broad range of operation, and some of them vary the steam temperature with the load on the system.

A review of the steam tables will reveal lower steam temperatures for pressures less than atmospheric or under vacuum conditions. Vacuum systems were developed to reduce the operating temperature of the steam. For example, steam at a vacuum of 15 inHg has a temperature of near 178°F. This enabled the radiators and heating coils to operate at temperatures lower than 212°F and provided better control of the heating system.

With the development of heating coils other than direct radiation, low-pressure steam systems were used in place of the vacuum systems. The pressure on these systems is normally in the range of 5 to 12 psig. These systems can provide low-pressure steam to other services such as domestic water heaters and converters. *Converter* is the name given to heat exchangers for converting steam to hot water for space heating systems that use coils, radiation, and convectors.

Steam traps are used to trap the steam in a heating coil but let the condensate through to return to the boilers or condensate pumps. Most steam traps used on low-pressure steam systems are thermostatic on small steam-heated equipment or float and thermostatic on larger equipment. These traps pass air as well as condensate through the use of thermostatic elements that expand and close when higher-temperature steam reaches the elements.

22.3.1 Configuration of low-pressure steam systems

The shape of the building in which the steam system is installed determines the equipment used in that system. The rate of return of the condensate determines the type of steam system that should be used. If the condensate can be returned rapidly to the boilers, low-pressure steam can be used without condensate or boiler feed pumps. One disadvantage of such direct return is the feeding of makeup water directly into the boiler. This may result in the injection of air and therefore oxygen directly into the boiler.

Tall buildings with boilers installed in the basement are candidates for such a system. The condensate passes through the steam traps and drops back into the boilers. As the steam system becomes broader in shape, the ability of the condensate to return rapidly to the boilers is reduced. This necessitates the use of condensate pumps to receive the condensate from the steam traps and return it to the boilers. If the steam system is extensive and the delay of the condensate return to the boilers is appreciable, a boiler feed system may be needed to provide storage space for the condensate as it is returned by the condensate systems. This prevents the undesirable condition of freshwater makeup entering the boiler because of the delay in the condensate returning to the boiler.

22.4 High-Pressure Steam Systems

High-pressure steam systems are used for (1) transporting steam over distances, (2) providing higher temperatures for heating equipment, and (3) high-pressure steam turbines. High-pressure steam is used in buildings such as hospitals that can be extensive in layout and require high temperatures for such services as sterilization.

Most high-pressure steam systems in the HVAC field operate with steam pressures from 40 to 110 psig. Many cast iron pipe fittings and valves are rated for 125- or 250-psig steam, so the steam operating pressure is kept below 125 psig. Steam at 110 psig or near 125 psia pressure has an evaporation temperature of 344°F, which provides adequate thermal difference for many heating services.

If steam is required for turbines that are driving chillers or other mechanical equipment, the steam may be generated at higher pressures or temperatures. In rare cases in HVAC systems, the steam may be *superheated,* which means that the steam is heated to a higher temperature than saturation temperature. If the steam is generated at 186 psig or 200 psia and passed through a superheater in the boiler so that the leaving steam temperature is 432°F instead of 382°F saturation temperature, it has 50°F of superheat. Instead of rating the steam in pounds of steam per hour from and at 212°F, it may be necessary to rate the capacity of the boiler in pounds of steam at a feedwater temperature to a steam pressure with so many degrees of superheat. It is important when specifying the capacity of a high-pressure boiler that the pounds of steam per hour be defined by feedwater temperature and operating steam pressure and temperature.

22.5   Steam Boilers

Steam heating boilers are rated in boiler horsepower or pounds of steam per hour. A boiler horsepower is equivalent to 33,475 Btu/h or 34.5 lb/h from feedwater at 212°F to steam at 212°F and 0 psig steam pressure.

Small high-pressure boilers are rated in boiler horsepower or pounds of steam per hour from and at 212°F. Most large high-pressure steam boilers are rated in just pounds of steam per hour at 212°F. High-pressure boilers that provide steam to steam turbines are rated in pounds of steam from a certain feedwater temperature to the actual operating pressure and temperature. As mentioned earlier, the boiler temperature may include superheat if large steam turbines are involved.

22.6   Condensate Systems for Steam Boilers

There are four distinct types of condensate systems for steam systems: (1) vacuum pumps, (2) condensate return systems, (3) boiler feed systems, and (4) deaerators. Each has its place in returning condensate to boilers.

22.6.1   Vacuum pumps

Vacuum pumps are used on vacuum steam systems to perform two services: (1) return the condensate to the boilers, a boiler feed system, or a deaerator and (2) maintain the desired vacuum on the steam system. Vacuum systems have been rated in thousands of square feet of radiation at a specified vacuum. Two different pumps were used on many vacuum systems: (1) to return the condensate and (2) to maintain the vacuum on the steam system. Few new vacuum systems are being used today as a result of the development of alternative methods of heating buildings. Vacuum pumps are still manufactured for existing and new systems.

22.6.2 Condensate return systems

Condensate return systems are designed to return condensate from the steam traps back to the boilers, a deaerator, or a boiler feed system. Condensate return systems (Fig. 22.1a), consist of a receiver, a float switch, and the condensate pump. The condensate fills the receiver, and at a preset level, the float switch starts the condensate pump. The pump removes the condensate from the receiver, and at a second, lower preset level, the float switch stops the pump. There is no other control procedure for these pumps to return the condensate to boilers. They are called *condensate return systems* because this is their principal duty. They do not include any storage volume for the condensate.

The condensate return system can be rated in square feet of radiation, pounds of condensate per hour, or MBH. Usually, this rating is specified at a discharge pressure such as 10 psig.

The pressure rating of the condensate system is determined by the type of condensate piping that is installed in the building. There are gravity returns where the condensate piping is pitched toward the boiler room and pressure returns where the condensate is pumped back to the boiler room by the condensate pumps.

If the condensate piping is of the gravity type, the condensate return system is only required to lift the condensate to the nearest return. Under this condition, the discharge pressure need be only the friction loss of the discharge piping and the static lift of the condensate to the gravity return pipe. The lift is the difference between the elevation of the condensate return system and the condensate pipe. The equation for the pressure is

$$\text{Discharge pressure} = \frac{h_f + \text{static lift}}{144/\gamma} \text{ psig} \qquad (22.1)$$

where h_f = pressure loss in feet for the discharge piping and valves
γ = specific weight of water at the condensate pump discharge temperature

If the condensate piping is of the pressure type, the discharge pressure must include the pressure loss of the piping between the condensate system and the point of delivery in the boiler room, the difference in elevation between the condensate return system and the point of delivery, and the pressure in either the boilers or the deaerator.

Equation 22.1 becomes

$$\text{Discharge pressure} = \frac{h_f + \text{static lift}}{144/\gamma} + I_p \text{ psig} \qquad (22.2)$$

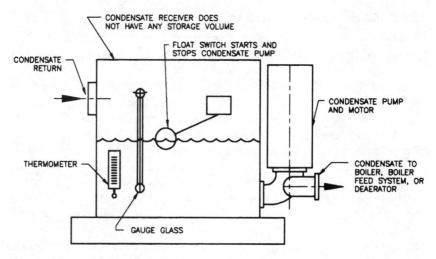

a. Condensate return system.

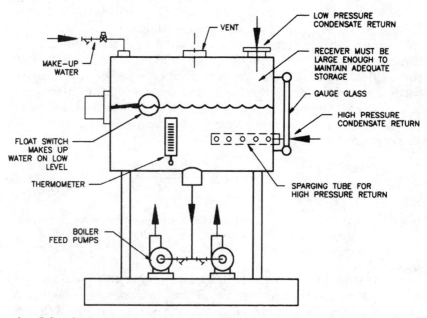

b. Boiler feed system.

Figure 22.1 Condensate and boiler feed systems.

where h_f = pressure loss in feet for the piping and valves between the condensate return system and the point of delivery in the boiler room

I_p = pressure in the boilers or the deaerator

If the condensate is delivered to a boiler feed system, the discharge pressure for the condensate return system is similar to Eq. 22.1, where h_f is the friction loss of the piping between the condensate system and the boiler feed system; likewise, the static lift is the difference between the elevation of the condensate system and the level of the condensate return connection on the boiler feed system.

22.6.3 Boiler feed systems

A boiler feed system (Fig. 22.1b) receives low- and/or high-pressure condensate from steam traps and holds that condensate until it is called for by the boilers. It has a much larger receiver or tank than a condensate return system, since it must provide storage of the condensate between periods of high and low steam production. The contemporary steam boilers with small water volumes are in greater need for condensate storage outside the boiler and therefore in the boiler feed system or deaerator.

The boiler feed system also maintains a minimum level in its receiver by means of a float valve or any other water level control system that adds makeup water to the receiver. This has a decided advantage over injection of makeup water directly into the boilers. Most of the air and oxygen is released in the boiler feed system receiver, not in the boilers. Boiler feed systems can be equipped with sparging tubes for injection of steam into the receiver water so that it can be maintained at 210°F. As demonstrated in Table 2.5, almost all the air is released from water at 0 psig pressure and 210°F temperature.

Unlike condensate return and vacuum pumping systems, the pumps on the boiler feed system are started and stopped or otherwise controlled by the boiler water level controller. These pumps are therefore called *boiler feed pumps*. When the water in the boiler drops to a preset level, a switch is closed in that controller, and the boiler feed pump is started; when the desired level in the boiler is reached, this switch opens, and the boiler feed pump stops. Few boilers that are fed by boiler feed systems are equipped with feedwater regulators; the pumps normally start and stop. Variable-speed pumps are seldom used on boiler feed systems because the system head curve for such pumps is so shallow or flat that very little energy is saved by varying the speed of the pump. On larger boilers, the boiler feed pump runs continuously with a feedwater regulator controlling the rate of flow of feedwater into the boiler.

Boiler feed systems are superior to condensate return systems in operation because the amount of air and oxygen fed into the boiler is greatly reduced. Boiler feed systems are used on low- and high-pressure steam systems that are relatively small and do not have strict requirements for oxygen content in the boiler feedwater. Large high-pressure boilers are almost always equipped with deaerators.

22.6.4 Deaerators

Deaerators (Fig. 22.2) are boiler feed systems with guaranteed air and oxygen content in the feedwater. This is accomplished by a number of different mechanical devices and arrangements. Deaerators accomplish the following duties:

1. Receive both low- and high-pressure condensate from the steam systems.

2. Maintain a minimum water level in the deaerator by adding make-up water.

3. Deaerate both condensate and makeup water to some guaranteed level of oxygen content such as 0.005 cc/liter.

4. Provide adequate boiler feedwater storage to pump the boilers

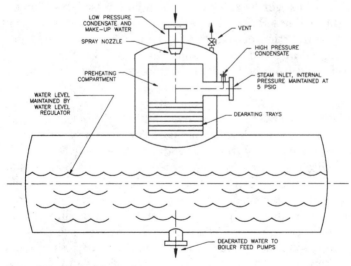

Figure 22.2 Typical deaerator.

under all steaming rates without excessive makeup water. The amount of makeup water should be limited to evaporation or leakage losses, and it should not include water needed to maintain level in the deaerator under heavy load swings on the steam system. Figure 22.2 describes a typical deaerator without a separate storage section. Some deaerators have a second storage section for larger volumes of low-pressure condensate; they are used on large low-pressure steam heating systems.

5. Maintain all the preceding duties without excessive steam loss up the vent pipe or pipes of the deaerator. Deaerators are usually sold under a guarantee of so much steam lost up the vent pipe per thousand pounds of water processed.

Deaerators are manufactured in a number of configurations, of tray or spray type and with or without the separate condensate sections. Manufacturers of deaerators with both a condensate section and a deaerated water section claim better performance for their units. Low-pressure condensate and makeup water are fed into the condensate section, while high-pressure condensate is returned to the deaerated water section. Other manufacturers claim better performance without the storage section.

The deaerator section can use a number of different mechanical devices for deaeration of the water; most deaerators are either spray or tray type. In spray-type deaerators, the condensate and makeup water from the condensate section are heated and sprayed over the water in the deaerator section, thus releasing any gases in the water. The gases, such as oxygen, pass up the vent pipe. In tray-type deaerators (Fig. 22.2), low-pressure condensate drops down over trays and steam passes through the trays, scrubbing the gases from the condensate. This configuration appears to be the preference where very low levels of oxygen are required.

Deaerators also can be classified as atmospheric or pressure type. Atmospheric deaerators operate with no steam pressure in the deaeration section, while pressure deaerators maintain a minimum steam pressure such as 5 psig in this section. Usually, the pressure type offers a higher level of deaeration than the atmospheric type. The question often raised is what is the necessary level of deaeration on many HVAC steam systems? Also, on many HVAC boilers the final removal of oxygen can be accomplished chemically with the feeding of sodium sulfite or similar chemicals into the feedwater. This increases the amount of surface blowdown required at the boiler. It is the decision of the steam system designer as to which is the best method for controlling corrosion in boilers and steam systems.

22.7   Design and Selection of Boiler Feed Pumps

Boiler feed pumping can be troublesome if some application problems are not addressed. Relatively high temperatures (200 to 230°F) at low NPSH available conditions and high discharge pressures require careful selection of boiler feed pumps.

22.7.1   Design of boiler feed pumps

Boiler feed pumps for HVAC boilers are almost always centrifugal of single-stage construction for low-pressure boilers and multistage for high-pressure boilers. Some small high-pressure process boilers use peripheral turbine-type pumps that have a relatively low efficiency.

Feed pumps for HVAC boilers usually are equipped with mechanical seals, not packing. The mechanical seal eliminates most of the leakage and evaporation at the stuffing boxes. The mechanical seals must be designed for the maximum temperature encountered with the hot condensate.

The boiler feed pump must have a low NPSH, since the condensate will be at or near evaporation temperature. Most boiler feed pumps have a required NPSH of less than 10 ft. The required NPSH must be evaluated at the maximum pumping rate that can be encountered in the boiler feeding cycle. The boiler feed pump may require a discharge pressure regulator that maintains a continuous discharge pressure on the boiler feed pump regardless of the boiler pressure. This reduces the ability of the pump to run at high capacities and high required NPSH.

Small low-pressure boiler feed pumps are normally single-suction, single-stage pumps flexibly coupled to standard electric motors. Small high-pressure boiler feed pumps can be peripheral turbine pumps, but for pump efficiency and continuous service, the multistage turbine pump is the preferred type. This pump should be designed for 250°F service and provided with flexibly coupled electric motors.

Large high-pressure boiler feed pumps can be multistage turbine or single-suction pumps. The multistage single-suction pump, often called a *multistage horizontal split case pump,* has been the standard pump of the medium-sized boiler. The multistage turbine pump has been developed to the point that it is often the most efficient, economical pump for this service.

22.7.2   Selection of boiler feed pumps

Boiler feed pumps must be selected for capacity, head, and NPSH available for each application. Each of these factors is critical in the selection of boiler feed pumps.

Boiler feed pump capacity. The capacity of boiler feed pumps must accommodate not only the feedwater requirements of the boilers but also the quality of the steam produced by the boilers as well as the blowdown requirements of the boiler treatment. *Quality* is the name given to wetness of the steam produced by the boiler. Pure steam is all gas and has no liquid content. Large boilers have water separators and scrubbers that enable the steam leaving the boiler to have very little water and are said to produce steam with close to 100 percent quality. Most of the steam boilers on HVAC steam systems have no steam separators and smaller steam releasing areas, so they do not produce high-quality steam. Because of this, a factor of 5 to 10 percent must be added to the boiler feedwater requirements to secure adequate pumping capacity. The following equation provides the method of calculating boiler feed pump capacity:

$$\text{Boiler feed pump gal/min} = \frac{\text{boiler lb steam/h}}{60 \cdot (\gamma/7.48) \cdot 0.90}$$

$$= \frac{\text{boiler lb steam/h}}{7.22 \cdot \gamma} \qquad (22.3)$$

where γ = specific weight of water at the feedwater temperature
0.90 = contingency for steam quality, leakage, etc.

For example, if a 10,000 lb/h boiler is being fed with 230°F feed water, from Table 2.4, $\gamma = 59.38$ lb/ft^3, and

$$\text{Pump gal/min} = \frac{10,000}{7.22 \cdot 59.38} = 23.3 \text{ gal/min}$$

Some designers will add a greater contingency than the 90 percent included in the preceding formula, particularly if peripheral turbine-type pumps are used as boiler feed pumps. The close tolerances in these boiler feed pumps can subject them to wear. Often, the boiler feed pump capacity is selected at double the boiler capacity for these pumps.

If the boiler capacity is provided in boiler horsepower, Eq. 22.3 becomes

$$\text{Boiler feed pump gal/min} = \frac{\text{boiler capacity (hp)} \cdot 34.5}{60 \cdot (\gamma/7.48) \cdot 0.90}$$

$$= \frac{4.78 \cdot \text{boiler capacity (hp)}}{\gamma} \qquad (22.4)$$

If the boiler feed pump operates continuously with the boiler feed rate controlled by a feedwater regulator, a circulating flow in the amount to keep the boiler feed pump from overheating must be added to the pump capacity. This can be computed by the boiler feed pump manufacturer and is often in the range of 5 to 20 gal/min.

Calculation of boiler feed pump head. Boiler feed pump head requires more care than other HVAC pumps due to the boiler pressure that must be included in the equation for the boiler feed head h_b:

$$h_b = P_f + P_d + \text{boiler pressure (psig)} \cdot \frac{144}{\gamma} \qquad (22.5)$$

where P_f = suction piping friction loss, ft
P_d = discharge piping friction loss, ft

The suction piping friction loss P_f must include (1) the exit loss from the deaerator or boiler feed system tank, (2) pipe and fitting losses, and (3) strainer losses if they exist. Suction piping friction must be calculated carefully, since it is also part of the NPSH calculation.

The discharge piping friction loss P_d must include (1) piping and fitting losses and (2) losses through feedwater regulators, if they exist.

The difference in elevation between the water in the boiler feed system or deaerator tank and the boiler is not figured in the equation for boiler feed pump head for most HVAC boilers. If for some reason the boilers are installed at a level different from that for the boiler feed system or deaerator, the elevation difference should be included in the boiler feed pump head equation.

NPSHA calculations for boiler feed pumps. NPSHA calculation is very important for boiler feed pumps because the feedwater temperature is close to the evaporation temperature. This means that the atmospheric pressure P_a and the vapor pressure P_{vp} in Eq. 6.8 cancel out; this equation reduces to the following:

$$\text{NPSHA} = P_z - P_f \qquad (22.6)$$

where P_z = static height of the water level in the boiler feed system or deaerator above the suction level of the boiler feed pump
P_f = friction loss in the suction piping

For example, if the maximum NPSHR for the pump at the peak flow rate is 8 ft and the friction loss in the suction piping is 2 ft, the working level of the deaerator must be 10 ft above the suction connection of the boiler feed pump. NPSHA ≥ NPSHR, so NPSHA must be greater than 8 ft.

$$\text{NPSHA} = 10 - 2 = 8 \text{ ft}$$

Some designers will prefer to have a safety factor in the NPSH available and will add several feet to Eq. 22.6.

22.8 Number of Boiler Feed Pumps

Boiler feed pumps are much like pumps for chillers or hot water boilers. If the boilers are utilizing on-off control of the boiler feed pumps, individual boiler feed pumps should be used with a standby pump, as described in Fig. 14.5b for chiller pumps.

If the boilers are equipped with feedwater regulators, central pumps can be used with one pump providing water to several boilers. The number of pumps installed depends on the characteristics of the actual installation, such as total number of boilers, size of the boilers, amount of standby capacity required, etc.

The selection of the number of boiler feed pumps is a process that requires careful consideration. Since most pumps on HVAC boilers are constant-speed pumps, it must be ensured that no pump operates at a minimum flow at low efficiency and high radial thrust when a boiler is operating at minimum load. The pumps should go from continuous operation into an on-off method of control when this minimum flow occurs.

22.9 Condensate Transfer Pumps for Deaerators

A special application of pumps exists on deaerators, which have a condensate section and a deaeration section. The condensate must be transferred from the condensate section to the deaerator section. These pumps comply with the equations illustrated earlier for boiler feed pumps except that the pressure in the deaerator section must be substituted for the boiler pressure.

22.10 Summary

Although condensate and boiler feed applications are small in capacity and motor horsepower on most installations, they are critical services for most buildings, and care should be given to their design and application. Calculation of NPSH available is a simple yet very necessary procedure for these pump installations

Installing and Operating HVAC Pumps

23

Instrumentation for HVAC Pumping Systems

23.1 Introduction

The ability to operate HVAC pumping systems efficiently depends on the quality of the instrumentation used to operate the pumps and their water systems. Digital electronics gives us tools by which we can secure efficient operation, but without accurate instrumentation, we have no idea what we are achieving in actual operation.

Instrumentation includes transmitters, indicators, and controllers. Transmitters measure values such as pump speed, temperature, differential temperature, pressure, differential pressure, level, and electrical characteristics such as voltage, amperage, kilowatts, and power factor. Indicators display similar operating values such as temperature, pressure, level, and electrical values. Controllers compute pumping system values such as wire-to-water efficiency, start and stop pumps, and control their speed if they are equipped with variable-speed drives. All instrumentation on HVAC installations where appreciable amounts of energy are consumed should be traced to the standards of the National Institute of Standards and Technology (NIST).

Following are some definitions and terms that may be helpful in the evaluation of instrumentation:

Accuracy. Degree of conformity of an indicated value to a recognized acceptable standard value.

Accuracy is usually expressed in percentage of span or of the measured rate. The preferable method is the latter, the measured rate.

Accuracy usually will be a combined accuracy including linearity, hysteresis, and repeatability.

There is no universally accepted procedure for calculating the overall accuracy of a system with a number of instruments, each with its individual accuracy. Only actual system calibration can reliably establish the accuracy of that system of instruments. Adding all the individual inaccuracies together will provide a very conservative and probably an excessive system inaccuracy. Another procedure is to use the instrument with the greatest inaccuracy as the system accuracy. This may be closer, but it is still poorer than actual evaluation of the system.

Analog signal.   A signal representing a variable such as pressure, temperature, level, etc. Usually, they are direct current of magnitude 4 to 20 mA, where 4 mA is the minimum point of the span and 20 mA is the maximum point of the span. If a control loop for pressure reads from 10 to 150 psig, the analog signal would read 4 mA at 10 psig and 20 mA at 150 psig.

Digital signal.   A discrete value at which an action is performed. A digital signal is a binary signal with two distinct states, on and off. The on or off state is usually determined by a condition such as on for failure and off for operational, this being typical for an alarm signal. An operational digital signal would be a pressure switch that opens above 75 psig and closes below 73 psig. The state of a digital signal is usually determined by an instrument signal, such as a voltage level, where 0 V is "off" and 5 V is "on."

Dead band.   The range through which an input can be varied without initiating an observable response.

Deviation.   Any departure from a desired value or expected value or pattern. For example, if the actual pressure is 101 psig and the set point is 100 psig, the deviation is 1 psig.

Overshoot.   A control term indicating the amount that the process exceeds the set point during a changing load on a system.

Process.   The actual value in a control loop. For example, if the actual pressure being measured is 101 psig, the process is 101 psig.

Response time.   The rate of interrogating a transmitter. This is very critical in the control of variable-speed pumps. If the pumps vary their speed continuously, there probably is a problem with the response time. On continuous service, changes in HVAC pump speed should never be audible or visual.

Set point. An input variable that sets the desired value of the controlled variable. For example, in the preceding control loop, if a pressure of 100 psig is desired, the set point is 100 psig.

P-I control (ANSI/IEEE Standard 100-1977). Proportional plus integral control systems. Control action in which the output is proportional to a linear combination of the input and the time integral of the input.

P-I-D control (ANSI/IEEE Standard 100-1977). Proportional plus integral plus derivative control systems. Control action in which the output is proportional to a linear combination of the input, the time integral of the input, and the time rate of change of the input.

Repeatability. The variation in outputs for an instrument or procedure for the same input.

Span. The range of an instrument; for example, if a pressure gauge reads from 10 to 150 psig, its span is 140 psig.

These are a few instrument terms that may be encountered in this book. Many excellent books exist on instrumentation and control.

This chapter reviews instrumentation for the operation of pumping systems; instrumentation for the testing of pumps and pumping systems requires a greater accuracy and is discussed in Chap. 24 on the testing of pumping systems.

23.2 Transmitters

Transmitters sense the actual values of an operating pumping system. Therefore, the accuracy of transmitters is of great importance in the operation of such pumping systems. Skimping on quality in these instruments is short-sighted economically. The recommended quality for HVAC water systems results from the tolerances and features listed hereafter for each type of transmitter.

23.2.1 Flowmeters

Flowmeters are some of the most misunderstood and misapplied instruments in this field. The result is questionable accuracy and reliability. Flowmeters can be classed into (1) head-loss instruments, (2) propeller instruments, and (3) electronic measurement instruments.

Head-loss meters. Head-loss flowmeters measure head loss to determine the flow. This class includes venturi, pitot tube, insertion tube, special valves, and any impact-measuring type of meter that uses friction or velocity head loss as a measurement of flow. This class of flowmeters is acceptable for measuring water systems that have a steady flow with very little variation. The reason for this is the fact that the velocity head of a water stream varies with the square of the flow (see Chap. 6). They are not acceptable for the measurement of flow in variable-volume water systems that have a flow range equivalent to a velocity of 1 to 10 ft/s. Such systems are said to have a turndown ratio of 10 to 1.

For example, an 8-in pipe carrying 1200 gal/min has a velocity head of 0.920 ft. At 400 gal/min, this velocity head has been reduced to 0.102 ft. This demonstrates that the accuracy of these meters drops off rapidly with broad changes in flow rate. Most head-loss meters are not used with turndown ratios much greater than 3 to 1. The development of electronic transmitters for these meters may extend their range somewhat.

Propeller meters.   Propeller meters use an insertion device that contains some type of rotating element that measures the flow. The speed of rotation of the propeller varies with the flow rate. These instruments have served the water industries well for years with greater flow ranges than the head-loss instruments. The actual flow range and accuracy for a specific meter should be certified by the manufacturer. These meters must not be of the magnetic type, since the use of magnets will cause welding spatter and other bits of metal in the water to cling to them, thus causing cleaning problems. Propeller meters can be full throated (across the entire pipe area) or insertion type.

Because of the moving parts of propeller meters, they should be calibrated often to ensure their accuracy. Impact of water-borne particles may damage them; also, the axles that the propellers turn on can wear due to loss of lubrication or dirt.

Electronic meters.   Great strides have been made in flow measurement through the development of electronic instruments. These meters emit an electronic signal across the water stream and use some method of evaluating the quality or quantity of that signal when it is received. These meters can be full throated, insertion type, or strapon, where they are attached to the exterior surface of the pipe.

Full-throated magnetic meters provide the greatest flow range, 1 to 30 ft/s, at probably the highest accuracies. Some manufacturers quote accuracies as high as ±0.5 percent from 1 to 30 ft/s velocity. Manufacturers of insertion magnetic meters offer similar claims for their instruments. These magnetic meters are so called because of the type of signal that is used to measure the flow; there is no existing magnetism in them that collects metallic debris.

A number of strap-on flowmeters are available with varied claims and results. Some strap-on meters that are supplied with carefully thought out electronics and proper application can approach the performance of full-throated magnetic meters.

23.2.2   Accuracy evaluation of flowmeters

The accuracy of flowmeters can be stated two ways: (1) as percent of full span or (2) as percent of rate. Percent of full span means that the accuracy stated must be taken as a percentage of the full scale of the instrument. For example, if a flowmeter has a maximum rate of 1000 gal/min and percent of full-scale accuracy of 1 percent, its accuracy is ±10 gal/min at any flow rate. At a 10:1 turndown, or 100 gal/min, the accuracy becomes ±10 percent. For this reason, flowmeters with accuracies quoted as percent of scale should not be used for most variable-volume HVAC water systems.

Flowmeters with accuracies quoted as percent of rate provide the proper instrumentation for most variable-volume HVAC water systems. If such a flowmeter has a maximum rate of 1000 gal/min and a quoted accuracy of ±1 percent of rate, at 100 gal/min, its flow variation is ±1 gal/min.

Some flowmeters are used for specific services where high accuracy is unnecessary but repeatability is important. Repeatability is necessary for flowmeter installations where a specific flow rate must be held in a water system. Flowmeter manufacturers may quote both accuracy and repeatability over various flow ranges. Others will include the repeatability as part of their accuracy statement.

23.2.3   Calibration of flowmeters

Flowmeters must have a calibration procedure that is independent of any field conditions other than the pressure loss or velocity head measured within the meters themselves. Flowmeters that use pump curves for calibration should be checked otherwise for their accuracy. There are two reasons for this: (1) pump manufacturers normally guarantee performance at only one or two points on their head-capacity curves, and (2) most pump head-capacity curves are dynamometer curves at a constant speed. If a certified pump curve has been run with the actual motor, the pump curve can be used for some calibration work. Figure 23.1 describes the difference between the two head-capacity curves, one being the constant-speed dynamometer curve and the other the actual head-capacity curve at the speed of an induction motor. The other problem is the inconsistency in the pump and motor industry to establish standard dynamometer speeds for pump curves. For example, dynamometer speeds for pump catalog curves

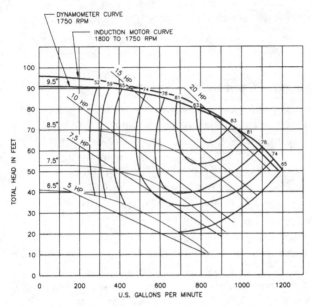

Figure 23.1 Speed variation in pump head-capacity curves.

for four-pole motors vary from 1750 to 1790 rev/min. A catalog curve for a particular pump may be 1750 rev/min, and it may be equipped with a motor that has a rated speed of 1770 rev/min at full load. The actual performance of a pump will vary greatly from its published catalog head-capacity curve. It is obvious that pump curves should not be used to calibrate flowmeters.

23.2.4 Installation of flowmeters

Flowmeters always should be installed in strict compliance with the instructions of the manufacturer. Most such companies require 10 pipe diameters in length of clear pipe ahead of the flow measurement point and 5 pipe diameters downstream. Often, it is difficult to get 15 diameters of length of straight pipe without valves, elbows, or other pipe accessories. If shorter lengths must be considered, some flowmeter manufacturers will advise the effect on the accuracy of the meter that is caused by the shorter distances. Compromising on the length of straight pipe around a flow sensor can have disastrous effects on transmitter accuracy and should not be done without the flow transmitter manufacturer's knowledge.

If there are two elbows upstream of the flowmeter that change the direction of flow twice in a short span of pipe, a torsional effect will be imparted to the water stream that may require straightening vanes or other devices to prevent a deterioration of flowmeter accuracy. The flowmeter manufacturer must be made aware of such specific installation details.

Magnetic-type flowmeters are not affected by dirt in the water, since they are used on sewage. Propeller-type flowmeters may be damaged by water-borne dirt, so care should be taken to ensure that such meters are protected from physical damage. Insertion-type flowmeters, which can be of several different types, should be equipped with hot tap assemblies that enable the flowmeter to be removed for inspection, cleaning, or repair without draining the pipe. Most manufacturers of insertion-type flowmeters have hot tap assemblies that include a gate valve and other piping accessories that enable removal of the transmitter from a pipe under water pressure.

23.2.5 Selection of type of flowmeter

It is obvious from the preceding discussion that there is a great difference in the cost of flowmeters. Before a flowmeter is selected, the decision must be made as to its primary use. Following are several uses for flowmeters:

1. Proof of flow in a pipe.
2. Approximate flow rate for rough calculations.
3. Precise flow rate for accurate calculations.

Some flowmeters are used strictly as flow switches to verify flows in a pipe. They may be selected instead of a flow switch due to their ruggedness and lack of any moving parts. Also, the flow may be varied as to where the switching function is to take place.

Other flowmeters are used for approximate calculations; high accuracy is not needed. Contrary to these applications, the development of on-line calculations of energy consumptions requires high accuracy and broad flow ranges. Typical values now being computed using precise flow rates are (1) pump performance, (2) pump wire-to-water efficiency, (3) system flow rates, (4) system British thermal units and tonnages, (5) kilowatt per ton calculations, and (6) coefficient of performance or energy efficiency ratio.

Flowmeters for precise calculations must have their control procedures checked to ensure that they can provide the desired results. This can become critical when pump speed or valve position is paced to the output of a flowmeter. Such use of flowmeter output should be avoided wherever possible.

Along with the preceding uses, the size of the pipe comes into the decision as to the type of flowmeter. For example, a small pipe such as 2-in diameter with a requirement that a flow of 50 gal/min be maintained in the pipe would be best suited for a head-loss or propeller-type flowmeter. A flowmeter for a 10-in pipe could be served by an insertion impeller meter if a flow range of 300 to 2000 gal/min is required with reasonable accuracy. If precise flow is required in this pipe for energy evaluations such as kilowatts per ton or wire-to-water efficiency, a full-throated magnetic meter may be the best selection. If a 36-in pipe with great variations in flow exists, an insertion-type magnetic meter may be cost-effective. The exact duties of a flowmeter should be established to ensure proper selection of the type of meter.

Flowmeters should be selected also with consideration for ruggedness, ease of service, frequency of cleaning, location of the installation, and general reputation of the manufacturer in providing an instrument that will function properly under the intended use.

23.2.6 Pressure and differential pressure transmitters

Pressure and differential pressure transmitters measure and transmit signals that provide the information on these values at specific points in an HVAC water system. In many cases, they are the most vital instrument in the operation of a pumping system or in reporting the quality of an entire water system's operation. They provide the information needed to calculate an individual pump's performance or that for a complete pumping system.

Pressure and differential pressure transmitters must be rugged for installation on most HVAC systems. This is due to the possibility of water hammer when systems are filled initially or when air collects in a system and suddenly is removed. Most manufacturers are providing casing design pressures as high as 1000 psig. Differential pressure transmitters should not require equalizing valves between the two pressure taps on the instrument. The diaphragm in the instrument must be rugged enough to withstand any differential pressure to which the instrument is subjected. The equalizing valve can be forgotten by the service personnel and the diaphragm destroyed when one of the two taps is closed inadvertently.

Accuracies in the range of ± 0.25 percent of calibrated span are quoted by manufacturers of this instrumentation. This accuracy includes the combined effects of linearity, hysteresis, and repeatability.

These transmitters should be located where they can be removed easily; a manual shutoff valve should be installed on all pressure-sensing lines for removal without draining the water system.

23.2.7 Temperature transmitters

Temperature transmitters are rated by accuracy, as are flow and pressure transmitters; usually, this accuracy is in the range ±0.25 percent of span. Temperature transmitters have an additional requirement, which is stability with respect to changes in ambient temperature and voltage changes in the power supply for the transmitter. Stability for a temperature transmitter should be around ±0.01 percent of span per degree Celsius change in ambient temperature and ±0.001 percent of span per volt change in line voltage.

Temperature transmitters should be installed where accessible and provided with stainless steel thermowells for ease of removal.

23.2.8 Level transmitters

Level transmitters are not used as often in HVAC water systems as other types of transmitters. Their installation is so specific that the accuracy, repeatability, and ruggedness cannot be defined but must meet the requirements of the installation. The supplier should demonstrate that the accuracy and structural capability of the instrument do meet that required for the actual application.

23.2.9 Watt transmitters

Watt transmitters are becoming more popular in the HVAC water industries due to the need to monitor power consumption of HVAC equipment including pumps. Watt transmitters consist of the transmitter itself and current transformers mounted on at least two of the three wires of a three-phase power supply.

Extreme care is required with the connections between the transformers and the transmitter. High voltage can exist between these wires if they are disconnected from the transmitter. The manufacturer's instructions should be followed carefully in installing and servicing these units. Only qualified electricians or instrument technicians should handle these transmitters.

A new type of watt transmitter is available with the current transformers as part of the instrument itself. This model reduces the danger of high voltage from the current transformer wires. These transmitters are available in sizes from 4 to 1000 A.

The accuracy of watt transmitters should be in the range of ±0.25 percent of span. This should be the total accuracy for the current transformers and the instrument.

Watt transmitters should be installed in power cubicles that are serviceable only by trained electricians. Extreme danger can result from touching the unconnected wires of the current transformers be-

cause of the very high voltages that are generated by these transformers.

23.3   Indicators

Indicators at one time were quite simple, most of them being of the
analog type found on pressure gauges. Today indicators can range
from such simple analog types to colored flat screens that include
switching and control functions.

Analog indicators include all types of gauges that use the dial face
of a gauge or a vertical bar that is calibrated to the span of the transmitter analog or process. Digital indicators are the alphanumeric type
that display the process in numbers and letters. There are a number
of different types of such indicators, such as (1) the seven-segment
neon number, (2) liquid crystal, (3) plasma, and (4) vacuum fluores-
cent. These indicators must be selected for the application conditions.
Some of them must be read directly in front; this limits their use
where the display must be read at angles from directly in front of the
instrument. Sunlight and interior lighting can adversely affect the
readability of some of these digital indicators.

23.4   Controllers

Controllers can range from simple switches to complex programmable
logic controllers with and without indication or operator interfaces.
This subject is so broad that only a synopsis of the controllers included in the HVAC water industries can be included in this chapter.

23.4.1   Simple controllers

Simple electric controllers include manual switches that can start or
stop HVAC pumps, or they can be switches that are set to open or
close an electric circuit at a discrete process value. Such switches are
operated by pressure, differential pressure, temperature, differential
temperature, level, flow, speed, volts, and amperes. Most of them
have an internal adjustment to control the set point at which the
switch either opens or closes an electric circuit.

Elementary mechanical controllers consist of pressure regulators,
pressure-relief valves, balance valves, and self-operated temperature
controllers. Pressure-relief valves protect water systems from excessive pressures. Care should be taken in the use of pressure regulators
and balance valves because of the possibilities of energy waste.

Other basic controllers are of the analog type such as pneumatically operated controllers where the output is directly proportional to control pressure or follows a known curve with respect to control pressure.

23.4.2   Digital controllers

Most HVAC systems are operated by some form of digital controller. There are a number of different digital controllers ranging from simple single-function controllers to complete universal controllers handling all functions of the interrogation and control of a HVAC water system.

Digital controllers can use either PI or PID control for the operation of HVAC water systems. The development of the control loop must recognize the rate of reaction of the water system and the needed rate of response. Too often HVAC control systems for pumps do not provide the necessary rate of response, and the result is hunting of the pump speed and unnecessary starting and stopping of pumps.

23.4.3   Quality of controllers

The quality and reliability of a controller for HVAC pump operations should be established by testing facilities such as Underwriters Laboratories or Electrical Testing Laboratories. The manufacturer of such controllers should be able to specify the levels of quality used in the development of a particular instrument. For example, the United States military standards MIL-STD-810D provides an excellent procedure for testing instrumentation and controllers for environmental stability, including temperature, humidity, and vibration.

23.5   Control Wiring

Control wiring should be designed and installed as recommended by the Instrument Society of America. Control wiring should be of the sizes recommended by this society; wire sizes between transmitters and pump controller are included in Table 10.7.

Of particular importance is keeping instrument and control wiring away from power wiring. Electromagnetic interference (EMI) and radio-frequency interference (RFI) should be kept within the levels recommended by the instrument and controller manufacturers that are involved on a specific project.

23.6   Control Valves

Great emphasis has been placed in this book on avoiding three-way control valves wherever possible because of the possibility of energy waste. There are applications of such control valves that are necessary and do not waste energy, such as changing the routing of water streams from chillers to boilers. Another application is the blending of two different water temperatures to produce a third temperature.

The quality of control valves and their actuators for these operations is very important. The valve itself should be of a metallurgy necessary for the application and should be the product of a manufacturer who can verify the quality of design and manufacture. The characteristic of a control valve, namely, flow versus valve position, should be known before that valve is applied to a specific application. For example, a butterfly valve is acceptable on some applications but not on others.

Valve actuators are available for HVAC water applications that can be either pneumatic or electronic. Pneumatic actuators with electronic pilots have proved to be very reliable for HVAC water systems. Electronic actuators are now available that have a quality that will provide years of continuous operation without continual service and repair.

23.7 Summation

Quality instrumentation becomes more of a necessity as attempts are made to improve the overall efficiency of HVAC water systems. Every designer of these systems should establish specifications and other requirements that ensure that the desired instrumentation is installed.

23.8 Bibliography

Bela G. Liptak, *Instrument Engineers Handbook,* Chilton Book Company, Radnor, Pa., 1982.
Charles L. Phillips and Royce D. Harbor, *Feedback Control Systems,* Prentice Hall, Englewood Cliffs, N.J., 1988.

24

Testing HVAC Centrifugal Pumps

24.1 Introduction

The continued need for a higher efficiency of operation of HVAC water systems necessitates the testing of HVAC centrifugal pumps or pumping systems for their actual performance. They must be tested for flow, head, and efficiency.

The Hydraulic Institute (HI) has produced a test standard for centrifugal pumps that has been approved by the National Standards Institute (ANSI). This standard is entitled ANSI/HI 1.6-1994, *Centrifugal Pump Tests*. This standard should be in the possession of designers of HVAC water systems and buyers of HVAC pumps or pumping systems. A general review of the scope of this standard is included herein.

24.2 Objective

The objective of this test standard is to provide uniform standards for the hydrostatic, hydraulic, and mechanical testing of pumps and for the recording of test results. In the HVAC industry, this standard allows designers to establish levels of quality and performance and enables pump and pumping system manufacturers to demonstrate that they have met those levels.

Vibration and acoustic standards may be required on some installations. Hydraulic Institute's Standard ANSI/HI 1.1-1.5-1994, *Centrifugal Pumps,* includes acceptable vibration levels for centrifugal pumps. Standard ANSI/HI 9.1-9.6-1994, *Pumps—General Guidelines,* includes procedures for the measurement of airborne sound.

24.3 Types of Tests

The ANSI/HI test standard provides for the following tests:

1. Hydrostatic test of the pressure for the liquid container of the pump, namely, the volute or bowl
2. Performance test to demonstrate the hydraulic and mechanical acceptability (Capacity, total head, power, and speed are measured.)
3. Net positive suction head (NPSH) required test
4. Priming time test of self-priming pumps

24.3.1 Hydrostatic test

The hydrostatic test demonstrates that the pump, when subjected to hydrostatic pressure, will not leak or fail structurally. Each part of a pump that contains liquid under pressure must be subjected to the following tests:

150 percent of the pressure that would occur in that part when the pump is operating at rated condition for the given application of the pump

125 percent of the pressure that would occur in that part when the pump is operating at rated speed for a given application but with the discharge valve closed

The conditions for the hydrostatic test are spelled out in detail for the pump manufacturer in the ANSI/HI 1.6-1994, *Centrifugal Pump Tests.*

24.3.2 Performance test

The performance test has two tolerance levels, namely, A and B. The level A test is for compliance with actual contract values that have been established by the purchaser of a pump. The level B test is for published performance in catalogs, CD-ROM, or diskettes such as pump curves for head capacity, efficiency, brake horsepower, and NPSH required. Normally, if a factory test is not specified, the manufacturer will ensure that its pumps will comply with level B tolerances. Other pump companies guarantee that their pumps will always meet the flow, head, and efficiency requirements of acceptance level A even if level A test performance is not specified. It is always in the best interests of the owner of the HVAC water system that the designer specify that all pumps comply with level A testing of ANSI/HI 1.6-1994, *Centrifugal Pumps Tests.*

Tables 24.1 and 24.2 are excerpts from this ANSI/HI test standard.

TABLE 24.1   Total Head and Efficiency Tolerances at Rated
Capacity and Speed

Head and flow ranges	Level A	Level B
Under 200 ft and 2999 gal/min	+8%, −0	+5%, −3%
Under 200 ft and over 3000 gal/min	+5%, −0	+5%, −3%
From 200 to 500 ft at any flow	+5%, −0	+5%, −3%
500 ft and over at any flow	+3%, −0	+3%, −0
Efficiency, % η_{p1} or η_{oa}		$\dfrac{100}{(D) - 0.2\%}$
rated flow and head		η_{p2}

TABLE 24.2   Capacity and Efficiency Tolerances at Rated
Total Head and Speed

	Level A	Level B
Capacity	+10%, −0%	+5%, −5%
Minimum efficiency	η_{p1} or η_{oa}	$\dfrac{100}{(120/\eta_{p2}) - 0.2\%}$

NOTE: η_{p1} is the contract efficiency specified by the designer of
the HVAC system; η_{p2} is the catalog efficiency on a published pump
curve.

The variations in head, capacity, and efficiencies shown in Tables
24.1 and 24.2 demonstrate the tolerances that pump manufacturers
must have to achieve reasonable production of pumps. The use of these
standards will ensure the accuracy and efficiency of HVAC pumps.
These efficiencies also demonstrate that pumps operating at 8 percent
of head higher than the published pump curve can have a pronounced
effect on the loading of the pump driver such as an electric motor.

For example, assume that a 1000 gal/min pump with a head of 100
ft is specified with a pump efficiency of 85 percent. The design brake
horsepower for this pump is 29.7 bhp. The actual pump is allowed to
have a pump head of 108 ft. Likewise, the pump could have a capacity
of 1100 gal/min at 100 ft of head. The actual pump horsepower in either case would therefore be 32.1 bhp, and this might overload a 30-
hp motor that could have been specified.

Two facts arise from this evaluation: (1) pumps of magnitude
should have certified tests to ensure closer compliance to the project
specifications, and (2) variable-speed pumps offer quick flexibility to
adjust to the actual operating conditions without overloading the
pump driver.

Manufacturers should verify the quality of their system head curve
and efficiency data by providing a pump performance diagram similar

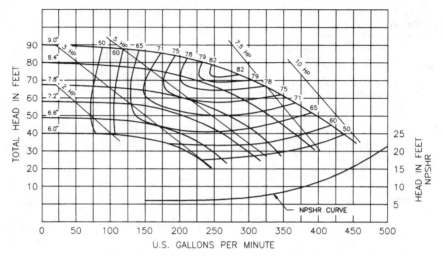

Figure 24.1   Standard system head curve diagram.

to Fig. 24.1. The manufacturer should certify that the data have been derived from tests conducted in accordance with ANSI/HI test standard 1.6-1994. Further, it should be stated whether the data are in compliance with acceptance level A or B as stated in paragraph 1.6.5.3 of this standard.

24.3.3   Net positive suction head (NPSH) required test

This test enables the pump manufacturer to develop the NPSH required curves that are included with the head-capacity curves for the pumps. Very detailed procedures are provided for the pump manufacturer to determine the NPSH required for each pump size at various flow rates. None of this information is of importance to the HVAC system designer; what is important to the designer is that the NPSH required curves provided for these pumps have been derived from NPSH required tests that have been carried out in accordance with the ANSI/HI test standard.

24.3.4   Priming time test for self-priming pumps

Although self-priming pumps are not used extensively in the HVAC field, when they are used, the priming time test enables the pump manufacturer to demonstrate to the HVAC system designer that the pump is truly self-priming.

Like the NPSH required test, detailed testing procedures are provided the pump manufacturer to verify the time required to prime each size pump at various capacities. The HVAC designer should specify that the priming time for self-priming pumps has been determined in accordance with the requirements of the ANSI/HI test standard.

24.4 Testing Procedures

The ANSI/HI test standard establishes the requirements for the pretest data requirements, test stand configuration, instrumentation, records, calculations, and reports. All these subjects are of importance to the manufacturer of pumps to ensure that its pumps comply with the levels of performance provided above. These matters are not of importance to the designer or purchaser of the pumps as long as the pump performance is ensured by certification that the pumps were tested in accordance with the standard as described above.

24.5 Field Testing of Pumps

Testing arrangements and instrumentation for testing are of importance to HVAC designers and pump buyers, as it is often recommended that pumps be tested in the field under questionable conditions. A survey of the ANSI/HI test standard will demonstrate the futility of trying to test a pump in the field without extensive and expensive test arrangements.

One significant problem with testing HVAC pumps in the field is achieving design flow without expensive piping arrangements. HVAC heating and cooling loads are so variable that it may be difficult to achieve design flow and head on the system when the test is scheduled to be completed. Also, so many HVAC pumps have future flow capacities designed into them that it may be years before design flow and head conditions are available. For example, a chilled water pump might be designed for a capacity of 1500 gal/min for future additions to the chilled water system even though the existing load is only 1000 gal/min. It would be difficult to certify the pump in the field for the 1500 gal/min condition.

24.6 Test Instrumentation

One of the difficulties in field testing of pumps is the required quality of the test instrumentation. Following are tables from ANSI/HI 1.6-1994, *Centrifugal Pump Tests,* that demonstrate the needed quality of instrumentation to achieve acceptable test results.

TABLE 24.3   Instrumentation Fluctuation and Accuracy

Value measured	Fluctuation of test readings, ±% of the values	Accuracy of the instrument, ±% of the values
Capacity	2.0	1.5
Differential head	2.0	1.0
Discharge head	2.0	0.5
Suction head	2.0	0.5
Pump power input	2.0	1.5
Pump speed	0.3	0.3

TABLE 24.4   Typical Recommended Instrument Calibration Intervals

Instrument	Interval	Instrument	Interval
Flowmeters		Power meters	
Venturi	*	kilowatt transducer	3 years
Turbine	1 year	Watt-amp-volt	1 year
Magnetic	1 year	Strain gauges	6 months
Pressure meters		Speed meters	
Bourdon tube	4 months	Tachometers	3 years
Manometers	Not required	Eddy current	10 years
Dead weight tester	1 year	Electronic	Not required

*Calibration not required unless critical dimensions have changed.

The instrumentation for testing pumps should be certified by an independent testing laboratory or consulting firm that specializes in instrument testing.

ANSI/HI 1.6-1994, *Centrifugal Pump Tests,* also provides a recommended instrument calibration interval; the test instrumentation must be calibrated periodically to ensure that the accuracies in Table 24.3 are maintained. Some typical intervals are listed in Table 24.4.

24.6.1   Installation of instrumentation

There are very specific requirements for the installation of this instrumentation in ANSI/HI 1.6-1994, *Centrifugal Pump Tests,* to ensure the accuracy of the test results. The location of the instrumentation, types of pressure taps, etc. are all defined in this test standard.

24.7   Test Reports and Records

One of the most significant parts of ANSI/HI 1.6-1994 is the organization of the data from pump or pumping system tests. A data form is provided to aid the manufacturer in the organization of the data resulting from the various test covered by this standard. Also, the calculations for developing pump or pumping system performance from these data are delineated completely in this test standard.

How accurate are the pump curves that are included in pump manufacturers' catalogs? Some manufacturers base their curves on acceptance level A, which allows -0, $+8$ percent variation in pump head at rated capacity and speed; others use level B, which allows -3, $+5$ percent variation in pump head at rated capacity and speed (see Table 24.1). It is not clear whether all pump manufacturers publish curves that are derived from tests in accordance with the ANSI/HI 1.6-1994. It is therefore recommended that consulting engineers, designers, and users of pumps specify that pump curves should state the level of acceptance for a proposed pump. Also, it is recommended that all pump curves be in the form shown in Fig. 24.1, which certifies the pump curves to be in accordance with the ANSI/HI Standard 1.6-1994.

Reiterating, no pump, however small, should be used in HVAC work without head-capacity and efficiency curves as shown in Fig. 24.1. The NPSH required curve is not necessary for small circulators unless they are taking suction from an open tank.

24.8 Summation

To ensure the quality of HVAC pumps, the designer of HVAC water systems should specify that all pumps be tested in accordance with the standards of the Hydraulic Institute. Small pumps should meet at least level B performance, while large pumps should be factory tested and certified to level A performance.

Pumps can be certified by the manufacturer of pumping systems, where the entire pumping system is tested for flow, head, power and efficiency. A wire-to-water efficiency test should be conducted on the entire pumping system. This is described in Chap. 26.

The ANSI/HI test standard is so complete for this testing and the interpretation of the results that there truly is a criterion now for complete evaluation of pumps or pumping systems for the HVAC industry.

24.9 Bibliography

ANSI/HI 1.1-1.5-1994, *Centrifugal Pumps,* Hydraulic Institute, Parsippany, N.J., 1994.
ANSI/HI 9.1-9.6-1994, *Pumps—General Guidelines,* Hydraulic Institute, Parsippany, N.J., 1994.
ANSI/HI 1.6-1994, *American National Standard for Centrifugal Pump Tests,* Hydraulic Institute, Parsippany, N.J., 1994.

25

Installing HVAC Pumps and Pumping Systems

25.1 Introduction

A pump's useful life is determined more by its installation than by its quality of manufacture. More pumps wear out because of improper installation and operation than from any other factors. As has been mentioned often in this book, HVAC pump duty is easy pump duty compared with other industries or applications of pumps.

HVAC pumps should not have unusual maintenance. The way to avoid maintenance is to install the pumps or pumping systems properly. Almost all pump and pumping system manufacturers have detailed installation instructions that should be followed carefully.

The basic effort in installing this equipment is to conclude with an installation that imposes no external forces on the pumps or pumping systems. This is achieved by setting the pump base properly and connecting the piping and electrical connections so that no forces are imposed on the finished assembly.

25.2 Preinstallation Procedures

Much agony and frustration would be eliminated if all pumps were checked carefully prior to installation. On many HVAC installations, there can be dozens of pumps of different sizes. Great care should be taken before setting any pumps on their foundations. Following are specific details that should be checked:

1. Each pump should be tagged with its project number and particular service, such as pump no. 1A, primary chilled water pump no. 1.

2. Each pump should be checked for model number, impeller diameter, and on double-suction pumps, the rotation of the impeller.

3. The pumps should not be set in place until the equipment room is completed to the point where the pumps will not be damaged by construction equipment, debris, or freezing. Some water may remain in the pump volutes from factory testing; drain plugs should be removed to prevent any freezing of water in the volutes.

4. Pumps should be lifted from rigging points only, not from the pumps or motors themselves on base-mounted pumps. On close-coupled pumps, the lifting points should be specified by the manufacturer.

5. The electric motors should be checked for horsepower rating and specified voltage.

6. On electric motor–driven pumps, the power supply should be checked for correct voltage as well as the variation in voltage from leg to leg on polyphase applications. A sizable variation in voltage in the three different phases can cause an appreciable reduction in power for an electric motor. See Table 7.1 for the amount of derating caused by a voltage imbalance between phases.

25.3 Pump and Pumping System Bases

It is quite obvious that the beginning of a successful pumping installation is providing the correct base for the pumps and their accessories. Much of the trouble with pumps results from improperly designed bases. The base begins the attachment of the pump to its environment, and that base must be designed to fit the conditions surrounding the pump. Pump bases can be solidly connected to the equipment room floor, or they can be floated on vibration-type bases. Various types of pump and pumping system bases are shown in Fig. 25.1.

This is the first question that must be asked: Is a vibration-type base required to prevent any vibrations originating in the pump or its driver from reaching the floor? If the pumping equipment is to be mounted on an upper floor with occupancy beneath it, very likely a vibration-type base will be required. There are cases where the vibration base is required even though the pump base is mounted on concrete that is poured on rock at the first elevation in a building. When the activity in the building is of a sensitive nature, such as experimental research, the vibration base prevents objectionable vibration and noise from being transmitted from the pumping system base through the building.

All bases should be designed to keep casual water away from any metal. Solid concrete bases with the pump and its metal base mount-

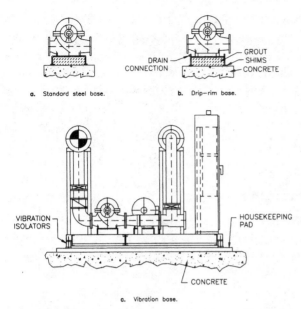

a. Standard steel base.

b. Drip-rim base.

DRAIN
CONNECTION

GROUT
SHIMS
CONCRETE

VIBRATION
ISOLATORS

HOUSEKEEPING
PAD

CONCRETE

c. Vibration base.

Figure 25.1 Pump bases.

ed on them do not require a housekeeping pad, but vibration-type bases should be mounted on a 2-in-high concrete pad that prevents water from reaching the vibration springs and pads.

Drip-rim bases were used extensively at one time; they were needed because packing, not mechanical seals, was installed in the stuffing boxes. The leakage from the packing was contained within the drip-rim base and piped to a drain. Today, most pump manufacturers provide a threaded drain hole below the mechanical seal that can accept a drain pipe. This provides a much cleaner appearance than the drip-rim base and eliminates standing water which can promote Legionnaires' disease.

Most pump companies provide an installation diagram for connecting the steel pump base to the concrete. These diagrams include the use of anchor bolts, shims, and nonshrink grout to properly install the pump on its concrete base. On large pumps, dowelling the pump and its driver to the steel base is a standard practice. These dowels are provided to ensure that the pump and driver are kept in alignment. This really is not necessary for most HVAC pump applications. Also, variable-speed pumps operate at reduced speed and radial thrust so that there seldom is any need to dowel them to the steel base. If dowelling is necessary to keep the pump in alignment with its driver, there may be something wrong with the installation, such as piping force on the pump connections.

25.3.1 Pump bases with seismic restraints

The persistence of earthquakes has necessitated the installation of many pumps with seismic restraints built into their bases and other supporting structures. The *Uniform Building Code* (UBC) has classified the United States into zones and has provided lateral force levels that can be used for the design of pump bases and supporting equipment. Manufacturers of equipment that controls vibration, shock, and sound have computer programs that can develop the needed seismic restraints for a specific application.

25.4 Connecting Piping to Pumps

There are two types of piping that impinge greatly on the efficiency and maintenance levels of pumps. These are (1) the fittings around the pump itself and (2) the header or system piping. Either of them can create energy waste and mechanical stress on the pump.

25.4.1 Pump fitting sizing

Pump fittings must be sized properly or there will be an appreciable waste of energy. The friction losses through the pump fittings should be totaled and should not exceed 8 to 10 ft on most installations at design flow. These losses include

1. Entrance loss from the header to the pump suction piping
2. Suction shutoff valve
3. Suction or discharge strainer
4. Suction piping
5. Suction elbow
6. Suction reducer
7. Discharge increaser
8. Discharge check valve
9. Discharge piping
10. Discharge shut off valve
11. Entrance loss into header from pump discharge piping

Not all pump installations will have all eleven of these losses, but each installation must be checked for them. It is obvious that undersizing these fittings can result in sizable and unexpected losses. Some fittings that absolutely must be avoided around pumps are bushings of any sort, reducing flanges, and balance valves wherever possible. Only as a last resort should a relief valve be installed on an HVAC pump!

25.4.2 Pump fitting arrangement

Suction elbows on pumps require some care to ensure that they are installed correctly. A double-suction pump with a suction elbow must have the elbow installed so that flow through the elbow is perpendicular to the pump shaft. If not, one side of the double-suction impeller will be loaded heavier than the other side, and the hydraulic and mechanical integrity of the pump will be impaired. This can be avoided by installing a suction diffuser on the pump. This is acceptable for most HVAC pumping systems but not for condenser pumps taking water from cooling towers.

Any pump taking a suction lift must have an eccentric reducer or long-sweep reducing elbow on its suction to avoid air pocketing at the pump suction.

At one time, the Hydraulic Institute required a specific number of lengths of pipe between the pump and a suction elbow. This is no longer required, since it has been proved that most volute-type pumps can operate with elbows mounted on their suctions without damage or loss of efficiency. In fact, many pumps, such as in-line and some petrochemical pumps, have the elbows integrated into their casings.

Hydraulic Institute standards are an excellent source for information on the proper arrangement of fittings around pumps. Many examples are provided by these standards that enhance the installation of pumps.

25.4.3 Expansion provisions at pumps

Expansion of piping in HVAC systems consists of thermal expansion in the piping and expansion due to changes in the building due to settling or other movements. Both these possibilities must be recognized in all HVAC systems.

One of the most damaging practices in the pump industry is not providing adequate thermal expansion provisions around pumps. This is not very important for chilled water pumps because they operate near 50°F and do not have much expansion in the piping. It is important for low-temperature hot water heating systems and particularly so for medium- and high-temperature water systems to have the correct expansion provisions.

Thermal expansion must be accounted for in both lateral and axial directions. Installation of the system piping must be such that excessive expansion does not occur at the pumps. Long runs of piping must include expansion provisions such as expansion joints, bellows assemblies, or offset piping. Chapter 43 of the ASHRAE *Applications Handbook* is an excellent source for evaluating complex piping systems to determine the type and location of expansion provisions. One common error in the pumping industry is to use flexible hose for pump connections; it can handle only lateral, not axial movement.

HVAC pumps must be installed without any thrust on the pump connections by the piping. After the installation is completed, the flange bolts on the pump connections should be removed. If the pipe flanges move at all, the piping should be redone. The worst practice is to have a gap between the pump and pipe flanges and then draw them together with the flange bolts. Other conditions that must be watched for are shifting of the pipe flange so that the flange holes do not line up or for the pipe flange to be at a slight angle from the pump flange. All these conditions create stresses in the pump casing and will, eventually, pull or push the pump out of alignment. *Do not use flexible joints to eliminate these misalignments of the pipe flanges.* The expansion joint and pump flanges must come together without any force or movement of any expansion provisions.

25.5 Electrical Provisions for Pumps

Electrical provisions for pumps consist of connections to pump motors and safety controls for pumps.

25.5.1 Electrical connections for pump motors

There are some simple rules for electrical connections for motors that, if followed, will result in trouble-free installations. Generally, the connection of power wiring to motor junction boxes should conform to the *National Electrical Code* and any special provisions for a specific application. The power connections should be bolted to the motor leads; wire nuts should not be used. The final connection to the motor must be flexible to allow some movement between the electrical power conduit and the motor.

A motor out of sight of or 50 ft from the electrical disconnecting means may require the installation of a manual switch at the motor. If there is a possibility that the motor can be stopped manually while in operation, this disconnection switch should be equipped with a control contact that opens the motor control circuit before the main power switch is disconnected.

Most HVAC pumps depend on the motor overloads to protect their motors. Larger motors should be equipped with temperature and vibration sensors that can be used to alarm the operators in event of high temperature or excessive vibration.

25.5.2 Safety controls for pumps

Small HVAC pumps are often installed without any safety controls. Several inspection agencies are now requiring differential pressure switches installed across the suction and discharge connections of the pumps. Such a switch proves that the pump has generated pressure; if this does not occur within a specific time period, the pump is stopped and an alarm signal is generated.

Larger pumps, like motors, should be equipped with temperature and vibration sensors that alarm the operator in the event of high temperature or excessive vibration.

25.6 Alignment of Pumps and Motors

Before attempting to start the pumps, the alignment of the pumps and their motors or other drivers should be checked carefully. All pump and flexible coupling manufacturers have alignment tolerances that must not be exceeded. Alignment should be checked for both off-set and angularity.

Alignment by straight edge across the coupling was acceptable at one time. Today, alignment by means of an Ames dial or equivalent should be accomplished for most HVAC pumps. Laser technology has been applied to machinery alignment and is now available for centrifugal pumps.

25.7 Initial Operation of Pumps

Chapter 27 is devoted to the pump operation, which pertains to the efficiency of that operation. Initial operation of pumps is part of the installation process, since it tells quickly the quality of the actual installation.

The first operation of a pump and its driver will indicate many things:

1. Unexpected noise will provide information on pump alignment, cavitation, or entrained air.

2. Pump speed on variable-speed pumps will provide information on the adequacy of the pump control.

3. Heat in the motor or pump can be an indicator of problems with alignment or improper operation.

4. If wire-to-water efficiency indication is provided, it will furnish information on the adequacy of the pump control system and whether the pumps, drivers, and pump fittings are of the quality and size expected.

25.7.1 Rotation of pumps

Many unnecessary service trips could be averted if the rotation of pumps was checked carefully at initial startup. A centrifugal pump will produce head with reverse rotation, but it will not perform to its design condition. Rotation can be checked easily; also, on three-phase power operations, the rotation can be changed by reversing two of the power leads.

25.8 Summation

Proper installation of HVAC pumps will provide the long useful life that these pumps should have without much maintenance. Finally, pumping systems should be observed carefully during their first hours of operation to ensure that the specifications have been met for that installation.

26

Factory-Assembled
Pumping Systems

26.1 Introduction

Traditionally, pumping systems have been field erected in the HVAC industry. The development of digitally controlled pumping systems along with computer-aided drafting and manufacturing has made economical the factory production of these systems.

Factory-assembled pumping systems have had the same natural development as other equipment in the HVAC field, such as chillers, boilers, and air-handling units, as well as condensate pumps, deaerators, and boiler feed systems.

26.2 Types of Factory-Assembled Pumping Systems

There are a number of different pumping systems available for almost any HVAC pumping application. The principal limitation for these systems is size. Systems can be manufactured to the size limits imposed by highway transportation or access into the building of final location. The principal applications in the HVAC field for factory-assembled equipment in addition to condensate return systems, boiler feed systems, and deaerators are for chilled water, hot water, and condenser water systems.

26.2.1 Chilled water systems

Pumping systems can be manufactured for chilled water in sizes from the simple primary pumping system (Fig. 26.1) to the large variable-

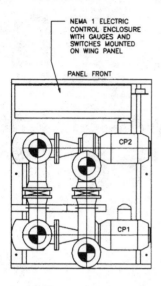

NEMA 1 ELECTRIC
CONTROL ENCLOSURE
WITH GAUGES AND
SWITCHES MOUNTED
ON WING PANEL

PANEL FRONT

CP2

CP1

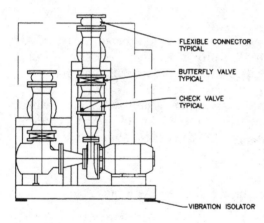

FLEXIBLE CONNECTOR
TYPICAL

BUTTERFLY VALVE
TYPICAL

CHECK VALVE
TYPICAL

VIBRATION ISOLATOR

Figure 26.1 Primary pumping system.

speed secondary pumping system (Fig. 26.2). Also, they can be simple systems or very complete ones of secondary pumps with expansion tanks and air separators (Fig. 26.3). Most HVAC pumping systems can utilize end-suction or double-suction pumps; larger systems are equipped with either horizontally or vertically mounted double-suction pumps. A unit with vertically mounted double-suction pumps is illustrated in Fig. 26.4. Axial flow pumps are seldom used on factory-assembled chilled water systems.

A popular factory-assembled system has been one with primary and secondary pumps, as shown in Fig. 26.5. This unit includes the primary chiller pumps and secondary distribution pumps as well as the

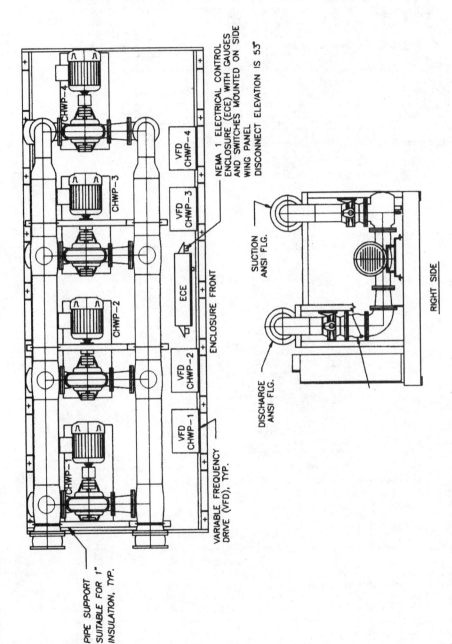

Figure 26.2 Large variable-speed secondary pumping system.

447

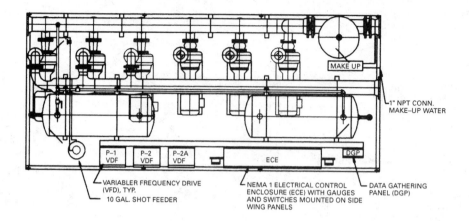

MAKE UP

1" NPT CONN.
MAKE-UP WATER

| P-1 VDF | P-2 VDF | P-2A VDF | | ECE | DGP |

VARIABLER FREQUENCY DRIVE
(VFD), TYP.

10 GAL. SHOT FEEDER

NEMA 1 ELECTRICAL CONTROL
ENCLOSURE (ECE) WITH GAUGES
AND SWITCHES MOUNTED ON SIDE
WING PANELS

DATA GATHERING
PANEL (DGP)

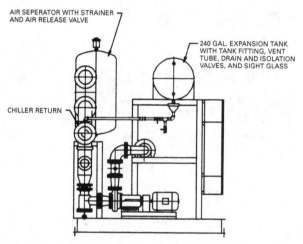

AIR SEPERATOR WITH STRAINER
AND AIR RELEASE VALVE

240 GAL. EXPANSION TANK
WITH TANK FITTING, VENT
TUBE, DRAIN AND ISOLATION
VALVES, AND SIGHT GLASS

CHILLER RETURN

Figure 26.3 Pumping system with expansion tank and heat exchanger.

chiller bypass; this ensures that the basic principles have been followed for good piping design for chillers.

26.2.2 Hot water systems

Factory-assembled hot water pumping systems are configured much the same as the chilled water systems described above. They are available with primary, secondary, or primary-secondary pumps. Some boilers do not require primary or boiler pumps, so only distribution pumps are required. In many cases, the hot water pumps are fur-

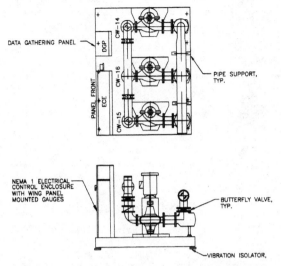

Figure 26.4 Pumping system with vertically mounted double-suction pumps.

nished on the same factory-assembled pumping system as the chilled water pumps.

Steam-heated hot water systems are available in a number of configurations, such as units with constant- or variable-speed pumps. Figure 26.6 describes a steam-heated unit with dual heat exchangers, variable-speed pumps, and a three-way valve for control of system temperature. This system offers many advantages over the traditional method of connecting steam-heated pumping systems. For example,

1. The control valve monitors steam pressure in the heat exchanger, not hot water system temperature. This ensures a positive pressure in the heat exchanger under all load variations in the hot water system.

2. With constant steam pressure in the heat exchanger, it is difficult for a vacuum to form in the heat exchanger under light loads.

3. The condensate flow through the steam trap is not erratic as with other steam-heated hot water systems. A steady flow of condensate through the steam trap ensures constant steam pressure in the heat exchanger.

4. Several different distribution loops with varying hot water supply temperatures can be served from one heat exchanger system by providing a three-way valve for each required temperature.

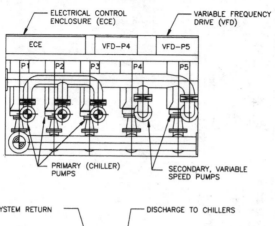

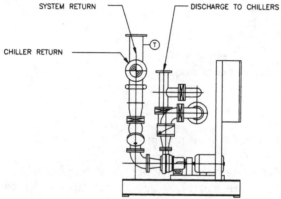

Figure 26.5 Primary-secondary pumping systems.

26.2.3 Condenser and cooling tower systems

Factory-assembled pumping systems for cooling tower and condenser water are similar to those for chilled water. They can range from constant-speed units like that in Fig. 26.1 to large pumping units similar to that in Fig. 26.2. Others can be primary-secondary as described in Fig. 26.5, where the primary pumps circulate the cooling towers and provide a suction pressure to the variable-speed condenser pumps.

The principal differences for cooling tower units are (1) the piping must be designed for higher pipe roughness and (2) the elevation of the suction header must be checked to ensure that its elevation is lower than the cooling tower sump.

Other special units for condenser water and cooling tower water are hot-well, cold-well systems and water source heat pump systems.

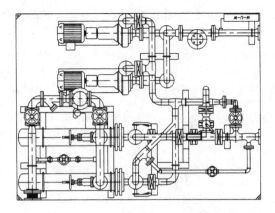

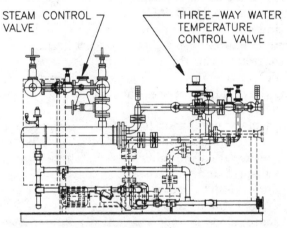

Figure 26.6 Steam-heated hot water system with three-way valve control.

26.3 Advantages of Factory-Assembled HVAC Pumping Systems

26.3.1 First cost

For most new installations, the factory-assembled pumping system offers the lowest-cost procedure for designing, building, and installing a pumping system. Fewer personnel hours are required than in designing a field installation and then actually installing the pumps, piping, and controls in the field. This, of course, is contingent on free access to the equipment room; many rehabilitation projects do not offer such access.

Factory-assembled pumping systems provide the production capabilities of a manufacturing plant and reduce the total cost of the whole process of design, manufacture, and installation of pumps. Following are some detailed processes that are enhanced by such systems:

1. The consulting engineer can use a catalog or drawing that provides a complete system of pumps, motors, piping, valving, controls, and switch gear all mounted on a single base. Figures 26.1 and 26.2 provide typical general arrangement drawings for such pumping systems. Figure 26.1 is for a simple primary pumping system, while Fig. 26.2 describes a large variable-speed secondary chilled water system. This system demonstrates the ability to computerize such complex systems for design and manufacture. This gives the pumping system manufacturer the same responsibility for detailed design of the pumping system as that of a manufacturer of a chiller or a boiler. All the contemporary electronic means of transmitting drawings can be utilized in providing these drawings to HVAC system designers.

The designer/consulting engineer can depend on the expertise of the pumping system manufacturer to provide a properly configured pumping system. The consultant can then devote more time to total evaluation of the building's environment and operation. This ensures that the client is getting the building services that are needed for the building to function as desired.

2. The contractor can place one order for a factory-assembled pumping system instead of hundreds of little orders that may be required to complete a field-installed pumping system. There is only one set of submittals to inspect and forward to the consulting engineer instead of many submittals. The contractor depends on only one source for the pumping system startup and service.

3. The owner saves space, since factory-assembled pumping systems usually reduce the space for the pumping equipment by around 30 to 50 percent. The owner's operating personnel have only one vendor to work with for service and adjustment after the consulting engineer and contractor have finished their work.

26.3.2 Time saving

Factory-assembled pumping systems offer savings in time during design, construction, and installation:

1. In the design phase, the pumping system manufacturer can offer drawings that enable the consulting engineer to select the complete system without becoming involved in the detailed design of the pump and piping details. On fast-track projects, this design procedure speeds up the layout of other piping in the equipment room.

2. On accelerated construction programs, the pumping system can be manufactured, tested, and prepared for installation long before the

site is ready to receive it. When the proper time occurs, the pumping system can be set in place. Much of the work delay due to coordination between various trades is eliminated.

3. Startup and commissioning are accelerated due to preplanning between the pumping system manufacturer and other control companies involved; the fact that one company is responsible for the operation of the complete pumping system also accelerates commissioning.

26.3.3 Unit responsibility

Factory-assembled pumping systems offer unit responsibility, as referred to earlier. This should be emphasized because of the great ease offered by this singular responsibility. Only one entity is involved with the startup of the pumping system. The consulting engineer, contractor, and owner have only one company to deal with in solving problems and getting the pumping system to operate as specified.

The pumping system manufacturer has so much at stake in terms of marketing that it is imperative that an ongoing relationship be established with the ultimate owner. Following are some of the duties that a pumping system manufacturer should carry out in commissioning a pumping system:

1. Verify the capacities of the system in gallons per minute and head and ensure that they conform to the requirements of the consulting engineer or owner.

2. Inspect the equipment room and ensure that it is adequate and that the ambient air is not hostile to the electronics of the pumping system.

3. Certify clearance around the system to ensure that adequate space is available for proper operation of the system.

4. Inspect electrical service and ventilation provisions to ensure that they are adequate.

5. Check all field piping attached to the system to ensure that it is properly supported and is not using the pumping system for support.

6. When the system is started, representatives of the owner, engineer, and contractor should be present to observe initial operation of the system.

7. The system should be run through all its capacities to verify that it is conforming to the specifications. Owing to the light loads that usually exist on HVAC systems at their initial operation, often it is impossible to achieve this in the field.

8. All alarm and safety circuits should be tested and verified to the owner's representative.

9. Complete operating instructions should be given to the operators of the system.

10. The service representative of the manufacturer of the system should establish a routine checkup procedure to ensure that the system is performing satisfactorily.

26.3.4 Less pump maintenance and repair

Factory-assembled pumping systems have demonstrated that there is less pump repair when the pump is assembled with a properly designed pumping system instead of mounting the pump in the field. This has been proved by years of operation of these factory-assembled pumping systems in the field. There are several reasons for this reduction in pump repair:

1. The pumping system manufacturer is totally responsible for the pump's environment.
    a. Suction and discharge piping is supported independently of the pump connections. There are no pipe forces on these connections.
    b. The pumps are mounted on a heavy structural steel base that does not shift as the building settles. This is not necessarily true for field-mounted pumps. There is little chance for the pump and motor to become misaligned after the pumping system is put into operation.
2. The pumping system manufacturer can evaluate the entire operating range of a pump and ensure that it does not operate at points of poor efficiency or at high radial or axial thrust.
3. The pumping system manufacturer should include the installation of differential pressure switches on all pumps to ensure that pump failure is alarmed and that the pump does not run continuously at a dangerous condition.

A properly installed factory-assembled pumping system on HVAC water systems should provide many years of service without significant repair or failure.

26.4 Components of a Factory-Assembled Pumping System

There is a variety of equipment that can be supplied with factory-assembled pumping systems. The principal components are listed below. Many special types of equipment are often furnished on these systems to meet the requirements of particular installations.

26.4.1 Pumps

Pumps for factory-assembled systems usually consist of single-suction and/or double-suction pumps. The number of pumps can vary from one to a dozen. The only limitation to the number of pumps is the ability to ship the pumping system.

The pumping systems can be a mixture of constant- and variable-speed pumps as long as constant- and variable-speed pumps are not designed to operate together in parallel.

Pump motor sizes vary from fractional horsepower up to 500 hp. Pumps with motors larger than this size are usually field installed.

26.4.2 Pumping system accessories

Following are standard accessories for these pumping systems:

1. Suction and discharge headers. Systems taking water from a sump and utilizing vertical turbine pumps are not equipped with suction headers.

2. Branch piping for each pump, including shutoff valves, check valves, and strainers. Some systems may include motorized valves instead of the check valves. Suction diffusers can be furnished in lieu of wye or basket strainers.

3. Chiller or boiler bypasses. These are furnished when both the primary and secondary pumps are supplied on one system.

4. All piping should be supported by structural steel to remove any stress from the pump flanges. Supports must be equipped with saddles and insulation spacers to reduce heat conductivity to and from the piping.

5. Hoists and rails can be furnished as an integral part of the pumping system for removing major components.

6. Heat exchangers, both shell and tube and plate and frame types. Steam-heated exchangers are furnished with steam control valves, steam traps, and water temperature control.

7. Expansion tanks with makeup water valves, including backflow preventers.

8. Mechanical air eliminators.

9. Chemical feeders and control.

10. Filters with control and piping.

11. Structural steel bases, standard and vibration types.

26.4.3 Electrical equipment

Factory-assembled pumping systems usually provide most of the electrical equipment required to integrate the pumping system into the rest of the water system. Following is a general listing of this equipment:

1. Switch gear
   a. Power supply can include main disconnect, watt transmitter, automatic transfer switch, and power feeders to variable-speed drives and magnetic starters on the pumping system.
   b. Combination starters for constant-speed pumps
   c. Variable-speed drives
2. Pump speed control and sequencing
3. Chiller or boiler sequencing
4. Instrumentation such as flowmeters, temperature transmitters, and pressure or differential pressure transmitters
5. Operator interfaces as well as information displays, alarms, and totalizers
6. Energy calculations such as Btu, Btu/h, tons, ton-hours, kW/ton, coefficient of performance, and wire-to-water efficiency
7. Automation control interfacing. This provides simple interconnecting with building management system manufacturers without special software and field coordination. These systems can become BACnet compatible when BACnet standards are completed.

26.4.4 Complete pump houses

The factory-assembled pumping system can be furnished in a pump house complete with the following equipment:

1. Base for installing the total assembly on a foundation in the field

2. Pumping system itself

3. Heating, ventilating, and air-conditioning equipment as required

4. Electrical switch gear for pumps as well as house lighting, heating, cooling, and ventilation equipment

The preceding is a brief synopsis of the equipment that can be furnished with a factory-assembled pumping system. Specific installations may require equipment not listed above.

On some applications, the installation of the pumps is away from the main building, or there is no room in the existing central energy

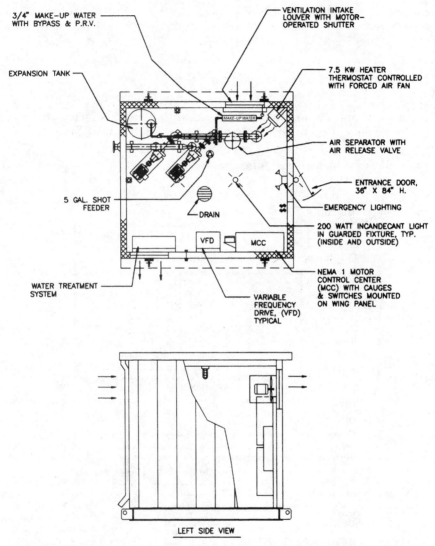

Figure 26.7   Large primary-secondary unit for cooling tower installation.

plant for the installation of new pumps. In such situations, it may be more economical to provide the pumps and switch gear in a separate house. A complete pumping system in its own house (Fig. 26.7) can be furnished with all wiring and internal piping completed and ready to be set on a field foundation. The only connections are the power supply, information interface, makeup water, and supply and return piping.

Special equipment such as boilers, chillers, electric generators, emergency power transfer switches, and water conditioning equipment can be furnished in these houses.

26.4.5 Control centers for existing pumps

Often, on retrofit projects or on an installation with limited access to the central energy plant, it is impossible to furnish a factory-assembled pumping system with pumps and piping all mounted on a structural steel base. In these cases, a complete control center (Fig. 26.8) can still be furnished that offers many of the advantages of the facto-

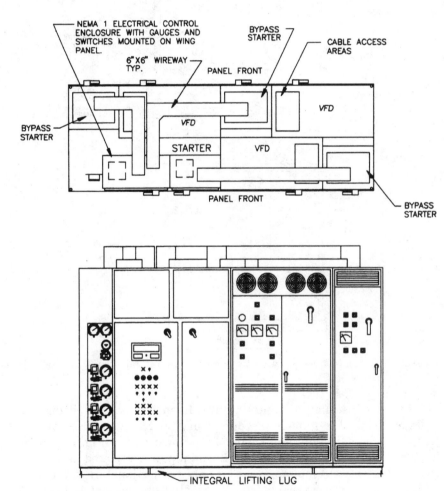

Figure 26.8 Variable-speed pumping control center.

ry-assembled pumping system. The supplier of these control centers should be able to furnish the support services and commissioning normally furnished by a manufacturer of pumping systems. The variable-speed drives for the pumps should be furnished by this manufacturer. This provides an assembly of the drives and pump controls that ensures the best fit between the pumps and controls. Also, factory wiring and assembly eliminate much of the field coordination that would be required when the variable-frequency drives are furnished by other suppliers.

The new procedures for controlling flow in chiller evaporators open a new field for the improvement of existing chilled water systems. An integrated control center providing all the necessary interfacing with chillers, chiller pumps, and any distribution pumps provides an economical and technical answer for achieving optimal operation of the chillers and pumps.

26.5   Testing Pumping Systems

The testing of factory-assembled pumping systems includes individual testing of the pumps, as described in Chap. 24. Also, the flow and friction loss tests of the entire pumping system should be completed to provide the designer or user of the pumping system with a total evaluation of the pumping system's performance. Test conditions equal to the initial and ultimate system flows should be performed at the factory test facility.

The flow test for a complete pumping system should include a wire-to-water efficiency evaluation, as described in Chap. 10 and Table 10.6. This enables the designer and ultimate user to have an overall evaluation of the pumping system before it leaves the manufacturer's plant. As with individual pumps, as described in Chap. 24, it is difficult to test pumping systems in the field due to the effort required to duplicate a factory test facility with its calibrated and properly installed instrumentation.

26.6   Summation

Factory-assembled pumping systems, like boilers and chillers, provide an economical answer for efficient pumping installations. They should be used where entrance clearances and economic conditions allow their application.

27

Operating HVAC Pumps

27.1 Introduction

Good operation of HVAC pumps consists of efficient performance without undue wear of the pumps, motors, and appurtenances. Efficient operation has been stressed throughout this book and will be summarized here; much of this chapter is found in other parts of the book. It is repeated here to emphasize the need for efficient operation, not just from an energy standpoint but also from a maintenance perspective. Further, a review will be made of how pumps fit into the overall energy conservation of HVAC water systems.

27.2 Checking for Efficient Selection of HVAC Pumps

Efficient operation begins with selecting the pump near the best efficiency point of its head-capacity curve. It is difficult to imagine how poorly pumps are often selected in the HVAC industry. Too many times pumps are selected far from the best efficiency point because one particular manufacturer does not have the proper sizes or types of pumps to meet a system's different pumping conditions. One pump manufacturer may have the best pump selections for the hot water pumps but not for the chilled and condenser pumps. On another project, one pump manufacturer may have the best selections for the single-suction pumps, while another may have better pumps for the double-suction applications.

The failure of the HVAC designer to realize that one pump manufacturer cannot truly meet all the various pump design conditions for a complex project results in poor selection of pumps. The designer should evaluate several manufacturers' pumps to ensure that the most efficient pump is selected for each pump duty. Too much emphasis is placed on having all the pumps from the same manufacturer.

In the past, great emphasis was placed on pump maintenance. Pump maintenance was high because of the use of constant-speed pumps and letting them operate at points of high radial thrust. Pump bearings and mechanical seals wore out rapidly and required continuous maintenance. Today, with proper selection and the use of variable-speed pumps, maintenance of HVAC pumps should be minimal. If an HVAC pump needs continuous maintenance, its application should be checked to ensure that the pump is operating near its best efficiency point. As has been mentioned throughout this book, HVAC pump duty is not difficult duty. Therefore, these pumps should run continuously without excessive wear. There should be little noise, no vibration, and no heat (except with hot water pumps). Evidence of any of these factors should be of concern and require inspection of the pumps to determine the cause. More on pump maintenance will be presented in Chap. 28.

Throughout this book, the emphasis has been on operating pumps near their best efficiency point to avoid radial thrust and resulting failure. Manufacturers of mechanical seals recognize this fact more than any other equipment manufacturer. A recent article in *Plant Engineering Magazine* estimated that 85 percent of mechanical seals replaced in pumps had failed and were not worn out. These failures were due to improper operation of the pumps, not the seals themselves. This article also pointed out the importance of operating pumps within ±20 percent of the flow at the best efficiency point. Further, this article provided additional information on extending the useful life of mechanical seals and centrifugal pumps.

27.3   Constant- or Variable-Speed Pumps

As was emphasized earlier, constant-speed pumps should be operated on constant-volume systems and variable-speed pumps should be operated on variable-volume systems. When investigating the operation of existing pumps, great concern should be given to ensuring that constant-speed pumps are not operated at the left side of the pump head-capacity curve. On most hot and chilled water systems, constant-speed pumps should not be operated on variable-volume systems. On light loads, this causes the pumps to run up their head-capacity curves and operate where the radial thrust is high. *Radial thrust wears casing rings, bearings, and mechanical seals!*

Variable-volume systems should be equipped with variable-speed pumps for energy conservation and prolonged useful life. This applies to most HVAC water systems, including hot, chilled, condenser, and cooling tower installations. The cost differential between variable- and constant-speed pumps has been dropping with the reduction in the cost of variable-speed drives. In fact, with the knowledge that we have today of properly designed variable-volume systems, a variable-speed system should be less expensive than a constant-volume, constant-speed system with the valving that is required to overcome the overpressure of the constant-speed pump.

27.4 Proper Selection and Operation of Variable-Speed Pumps

Contrary to many who look at any pump as a simple thing that just produces pressure, the proper selection and operation of variable-speed pumps provide great savings in energy and maintenance. Following is a brief resumé on these two important subjects.

27.4.1 Selection

1. Select the pumps just to the right of the best efficiency point.
2. Determine the optimal number of pumps. Remember that three pumps, each with a capacity equal to 50 percent of system capacity, is often the best number of pumps.
3. Determine the system head area of the proposed HVAC water system, from minimum to maximum load.
4. Run a wire-to-water efficiency calculation from minimum to maximum system flow to determine the acceptability of the pump selection.

27.4.2 Operation

1. Verify the actual maximum and minimum flow in gallons per minute of the system. So often systems are designed for future loads; if so, the pumping program should be adjusted to the initial operation.
2. Confirm the system heads in feet at both the maximum and minimum flow rates. They should be taken into consideration when adjusting the initial pump operation.
3. If a wire-to-water efficiency calculation has been completed and wire-to-water efficiency indication and control are available, com-

pare the actual operation with with that projected in the calculations. If a calculation has not been made, depending on the instrumentation available, the following equation can be used to determine the efficiency of the system operation.

$$\text{Wire-to-water efficiency} = \frac{\text{flow (gal/min)} \cdot \text{pump head}}{53.08 \cdot \text{pump kW}} \quad (10.2)$$

The pump flow can be measured from a flowmeter, the pump head from a differential pressure transmitter as installed in Fig. 10.8, and the pump kilowatts from a watt transmitter.

The control of variable-speed pumps includes programming both the number of pumps that should be operating and the speed of the pumps. Too often only speed is considered in the operation of variable-speed pumps.

27.5   Control Signals for Speed Control

Various methods of speed control have been tried for HVAC variable-speed pumps. The most successful has been the differential pressure transmitter for hot and chilled water distribution systems, as well as for condenser water systems. In the latter case, the differential pressure transmitter measures the lift pressure of the chiller.

Controlling pump speed with temperature, differential temperature, level, or flow signals can create problems because of the rate of response of the system itself. These signals can be used on systems that do not change such conditions such as temperature, level, or flow rate rapidly.

27.6   Sequencing and Alternation

Two of the most neglected control procedures for HVAC pumps are proper sequencing and alternation of pumps operating in parallel. Here again, with digital electronics, it is so easy to write the correct program for these two procedures.

27.6.1   Sequencing

Pump sequencing for variable-speed pumps has been discussed at several points in this book. The point at which this takes place depends on the load characteristics of the system. This is operational sequencing; sequencing under failure of an operating pump has not been discussed.

In the past, when a pump failed, an alarm was sounded, and the standby pump did not start until the system conditions detected a reduction in flow or pressure. With digital electronics, the program can be written to start the standby pump immediately on failure of a pump and not waiting for system conditions to deteriorate. Also, an alarm should be initiated at pump failure. These procedures protect the system and give the operator earlier notice that a pump has failed.

27.6.2 Alternation

Pump alternation is the selection of lead and lag pumps. For example, if three pumps are operating in parallel, there are several sequences, such as no. 1, no. 2, and no. 3 or no. 3, no. 2, and no. 1, etc. There are a number of different methods of sequencing pumps. Following is a description of most of them:

1. *Manual:* The operator selects the lead pump and the sequence of the lag pumps.

2. *Duty alternation:* Whenever a pump is stopped, the program indexes down one pump so that there is an automatic change in the lead pump. This is called *first on, first off,* which means that the lead pump is stopped instead of the last pump started.

3. *Timed alternation:* With this method, the pumps are alternated with a time clock, such as daily or weekly.

4. *Equal run time:* The pumps are equipped with elapsed-time circuits, and the pumps are sequenced so that equal run times are achieved for all the pumps.

The three automatic alternation methods listed above are designed to achieve equal wear on the pumps. Alternation in HVAC pumps is not as an important issue as it was in the past because there should be little wear in them. Also, it has been found that "equal wear" is a dangerous practice. If pumps are to wear, it should be known which pump should wear out first. If all the pumps have equal wear, they all should wear out at the same time!

None of these automatic alternation procedures should be used on any HVAC pumps, even those on condensate return systems. *Manual alternation* is the best procedure for operating HVAC pumps.

1. The operator should change the lead pump every month or so and should be there when the pump that has been at rest is brought on the line. This provides observation of the pump as it comes up to

speed. Any unexpected noise, vibration, or heat can be detected immediately, not after the pump has been operating for some time.

2. Larger pumps or those supplying critical processes should be equipped with elapsed-time meters, and the operator should manually select the pumps so that a difference of 2000 to 4000 hours is maintained between the pump running times. If wear is occurring, the pump with the most hours should be the first to require maintenance.

27.7 Summary

Too often HVAC pumps are looked on as simple devices that require no thought about their operation. If you need some more flow or pressure, just turn on another pump. It should be remembered that under load-shedding procedures such as use of energy storage, the pumps and fans of an HVAC system comprise the remaining load. In the case of chiller operation, the pump horsepower can equal that of the chillers under reduced-load conditions. Pump energy consumption should be given serious consideration under any load condition.

Typical wire-to-water efficiencies are provided throughout this book. Actual wire-to-water efficiencies for operating HVAC systems should be in the same general ranges.

27.8 Bibliography

William "Doc" Burke, Extending seal and pump maintenance intervals, *Plant Engineering Magazine,* July 10, 1995, p. 83.
American National Standard for Centrifugal Pumps, ANSI/HI 1.1-15-1994, Hydraulic Institute, N.J., 1994.

28

Maintaining HVAC Pumps

28.1 Introduction

If all HVAC pumps were installed and operated properly, this would be the shortest chapter in this book. Also, anyone reading the previous chapters may wonder if HVAC pumps require any maintenance. Compared with other pump fields such as industrial pumping, very little repair should be required by HVAC pumps. Actual cases exist where variable-speed pumps have operated on chilled water for over 20 years with no maintenance other than changing the mechanical seals that were damaged by the chemistry of the water.

There are certain maintenance procedures that should be performed and which are outlined by most pump manufacturers. They are provided with the pumps in the form of installation, operation, and maintenance instructions.

Pump maintenance should consist of (1) sustaining the performance of the pump, its driver, and accessories at a high level of efficiency and (2) providing the lubrication and servicing specified in the pump manufacturer's instructions.

28.2 Maintaining the Pumping Equipment at High Efficiency

So often pump maintenance means just making sure that the pump is pumping water without concern for the efficiency of its operation. There are a great many procedures that can be followed to ensure that the pump or pumping system is operating at the expected efficiency.

The ability to check a pump or pumping system's performance depends on the instrumentation that was provided with the pumping system. This instrumentation can vary from nothing to overall wire-to-water efficiency indication and control. It can be expected that a pumping system without instrumentation does not have a very high continuing wire-to-water efficiency. Reiterating, wire-to-water efficiency is not an esoteric thing that is hard to understand. It is merely the ratio of energy imparted to the water divided by the energy input to the pump driver or variable-speed drive for variable-speed pumps.

If the installation is not equipped with wire-to-water indication but has a flowmeter that registers in gallons per minute, wire-to-water efficiency can be developed by installing a watt meter on the power supply to the system along with a differential pressure indicator across the pumping system connections, as shown in Fig. 10.8. If the watt transmitter reads in kilowatts and the differential pressure indicator in feet, Eq. 10.2 provides the wire-to-water efficiency, and it is repeated here:

$$\text{Wire-to-water efficiency} = \frac{\text{flow (gal/min)} \cdot \text{pump head}}{53.08 \cdot \text{pump kW}} \quad (10.2)$$

For example, if the flowmeter is reading 600 gal/min, the differential pressure indicator 80 ft, and the watt transmitter 14 kW, then

$$\text{Wire-to-water efficiency} = \frac{600 \cdot 80}{53.08 \cdot 14} = 64.6\%$$

Periodic readings can be taken from this instrumentation to ensure that an adequate pumping efficiency is being maintained. Obviously, with today's digital computer technology, it would be better to read the three values of flow, head, and wattage as analogs and input them into a computer where the wire-to-water efficiency could be read continuously.

The advantage of the wire-to-water efficiency technology is that the single value of efficiency can be watched or recorded to provide a continuous record of pumping efficiency. Any decay in this efficiency indicates one of the following problems:

1. The wrong number of pumps may be in operation. Too many or too few pumps will cause operation at low wire-to-water efficiency.

2. Dirt or debris may be in a pump impeller, reducing the efficiency of that pump.

3. A pump's casing rings may have opened, allowing greater bypassing and therefore creating poor pump efficiency.

4. The pump may have air in it, or it may be cavitating.

5. A pump isolation valve may be partially closed, or a check valve may not be operating properly.

6. If the pump is a variable-speed pump, there may be problems with the variable-speed drive, although this has seldom been the case in actual operation.

Thus wire-to-water efficiency can be a watchdog on pumping system operation.

28.3   Checking the Pump Itself

There are several procedures that can be completed to determine a pump's condition. The first check is to listen to the pump. If it is noisy, it could have air in it, or it could be cavitating. Moving the discharge isolating valve to 90 percent closed will differentiate between the two. If the noise persists with this valve nearly closed, the problem is air; if the noise goes away with the valve nearly closed, the problem is probably cavitation. As indicated before, efforts should be made to ensure that there is no air entrainment. If the pump is cavitating, the net positive suction head (NPSH) required by the pump should be checked, and the NPSH available should be recalculated. If there is no flowmeter to verify flow, cavitation may be developing by the pump running out to the right-hand part of its head-capacity curve where the NPSH required is high.

Suction and discharge gauges are invaluable in checking pump performance. If the pump is not noisy, the pump should achieve shutoff head if it is in good condition. The casing wear rings can be checked by again moving the discharge isolation valve to around 90 percent closed. If the pump produces the shutoff head that is indicated on its head-capacity curve, the casing rings probably are in good shape. A second check on the casing rings is to operate the pump at design flow and verify that the design head has been achieved.

If the wire-to-water efficiency is low or the pump is not achieving design flow, the interior of the pump should be checked for dirt or debris. This is very easy to do with a double-suction pump with an axially split casing. It is not as easy with other types of pumps.

Other parameters to check are the pump's speed and the amperage or kilowatt input to the pump motor. If the pump's speed meets the specification, the power input to the motor will indicate the amount of energy being consumed by the pump. If there is an accurate flowmeter that can be used to check the pump's performance, a total diagnosis can be made of the pump through use of the pump gauges, power input, and the flowmeter.

28.4   Checking the Pump Installation

As mentioned earlier, the installation of the pump has much to do with its maintenance. Following are some conditions that should be checked:

1. Is the base sturdy and supporting the pump and its driver properly without vibration? Is it strong enough that the piping is not disturbed? Does the pump or its driver require periodic realignment?

2. Are any vibration isolators on the pump or base operating properly? Have the isolators been selected properly, and are any axial thrusts being imposed on isolators that are not designed for such thrusts?

3. Are ambient conditions conducive to good pump, motor, and switch-gear operation, namely, proper air temperature, humidity, and cleanliness?

4. Are any water leaks apparent? If so, they should be repaired as soon as possible due to the possibility of resulting in Legionnaires' disease. Likewise, leakage from packing or mechanical seals should be piped to floor drains, not run in open gutters.

5. Has the water being pumped been checked for proper chemistry? Bad water is the source of many mechanical seal failures. Are necessary filters, air vents, and chemical feeders installed to prevent erosion and corrosion?

28.5   Checking the Electrical Conduit and Piping Installation

The attachment of electrical conduit to the pump motor obviously should not impair the pump movement of any vibration isolators. A flexible connection should be provided on the electrical conduit that will eliminate any such interference and any abrading of the conduit or fittings.

The piping should not impose any forces on the pump connections. On flanged pumps, this is easy to check by closing the shutoff valves on the pump suction and discharge and removing the flange bolts. It the mating pipe flanges gap, shift, or move to an angular position, the flexible connectors on the piping should be reevaluated or the pipe hangers should be adjusted. Forces imposed on the pump flanges by the piping can accelerate the wear of bearings, sleeves, and mechanical seals.

28.6   Maintenance Observations and Scheduling

Although HVAC pumps should be relatively free of maintenance, continuous evaluation of parameters such as wire-to-water efficiency guarantees that a pumping system is operating at optimal efficiency. Data on wire-to-water efficiency can be fed into computer memory and recalled to verify if current operation is similar to that when the system was commissioned. Load increases due to system changes may dictate the need to reset the points of addition and subtraction of pumps. Wire-to-water efficiency degradation indicates systems changes, improper operation, or dirty equipment.

Many rigid, real-time maintenance schedules are unnecessary when operational parameters such as wire-to-water efficiency are used to provide a running evaluation of the pumping equipment. With such information on a computer screen, it will advise the operator when pumps should be serviced or repaired. Usually, it is unnecessary to periodically disassemble an HVAC pump on some theoretical maintenance schedule when computer data on the pump's operation is available to the operator.

28.7 Summation

As with any mechanical equipment, good housekeeping always promotes longer life for pumps, motors, and electrical switchgear. The preceding practices, along with the pump operation recommended in Chap. 28, should provide long pump and motor operation with a minimum of maintenance.

The pump or pumping system manufacturer's installation, operation, and maintenance instructions should be reread to ensure compliance. Are these instructions located at a known place in the maintenance offices?

Noise, heat, vibration, and continual speed changes have no place in a well-designed and properly controlled pumping system. The quality of pumping equipment and operational procedures available today should provide HVAC pumping systems that are relatively free of maintenance.

29

Retrofitting Existing HVAC Water Systems

29.1 Introduction

There are so many HVAC water systems in operation and designed to the old constant-speed mechanical control procedures that it is important that a chapter of this book be devoted to retrofit of these existing systems. Some of this chapter will be a reprise of other chapters and will utilize the information provided in them.

There have been earlier discussions of energy waste in existing systems, e.g., the reference to the system with balance valves on pump discharges that wasted over 900,000 kWh per year. Any water system with a preponderance of balance valves should be studied for energy savings by eliminating them through the use of the various system analyses offered herein.

29.2 System Evaluation

This chapter will try to develop a systematic evaluation of existing systems in the hope that the systems with the greatest potential for savings in energy and maintenance will be highlighted.

29.2.1 Evaluating existing systems

Water systems will be evaluated first by the number of mechanical devices used to control water flow and pressure in these older systems. Most of these systems will be found in the distribution of chilled and hot water; chilled water is usually the best candidate because of the higher flows, heads, and energy consumption.

The existence of the following mechanical equipment in a system will point the way to possibilities for energy savings:

1. Pressure regulators on pump discharges
2. Manual or automatic balance valves on pump discharges
3. Pressure-reducing valves anywhere in a system
4. Balance valves on zones in heating and cooling systems
5. Crossover bridges that equalize pump pressure
6. Circulating pumps in series with heating or cooling coils.

All these devices can waste energy, and their existence is an indicator that the system should be evaluated.

29.2.2 Graphic description of flow in an existing system

Figure 29.1 describes the use of the valves and devices listed above to regulate flow and pressure to the water system. The shaded area of this figure describes the energy that can be saved through proper sys-

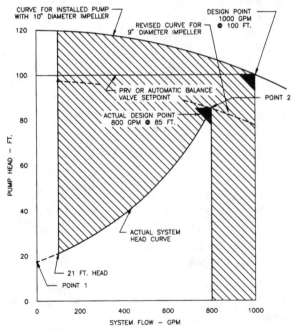

Figure 29.1 Evaluation of an existing pump installation.

tem analysis of a chilled water system and replacement of these mechanical devices with proper pump control.

The system head curve in Fig. 29.1 is typical for a variable-volume hot or chilled water system. The perfect pumping and control system would be one that tracks this curve at any load on the water system; in other words, there would be no overpressure at all.

The procedure for evaluating existing hot and chilled water systems can best be demonstrated by setting up a typical system, generating the system head curve, and establishing the ways that energy can be saved. First, a typical chilled water system should be established. Assume

1. The system is served by only one pump. This simplifies the procedures that follow. (Actual systems would normally be equipped with two 500 gal/min pumps.) Original design for the pump was 1000 gal/min at 100 ft with a 10-in-diameter impeller operating at 1770 rev/min.

2. A flowmeter is available to read actual system flow in gallons per minute. (*Note:* If a flowmeter is not installed, an insertion-type flowmeter could be hot tapped into the piping without draining the system.) A flowmeter is an essential instrument for monitoring system performance and should be part of the permanent instrumentation.

3. Two-way valves are installed on all cooling coils.

The system head curve can be calculated by

1. Determining the maximum pressure loss of the cooling coils, their control valves, and associated piping. Usually this loss is in the range of 15 to 30 ft. This is point 1 in Fig. 29.1, or 20 ft, and is at a no-flow condition for the system.

2. Developing the maximum flow and head of the system (point 2). This can be achieved by reading the flowmeter for maximum flow under full-load conditions. Suction and discharge pressures at the pump headers can be read to secure the head of the system at full load. Existing pressure gauges should be checked for accuracy. If pressure gauges are not installed, new calibrated pressure gauges should be installed. On larger systems where wire-to-water efficiency is to be checked, a differential pressure transmitter should be installed, as shown in Fig. 10.8. The flow and head of point 2 are assumed to be 800 gal/min and 85 ft, as shown in Fig. 29.1.

3. Determine the minimum flow for the system. In this case, it is 100 gal/min.

4. If it is difficult to accomplish the measurements of item 2, a recalculation of the system flow and head loss can be made based on actual operation of the system and known cooling loads.

5. Once points 1 and 2 are known, an approximate system head curve can be developed from Eq. 9.3:

$$H_a = H_1 + \left(\frac{Q_a}{Q_2}\right)^{1.9} \cdot (H_2 - H_1) \tag{9.3}$$

where H_a = head at any point a on the system head curve
$\quad H_1$ = head at point 1 on the curve (the maximum pressure loss for any coil, its valve, and piping)
$\quad H_2$ = head at point 2 on the curve (the maximum head of the water system)
$\quad Q_a$ = flow at any point a on the system head curve
$\quad Q_2$ = flow at point 2 (the maximum flow in the water system)

Note: The exponent 1.9 follows the Darcy-Wiesbach equation closer than does an exponent of 2.0.

Although the system head curve thus generated does not represent the system head area of a water system as developed in Chap. 9, it can be used as recommended in Chap. 9 to develop the approximate system head area.

29.2.3 Evaluation of existing pumps and motors

Just because HVAC pumps look dirty and beat up on the outside does not mean they should be replaced. As mentioned so often, *HVAC pump duty is easy duty.* Before considering the replacement of existing pumps, the pumps should be opened and the impellers and casing rings should be inspected. The inside diameter of the casing rings should checked to ensure that they are near their original diameter. If not, they should be replaced. Their cost is much less than the cost of new pumps!

The great advances that have been made in electric motors requires that existing motors on pumps be checked very carefully. The principal motor manufacturers and other companies have programs, including actual testing, that provide accurate checking of these motors. The amortization evaluation can be provided to demonstrate the value of replacing existing motors on an energy basis.

Do not change existing electric motors just because the pumps are being changed to variable speed! Most existing motors will operate very well with contemporary variable-speed drives. This includes old

T-frame, U-frame, and even wound rotor motors converted to induction type by shorting the rotor windings. The variable-speed drive manufacturers can verify the suitability of these old motors operating with their drives. The question is usually energy, not mechanical conditions, that favors the installation of new motors.

It must be remembered that variable-speed drives are current-rated devices, so their selection for existing motors must take into consideration the full-load ampere rating of these motors.

29.2.4   Evaluation of the number of pumps

Too often existing HVAC water systems, particularly chilled water systems, are equipped with oversized pumps. In many cases, the systems are equipped with pumps with 100 percent system capacity. Any evaluation of existing pumps must include a review of the system's requirements. The overall wire-to-water efficiency can be improved in many cases by replacing these larger pumps with three 50 percent capacity pumps. The wire-to-water efficiency evaluation that has been provided in this book is an excellent procedure to determine the value of changing to multiple smaller pumps.

29.2.5   Control of existing pumps

This is one of the most overlooked tasks and sometimes the most difficult task in evaluating an existing system. The type of pump control must be determined, from manual to the most complete control such as wire-to-water efficiency.

Manual control of HVAC pumps is a holdover from the past and usually indicates possibilities of improving energy conservation through programming of the pump to match the system's requirements. Constant-speed pumps should be programmed on and off to match the system head curve or area as closely as possible. This will reduce the overpressure at lower loads on the water systems.

Variable-speed pumps must be programmed on and off as well and have their speed controlled to achieve the highest wire-to-water efficiency. Most of this work has been covered in previous chapters. Existing systems, when retrofitted to variable speed, can be every bit as efficient as new systems with the correct approach to pump sizing and control.

Following are some system characteristics that should be observed with the proper control of variable-speed pumps:

1. There should be very little overpressure on the system at minimum to 50 percent of total load on the water system.

2. There should be no discernible fluctuation in pump speed. If an operator can hear or see HVAC variable-speed pumps change their speed, there are pump control problems. The rate of response for the control system may be too slow.

3. All equal-sized pumps should be operating within 15 rev/min of each other. If not, there again are control problems.

4. The controlling differential pressure transmitter should hold the differential pressure at this transmitter to within ±½ to 1 ft of head. Fluctuation of the pressure beyond 1 ft from the set point indicates insufficient control.

29.3   Evaluation of Existing Procedures

Once the system head curve or area is developed, a figure similar to Fig. 29.1 can be developed for the system. If the shaded area is appreciable, there exist possibilities for energy savings. The engineers for system design and operation should then continue their evaluation of the water system to determine actual energy savings by several procedures; some systems may require several such procedures:

1. Eliminate some or all of the existing mechanical devices such as balancing or pressure-reducing valves.

2. Trim existing pump impellers.

3. Change to variable-speed pumps.

4. Determine if the pump programming is correct. Are the right number of pumps being operated at all system loads?

The amount of overpressure that is occurring at the minimum flow point will help determine what procedure should be used. In Fig. 29.1, this overpressure is $120-21$, or 99 ft. The existing balance or pressure-regulating valves may have been used to control this overpressure. Removing them may cause operating problems without correcting the pumps as well.

29.3.1   Trimming the pump impeller

The first step that should be considered is to trim the pump impeller to provide the maximum flow and head of 800 gal/min at 85 ft of head. The dotted curve in Fig. 29.1 describes the resulting head-capacity curve for the pump.

Care should be taken in trimming the pump impeller. The affinity laws cannot be taken for granted in trimming this impeller, since the system contains 20 ft of constant head across its cooling coils and their control valves and associated piping. Equation 10.1 should be used to compute the impeller diameter. This is done by finding the equivalent operating point on the present pump curve.

$$Q_3 = \sqrt{\frac{Q_2{}^2}{H_2} \cdot H_3} \tag{10.1}$$

where Q_2 and H_2 = maximum flow and head conditions of 800 gal/min and 85 ft

Q_3 and H_3 = unknown equivalent points on the pump head-capacity curve

By trial and error, values of H_3 are inserted in the preceding equation until Q_3 and H_3 fall on the known pump head-capacity curve for the 10-in-diameter impeller operating at 1770 rev/min. The equivalent operating point, after several trials, is determined from the equation to be $Q_3 = 889$ gal/min and $H_3 = 105$ ft.

Since the diameter varies directly with the flow, the new diameter is $800 \div 889 \times 10$, or 9.0 in. If the affinity laws had been used directly without recognition of the system head curve, the impeller diameter would be $800 \div 1000 \times 10$, or 8.0 in. The impeller would therefore be undercut and would not have sustained the desired point of 800 gal/min at 85 ft of head.

Equation 10.1 should always be used for trimming impellers, not the ratio of flows in gallons per minute from the affinity laws themselves. Also, it must be remembered that the trimmed impeller must be replaced if the original design condition of 1000 gal/min at 100 ft is ever reached by the system in the future.

Trimming the impeller reduces the pressure from 120 ft at shutoff to 98 ft. At minimum flow of 100 gal/min, the head will be 98 ft and the overpressure $98 - 21$, or 77 ft. This may still be too much overpressure for the coil control valves. The trimmed impeller has saved energy, but much is still to be saved by the use of a variable-speed pump. All the shaded area below the dotted curve for the trimmed pump can be saved by changing the original 10-in-diameter impeller to variable speed.

29.3.2 Changing to a variable-speed pump

The trimming of the impeller requires some careful calculations to ensure that the correct impeller diameter is selected. The variable-speed pump requires none of these calculations. Instead of cutting the impeller, the pump will run at a lesser speed to maintain the desired condition of 800 gal/min at 85 ft. At this point, from the preceding calculations, the pump speed will be 800 ÷ 889 × 1770, or 1593 rev/min.

There are several advantages for the variable-speed pump over trimming the impeller.

1. The original 10-in impeller can be retained.

2. Since the 10-in impeller is retained, future conditions of 1000 gal/min at 100 ft can be achieved.

3. No overpressure will occur at any load on the system because the pump will merely slow down to the needed speed to maintain the head required by the system head curve.

4. If the system has a broad system head area that is difficult to compute, the variable-speed pump, if properly controlled, will automatically adjust to this variable-head condition.

5. All the energy shown by the shaded area in Fig. 29.1 will be saved. Additional energy will be saved because the wire-to-water efficiency of the variable-speed pump will be much greater than that for the constant-speed pump.

It is up to the designer to run the calculations and determine the energy and maintenance savings for the variable-speed pump as compared with a pump with a trimmed impeller.

Many public utilities in the United States are offering rebates for programs aimed at energy savings. Often, the installation of variable-speed drives qualifies the owner for these rebates. These rebates must be factored into the costs for the conversion to the variable-speed pump.

There are other conditions in the system that must be considered, whether the pump impeller is trimmed or the pump is converted to variable speed. For example,

1. All three-way valves on heating or cooling coils should be converted to two-way valves. Three-way valves should not be converted to two-way valves by shutting off their bypasses without a very careful evaluation of the maximum pump head that can be imposed on them. The valves should be checked to ensure that they will operate satisfactorily as two-way valves.

2. Differential pressure-control transmitters should be installed in accordance with Chap. 8 to control the variable-speed pump.

3. Adequate instrumentation should be provided in the new system to ensure that projected energy savings are achieved.

29.3.3   Control and drives for modified systems

When variable-speed drives are added to existing pumps on a water system, some care must be taken in evaluating the electrical power installation. For example,

1. A study of the power feeds to each motor should be made to ensure that they are adequate physically for any new motors. An appreciable savings can be made if they can be reused, particularly if they are buried in the cement slab.

2. Existing magnetic starters should be evaluated for use as emergency standby in event of variable-speed drive failure. This could eliminate the need for new standby starters.

3. Since there is some power loss in the variable-speed drives with their approximate efficiency of 95 percent, the total power consumption of the drives should be checked against the capacity of the existing power supply. The maximum ampere rating for the drives should be considered in this evaluation.

A completely integrated control and power center as described in Fig. 26.8 should be considered for existing pumps. This provides a concentrated location of all the switchgear and controls needed for proper functioning of the variable-speed pumps.

29.4   System Configuration Modification

Too often boilers or chillers are added because there is a lack of sufficient hot or chilled water. Inefficient water distribution systems can be the cause, not boiler or chiller capacity. Before such equipment is added,

1. The heating and cooling loads imposed on these systems should be recalculated to verify that actual boiler or chiller capacity is truly needed.

2. The return water temperatures of these systems should be checked to see how close they are to the design return temperatures. For example, assume that a chilled water system was designed for 44°F supply and 56°F return temperatures, and the actual return temperature is 52°F. The pumping rate from Eq. 9.2 should be $24 \div 12$, or 2 gal/min per ton; instead, it is $24 \div 8$, or 3 gal/min per ton. This indicates that work should be done to improve the water distribution system before additional chiller capacity is considered.

Correcting or changing central pumps may be an easy task compared with changing system piping to achieve the best possible arrangement with the lowest energy consumption. Many existing systems for hot and chilled water have three-way valves and are therefore constant-volume systems that may have great waste in pumping energy. Changing the three-way valves to two-way valves makes the system variable volume, not constant volume. Boilers and chillers that require constant flow through them must be taken care of with the proper type and location of a bypass.

29.4.1 Changing constant-volume systems to variable volume

Existing systems that are changed from constant volume with three-way coil control valves to variable volume with two-way coil valves must accommodate chillers or boilers that may require constant flow through them. One procedure is to install new primary pumps for the chillers and make the existing constant-speed pumps into secondary pumps by converting them to variable speed. This procedure is often difficult due to lack of space for the new pumps and the amount of repiping that must be done to produce a primary-secondary pumping system.

An alternative design for making the system variable volume is to convert the existing pumps to variable speed and provide the means to ensure adequate flow through the chillers or boilers. This is done by adding a bypass with a flowmeter and control valve in it, as described in Fig. 15.1b. This works most effectively on systems with three or more chillers. All the coils should be equipped with two-way control valves, and differential pressure transmitters should be located as recommended in Chap. 10.

Following is a description of the operation of this system:

1. The speed of the chiller pumps is controlled by the differential pressure transmitters, maintaining the desired differential pressure required by the cooling coils.

2. Flowmeter FT1 measures the actual flow to the chilled water system.

3. The system flow is compared with the required flow for the chillers. The difference is made up through the bypass and is monitored by flowmeter FT2. This flowmeter controls the bypass valve to maintain this desired flow in the bypass. If it is acceptable to operate a chiller at less than 100 percent flow through the evaporator, this can be accommodated by changing the set points in the chiller sequencer.

4. Any size of chiller can be set in any position in the chiller sequencing control. The chillers are rated in gallons per minute; the actual flow to the system determines the number of chillers that should be in operation.

Contemporary chillers with PID control of their chilled water temperature may allow variable flow in their evaporators. The piping arrangement in Fig. 15.1b can be utilized; the only difference is that the flow set point for the bypass is reduced to the minimum allowable flow for the evaporator. This procedure has some significant advantages over primary-secondary pumping because space, first cost, and possibly energy are usually saved with such installations.

29.4.2 Changing pumped coil systems to variable-speed pumping

Some existing systems may have been designed with coil pumps with bypasses, as shown in Figs. 8.2b and 8.3b. This is a very inefficient arrangement because of the piping losses and the efficiency of the pump and motor themselves. So often such pumps are small circulators with very low wire-to-shaft efficiencies for the pump and motor. Such pumped coil systems can be converted to variable speed by installing two-way valves on the coils and using variable-speed pumps at a central or building location. The work in redesign and retrofit is usually worthwhile in energy savings. To verify this effort, the energy consumption of the small circulators should be checked along with their maintenance record.

Pumped coils with circulators in the bypass, as described in Fig. 8.4, should not be confused with the preceding pumped coil systems. If the circulator is in the bypass, it was installed there for freeze protection or for eliminating laminar flow in the coil.

29.4.3 Variable-volume systems with bypass valves

Many chilled and hot water systems were installed with constant-speed pumps and utilized bypass valves to accommodate the chiller flow and to achieve some form of variable volume. The problem with this type of system, in terms of energy, is that all the water must be pumped at a head equal to or greater than the design head. The pump or pumps are forced to run up the pump head-capacity curve, which increases the overpressure on the system. Also, this increases the wear on the pumps, since they are forced to operate with high radial thrusts (see Fig. 6.9a).

These bypass valve systems can be changed to variable-volume, low-head operations by eliminating the bypass valve and installing a bypass with a flowmeter and a regulating valve, as shown in Fig. 15.1*b*.

29.4.4 Adding variable-volume systems to existing systems

The retrofit of existing systems, particularly large, multiple-building installations, can be prohibitive in cost to convert to contemporary pumping and control systems. These systems require a great amount of thought and evaluation to achieve the most economical plan of operation. This does not mean that additions to these older systems cannot be made that utilize efficient variable-speed pumping.

There are several rules that can be followed that may help the planning for such complex installations:

1. Do not install variable-speed pumps centrally with constant-speed pumps operating on their discharges unless the constant-speed pumps are located in chiller or building bypasses.

2. Variable-speed pumps can be installed in the discharge piping from existing constant-speed pumps if (a) the variable-speed pumps are controlled properly with differential pressure controllers, and (b) a bypass is provided around the variable-speed pumps to accommodate the constant-speed pumps under reduced load conditions (see Fig. 15.4*c*).

29.4.5 Converting to distributed pumping

Some existing campus systems that contain tertiary pumping can be converted to distributed pumping by eliminating existing secondary pumps. This requires extensive evaluation, but it may pay off in great savings in energy and may eliminate the need for additional chillers that are required because of greater system flows but not because of additional tonnage on the system.

If the coil control valves are to be converted at the same time from three-way to two-way design, a sizable amount of money may be saved by the reduction in maximum differential pressure across these valves that is achieved by distributed pumping.

29.4.6 Combining multiple-chiller plants

There are a number of small colleges and other campus-type installations that have chillers in individual buildings with a total tonnage much greater than that required by the entire system. By combining the chiller plants into one system, better standby capacity will be provided, and the running chillers will operate at a much higher efficiency. *A study should be made to consider combining the chillers, as was described in Fig. 14.7.* Appreciable energy savings may be achieved, along with greater flexibility in the selection of operating chillers.

29.5 Summation

A great amount of energy can be saved through the use of contemporary digital electronic procedures on existing HVAC water systems. The preceding examples cover only a small part of all the existing chilled and hot water systems that should be retrofitted. Reiterating, systems should be evaluated if they depend on any of the mechanical devices listed earlier to control system pressure or to move water through the various components of the system.

At present, the program for changing of refrigerants in chillers offers opportunities to update and improve the pumping efficiency of these systems. When an analysis is made of the pumping programs, it is often found that reprogramming the chillers will achieve a lower kilowatt per ton figure that yields great energy savings. Older chiller installations accommodated chilled water pumping systems that reduced the efficiency of the chillers as well as the pumping. The effort to improve overall system efficiency should, therefore, include evaluation of the chiller operation as well as that of the pumping system.

30

Summary of HVAC Energy Evaluations

30.1 Review of Efficiencies

Much has been offered here in the design and operation of HVAC water systems. Configuration of these systems as well as methods of operation that have been provided should aid in achieving optimal energy consumption for them. Although no final design can be offered for a specific system, a number of equations have been delineated for the various efficiencies of an HVAC water system. These equations have been kept at the level of algebra, not higher mathematics, in order to render easily the conversion of this information to personal computer programs that fit the language and format most familiar to the reader. As mentioned in Chap. 1, computer programs are now available that will accommodate almost any equation.

It is obvious that this book is but a guidepost in the endeavor to produce more efficient HVAC water systems. Some of the efficiency equations included are for equipment, whereas others are for part or all of a water system. The individual equipment efficiencies are for boilers, chillers, motors, pumps, and variable-speed drives. All the actual efficiencies for this equipment are determined, to a large extent, by their manufacturers.

30.2 System Efficiencies

Two efficiencies are included that are composites of equipment efficiencies; these are wire-to-water efficiency for pumping systems and overall kilowatts per ton for chilled water plants. The equations are repeated here:

$$\text{Wire-to-water efficiency} = \frac{Q \cdot H}{53.08 \cdot \text{kW}} \qquad (10.2)$$

$$\text{Chiller plant kW/ton} = \frac{\Sigma\text{equipment kW}}{\text{tons of cooling on system}} \qquad (14.4)$$

One other system efficiency is for the use of energy in a hot or chilled water system, and this is explained in Chap. 8:

$$\text{System efficiency } WS_\eta = \frac{K_e \cdot 100\%}{K_i} \qquad (8.3)$$

Alternate pumping efficiencies that may be easier to use in system analysis are as follows. For chilled water:

$$\text{kW/100 tons} = \frac{0.452 \cdot H}{P_\eta \cdot E\eta \cdot \Delta T \,(°\text{F})} \qquad (8.4)$$

or:

$$\text{kW/100 tons} = \frac{2400 \cdot \Sigma\text{pump kW}}{\text{gal/min} \cdot \Delta T \,(°\text{F})} \qquad (8.5)$$

For hot water:

$$\text{kW/1000 mbh} = \frac{23.48 \cdot H}{P_\eta \cdot E_\eta \cdot \Delta T \,(°\text{F}) \cdot \gamma} \qquad (8.6)$$

or

$$\text{kW/1000 mbh} = \frac{124{,}700 \cdot \Sigma\text{pump kW}}{\text{gal/min} \cdot \Delta T \,(°\text{F}) \cdot \gamma} \qquad (8.7)$$

The actual equations that should be used on a specific installation depend on the instrumentation that is available to provide the needed analog signals.

30.3 Purpose of Efficiencies

All these equations have two purposes: (1) to aid the designer in the development of efficient water systems and (2) to provide the manager of HVAC systems with operating tools that ensure optimal system operation.

The most important efficiency in this book for chilled water systems is that for chiller plant kilowatts per ton (Eq. 14.4). This efficiency gives plant operators and managers an overall insight as to how efficient the plant operation is. The cost of this instrumentation is relatively small for medium and large chiller plants. More can be gained from this efficiency than from any other system or equipment efficiency for chilled water operations.

Not all these equations will be useful for every HVAC water system. The size and economics of each application will determine their value. Systems with high energy costs will justify their use more than systems with low energy costs. It is in the province of the designer to determine which of them will aid in the design and operation of specific HVAC water systems. Obviously, the first cost of the instrumentation and electronics will have a bearing on which efficiencies are utilized for a particular installation.

Generally, it is not feasible to have one overall efficiency for an HVAC water system, since there are several uses of energy in such a system. At this date, there does not appear to be any interest in such an efficiency. Also, some systems use several sources of energy, which makes difficult the development of an overall efficiency. The objective of the designer now should be to determine which of these expressions of efficiency will establish and sustain the operating efficiency of HVAC water systems.

30.4 Sustained System and Equipment Efficiencies

Sustaining operating efficiency is as important as the original equipment or system efficiencies. In the past, the difficulty in maintaining analog instrumentation often resulted in loss of the equipment that indicated efficiency of operation. Sometimes the instrumentation was not operator friendly, so it was neglected and not maintained. Contemporary digital instrumentation is relatively maintenance-free; with proper instructions at the beginning, this instrumentation should provide continued indication of operating conditions for the operators and managers of HVAC energy plants. The great amount of data storage capacity of contemporary computer systems enables operators to maintain detailed records on operating parameters such as coefficient of performance, kilowatts per ton, and wire-to-water efficiency; sustained high efficiencies are now achievable through the interpretation of this information.

30.5 Summary

The emphasis in this book has been on making available the various expressions of efficiency in a form useable to the designer and system operator. The elimination of older types of water system flow control such as balance valves, pressure-regulating valves, and most crossover bridges should be urged continuously to direct the designer toward the use of digital processes that aid in the rapid design of high-quality HVAC water systems. It is obvious from all these simple algebraic equations that they are tools that can be inserted easily into computer programs to aid the efficient design and operation of HVAC water systems.

Every energy-consuming device that is proposed for an HVAC water system should be evaluated to see if it can be eliminated by changing the water system design. The information, software, design expertise, and equipment are now available to achieve highly efficient water systems for the HVAC industry.

Abbreviations

Following are the abbreviations used in this book. An attempt has been made to use the standards of the pumping industry where possible.

ASCII	American Standard Code for Information Interchange
ASHRAE	American Society of Heating, Refrigerating and Air-Conditioning Engineers
ASME	American Society of Mechanical Engineers
ASTM	American Society for Testing and Materials
ANSI	American National Standards Institute
ac	Alternating electric current
bhp	Brake horsepower required by a pump
bhp_{bv}	Brake horsepower consumed by a balance valve
bhphr	Energy consumed by a pump operating at 1 bhp for 1 hour
Btu	British thermal unit (energy equal to 778.2 ft · lb or the amount of heat required to raise 1 lb of water 1°F at 60°F)
BAS	Building automation system
Binary	Any process with two possible conditions, i.e., a digital output or a two-position switch
C	(1) Williams and Hazen constant or (2) temperature, degrees Celsius
cfs	Flow, in ft³/s
CAD	Computer-aided design
COP	Coefficient of performance, work done divided by work applied
cP	Centipoise, a measure of absolute viscosity
c_p	Specific heat of water at constant pressure
γ	Specific weight of water, in lb/ft³

D	Inside diameter of pipe in feet
d	Inside diameter of pipe in inches or diameter of pump impeller in inches
dc	Direct electric current
ϵ	Roughness parameter
$E\eta$	Electrical efficiency of (1) a motor or (2) a motor and a variable-speed drive
EDR	Equivalent direct radiation
EER	Energy efficiency ratio
EFKW	Equivalent gas fuel in cubic feet of gas per kilowatt
EMI	Electromagnetic interference
EMS	Energy management system
ETL	Electrical Testing Laboratories
f	Colebrook friction factor
fps	Velocity of water in feet per second
ft	foot of measure, an English unit; in this book 12 in in length. Also, 1 ft of pressure (0.433 lb/in^2 at 60°F)
F	Temperature, degrees Fahrenheit
gal/min	U.S. gallons per minute
g	Universal acceleration constant, 32.17 ft/s^2 at sea level
HVAC	Heating, ventilating, and air-conditioning
H	Total head of a system in feet of water
h	Total head of a pump
HI	Hydraulic Institute
hp_v	Horsepower consumption for part of a water system
H_v	Pressure drop in feet through a part of a water system such as a balance valve
H_{ef}	Heating effect, percent, for chiller kilowatts
h_f	Friction head in feet of water
h_g	Gauge pressure in feet of water
Hz	Hertz (cycles per second)
h_v	Velocity head in feet of water
IEEE	Institute of Electrical and Electronics Engineers
IKW	Input kilowatts to a water system
I-P	Inch-pound (English) system of units as opposed to SI
K	Resistance coefficient for valves and pipe fittings
K_e	Useful energy of a water system
K_i	Energy input to a water system

kW	1000 watts
kWh	Kilowatthour
kWh$_{bv}$	kWh lost through a balance valve
L	(1) Length of pipe in hundreds of feet or (2) liters of volume
lb	One pound of weight
m	Meters of length
mA	Milliampere signal
MBH	Thousands of Btu transferred per hour
min	One minute of time
NEMA	National Association of Electrical Manufacturers
NIST	National Institute of Standards and Technology
NPSHA	Net positive suction head available
NPSHR	Net positive suction head required
N_s	Specific speed of a pump
η	Subscript for any expression of efficiency
psia	Absolute pressure in pounds per square inch
psig	Gauge pressure in pounds per square inch
P_a	Atmospheric pressure in feet of water
ΔP	Differential pressure in feet or pounds per square foot
P_e	Atmospheric pressure, in lb/in^2
$P\eta$	Pump efficiency as a decimal
P_f	Friction of suction pipe in NPSH equations
PF	Power factor in electrical operations
pH	Index of acidity or alkalinity of a fluid, where 7.0 is neutral
PkW	Electrical power required by (1) a pump and motor or (2) a pump, motor, and variable-speed drive
PRV	Pressure reducing valve
P_s	Static head in NPSH equations
P_{vp}	Vapor pressure of water in feet
Q	Pump or system flow in U.S. gallons per minute
Q_v	Flow in gallons per minute through a part of a water system such as a balance valve
R	Reynolds number
RFI	Radio-frequency interference
rev/min	Speed in revolutions per minute
S'	Pump speed, in rev/min
s	Specific gravity of a liquid
Ton	Time rate of cooling equal to 12,000 Btu/h

Ton-hour	Quantity of thermal energy in tons absorbed or rejected in 1 hour
ΔT	Differential temperature, in °F
μ	Absolute viscosity in centipoise or lb · ft/s^2
μ	Kinematic viscosity in centistokes or ft^2/s
UBC	*Uniform Building Code*
V	Velocity of water in feet per second
V_s	Volume of a water system in U.S. gallons
VFD	Variable-frequency drive
VHS	Vertical hollow-shaft motor
VSS	Vertical solid-shaft motor
ω	Specific volume, in ft^3/lb
whp	Water horsepower (useful work to a water system)
WK^2	Rotational inertia of an electric motor rotor
WR^2	Rotational inertia of a pump rotor or any driven member
$WS\eta$	Water system efficiency
WWE	Wire-to-water efficiency in percent
w_2	Centrifugal force vector for a pump impeller
Z	Static head of a water system

Terms and Nomenclature

The following list of terms and nomenclature has been compiled from this book as well as from field experience with pumps and pumping systems.

There are two very complete lists of terms and nomenclature that provide much additional information. These are

American National Standard for Centrifugal Pumps, ANSI/H11.15-1994, Hydraulic Institute, Parsippany, N.J., 1994.
ASHRAE Terminology of Heating, Ventilation, Air Conditioning, and Refrigeration, ASHRAE, Atlanta, Georgia.

Absolute viscosity   Absolute or dynamic viscosity is the measurement of the force per unit area required to produce a unit velocity between two parallel planes of unit area that are a unit distance apart.

Absorption chiller   A chiller that utilizes thermal energy and absorption to produce chilled water.

Accuracy   Degree of conformity of an indicated value to a recognized acceptable standard value.

Affinity laws   The basic laws that provide the relationships between pump speed, impeller diameter, flow, head, and power required for centrifugal pumps and fans.

Air entrainment   The inclusion of air in a body of water.

Assembly, rotor   The rotating assembly of either a pump or a motor.

Assembly, stator   The fixed assembly of an electric motor.

Alternation   The procedure used in selecting lead and lag pumps for pumps operating in parallel.

Axial-flow pump   A centrifugal pump in which the water flows parallel to the pump shaft.

Backflow preventer   A special double check valve assembly that prevents flow of nonpotable water back into domestic water systems. These systems must bear the approval of an inspection agency such as Underwriters Laboratories.

Backpressure regulator   A pressure regulator that senses the pressure in a water system and maintains that pressure at a preset point by flowing water from that system.

Barrel or can   An assembly for conducting water into an axial-flow pump.

Base or baseplate   The metal member that supports a pump and its driver.

Bearing, inboard   The bearing nearest the coupling of a double-suction pump.

Bearing, outboard   The bearing farthest from the coupling of a double-suction pump.

Bearing, thrust   The bearing that withstands the linear thrust of a volute-type pump.

Bearing, radial   The bearing that withstands the radial thrust of an end-suction pump.

Bell, suction   The inlet section that directs the flow into an axial-flow pump.

Balance valve   A valve used to eliminate excess pressure in a water circuit.

Boiler horsepower   A method of rating boilers where one boiler horsepower is equal to 33,475 Btu/h or 34.5 pounds of steam from feedwater at 212°F to 0 psig steam at 212°F.

Bowl   The enclosure in which the impeller of an axial-flow pump rotates.

Bracket (frame)   The metal support for a flexibly coupled end-suction pump and its shaft.

Brake horsepower   The work applied to a pump shaft.

Bridge, crossover   A piping arrangement designed to balance external pressures on a portion of a hot or chilled water system.

Bushing, stuffing box   A replaceable sleeve or ring placed in the end of the stuffing box.

Bushing, throttle   A stationary ring or sleeve in the gland of a mechanical seal subassembly to restrict leakage in the event of seal failure.

Bypassing   The passage of water between the impeller and the casing ring. The amount of bypassing determines the efficiency of the pump's operation.

Cap, bearing   The removable cover for a bearing.

Carry-out   A pump is said to have "carried out" when the pump is operating out at the right-hand end of its head-capacity curve at a very poor efficiency.

Casing   The stationary body of a pump that houses the impeller (same as the bowl for an axial-flow pump).

Centrifugal pump   Any pump that uses the centrifugal force of the included liquid to impart pressure to that liquid.

Churn   A pump is said to be in "churn" when it is operating at the no-flow or shutoff head condition.

Circulator   A name given to a small in-line pump.

Closed system   A water system that has all its parts under a pressure greater than atmospheric pressure.

Condenser   A heat exchanger that cools hot gas refrigerant.

Coupling, shaft   The assembly connecting the pump shaft to the driver shaft.

Coupling, spacer type   A shaft coupling with a removable spacer that allows back pullout of the bracket or mechanical seal of an end-suction pump or removal of the seal of an axial-flow pump without removal of the motor.

Cup, grease   A container for the grease of a grease-lubricated pump.

Cyclone separator   A mechanical device that uses centrifugal force to eliminate solid material from a liquid stream.

Design load   The computed load on an HVAC water system in Btu/h or tons of cooling.

Diffuser   The piece adjacent to the impeller exit on an axial-flow pump.

Distributed pumping   A procedure for locating distribution pumps in buildings in lieu of in the central plant.

Distribution system   The part of a chilled or hot water system that moves the water from the point of generation to the point of use.

District heating or cooling   The name applied to centralized heating or cooling.

Double-suction pump   A volute-type centrifugal pump that has a double-suction impeller located between inboard and outboard bearings.

Driver   The motor, engine, or turbine that operates a pump.

Economizer   A mechanical or control device that increases the utilization of energy.

Eductor   A device that uses one liquid stream to move another liquid stream.

Elbow, discharge   The pipe elbow that may be attached to a pump discharge.

Elbow, suction   The pipe elbow that may be attached to a pump suction.

End bell or cover   The ends of an electric motor.

End-suction pump   A volute-type single-suction centrifugal pump that receives the water at the end of the volute.

Evaporator   A heat exchanger that cools a liquid with cold refrigerant.

Evaporative cooler   A mechanical device that lowers a fluid temperature by evaporating moisture.

Frame   See *Bracket.*

Gland   A follower that compresses packing in a stuffing box or retains the stationary element of a mechanical seal.

Guard, coupling   A protective shield over a shaft coupling.

Head, discharge   The top assembly of an axial-flow pump that supports the driver and the pump assembly through which the water leaves the pump.

Housing, bearing   The body in which a bearing is mounted.

Housing, stator   The body in which an electric motor stator is mounted.

Impeller   The bladed member of a pump that imparts the principal forces to the liquid.

Impeller nut   The nut that holds the impeller on the pump shaft.

In-line pump   A pump mounted directly in the pipeline.

Key, impeller   A metal device that prevents rotation of the impeller on the pump shaft.

Kinematic viscosity   Absolute or dynamic viscosity divided by the specific gravity of the liquid.

Laminar flow   Streamline flow in a fluid stream with Reynolds numbers less than 3000.

Lateral setting   Often just called the "lateral," this is the clearance between the impeller and bowl on axial-flow pumps. It can be adjusted to the desired clearance that will provide the greatest efficiency.

Mixed-flow pump   An axial-flow pump with impellers of high specific speed for high-volume and moderate-head applications.

Open system   A water system that has part of its system exposed to atmospheric pressure such as a cooling tower on a condenser water system.

Overshoot   A control term indicating the amount that the process exceeds the set point.

Packing   The flexible material used to provide a seal around the shaft in the stuffing box.

Pipe, column   The vertical support pipe of axial-flow pumps.

Pressure gradient   A graphic representation of the pressure in parts of a water system.

Process   The actual value such as flow or pressure being maintained in a control loop.

Positive-displacement pump   A pump that moves a liquid by means of a continuously moving cavity.

Power factor The ratio of kilowatts to kilovolt-amperes for an electric device.

Primary A piping circuit for hot and chilled water that includes all the parts of those systems.

Primary-secondary A piping circuit for hot and chilled water systems separating the chillers and boilers from the distribution system. It is also used to separate cooling towers from chillers or process load.

Primary-secondary-tertiary A primary-secondary circuit with a tertiary third set of pumps in the buildings or zones

Propeller See *Impeller.*

Propeller pump A very high specific speed pump with an open impeller for high-volume, low-head applications.

Reciprocating pump A positive-displacement pump that has a piston that moves back and forth in a cylinder to achieve the pumping.

Regenerative turbine A pump that moves the liquid peripherally around the impeller, which is closely fitted to the casing.

Repeatability The variation in outputs for an instrument or procedure for the same input.

Retainer, bearing A bracket used to support an open-line-shaft bearing on axial-flow pumps.

Reynolds number A dimensionless number named after its originator to evaluate the characteristics of a fluid stream (see App. C).

Ring, bowl A replaceable wearing ring installed in the bowl of an axial-flow pump to provide a running fit with the impeller.

Ring, casing A replaceable wearing ring installed in the casing of a volute-type pump to provide a running fit with the impeller.

Ring, impeller A replaceable wearing ring installed on an impeller to provide a running fit with a casing ring.

Ring, lantern An annular ring in a stuffing box to provide a liquid seal around the shaft and to lubricate the packing.

Rotor The rotating assembly of a pump or an electric motor.

Seal, mechanical A device to prevent the flow of liquid around a shaft.

Seal, rotating element That part of a mechanical seal which rotates with the pump shaft.

Seal, stationary element That part of a mechanical seal which remains stationary in the pump.

Set point The desired value to be maintained in a control loop.

Shaft The cylindrical element of any pump that transmits power from the pump driver to the impeller.

Shaft, enclosed line The type of axial-flow pump construction where an inner tube separates the shaft and shaft bearings from the liquid being pumped; the tube is filled with grease, oil, or clean water.

Shaft, open line The type of axial-flow pump construction where the shaft and shaft bearings are immersed in the liquid being pumped.

Shaft, two-piece The top drive shaft of an axial-flow pump that enables removal of the driver without disassembling other parts of the pump.

Shutoff head The pump head at the no-flow or churn condition.

Sleeve, shaft A cylindrical piece fitted over the shaft to protect the shaft from wear or to locate the impeller on the shaft.

Slinger A disk rotating with the shaft to prevent migration of water down that shaft.

Sparger A device that reduces noise and vibration when steam is introduced into water.

Specific heat c_p The amount of heat required to raise 1 lb of water 1°F at constant pressure.

Specific speed A design speed expressing a relationship between actual pump speed, flow, and head.

Static pressure The pressure at the bottom of a water system when at rest.

Steam quality The amount of moisture in steam expressed as a percentage of the weight of the vapor and water mixture.

Stuffing box The portion of the pump casing that houses either the packing or the mechanical seal.

Submergence The depth at which an axial-flow pump suction must be immersed for satisfactory operation without cavitation.

Sump A container used to receive water from a cooling tower or other processes.

System head area A graphic figure that describes the flow-head relationships for a water system.

System head curve The curve that defines the flow-head relationships for a water system that is uniformly loaded.

Torque A turning force acting through a radius and measured in pound-feet.

Turbine pump An axial-flow pump of moderate to high specific speed with either open or closed impellers.

Turbulent flow Nonstreamlined flow in a fluid stream with a Reynolds number greater than 4000.

Velocity head The kinetic energy of water flowing in a pipe, conduit, or channel; equal to $V^2/2g$.

Volute pump A pump in which the liquid is collected peripherally from the impeller and directed toward a discharge connection.

Vortexing   The entrainment of air in water through the rotation of that body of water.

Water horsepower   The useful power transmitted to a water system.

Wire-to-water efficiency   The efficiency for a pumping system that is derived by dividing the work done on a water stream by that pumping system by the work applied to that system.

Zone   A part of the distribution of a hot or chilled water system.

Glossary of Equations

Following are the equations of this book by chapter and page.

Chapter 2

Pressure relationships:

$$\text{psia} = \text{psig} + P_e \qquad (2.1)$$

where psia = absolute pressure, lb/in^2
psig = gauge pressure, lb/in^2
P_e = atmospheric pressure, lb/in^2

Thermal equivalents for this book:

$$1 \text{ Btu} = 778.2 \text{ ft} \cdot \text{lb}$$

$$1 \text{ bhp} = 33,000 \text{ ft} \cdot \text{lb/min}$$

$$1 \text{ bhph} = 2544 \text{ Btu/h}$$

$$= 0.746 \text{ kWh}$$

$$1 \text{ kWh} = 1.341 \text{ bhp}$$

$$= 3412.0 \text{ Btu/h}$$

Viscosity equivalents:

$$\upsilon = 6.7197 \cdot 10^{-4} \cdot \left(\frac{\mu}{\gamma}\right) \qquad (2.2)$$

where v = kinematic viscosity, in ft²/s
   μ = absolute viscosity, in cP
   γ = specific weight, in lb/ft

$$v \text{ (ft}^2\text{/s)} = 1.0764 \cdot 10^{-5} \cdot v \text{ (cs)} \tag{2.3}$$

Pump gallons per minute for a glycol solution:

$$\text{Pump gal/min} = \frac{0.125 \cdot \text{Btu/h}}{c_p \cdot \Delta T \text{ (°F)} \cdot \gamma \cdot s} \tag{2.4}$$

where c_p = specific heat of liquid at constant pressure
   γ = specific weight of water, lb/ft³
   s = specific gravity of the solution

Chapter 3

Hydraulic radius of pipe:

$$\text{Hydraulic radius} = \frac{\text{area}}{\text{inside circumference}} = \frac{d}{4} \tag{3.1}$$

where d is the inside diameter of the pipe in inches.

The Bernoulli theorem:

$$H = Z + h_g + h_v \tag{3.2}$$

where H = total system head, ft
   Z = static head, ft
   h_g = system pressure, ft
   h_v = $V^2/2g$, velocity head, ft

Darcy-Weisbach equation:

$$H_f = f \cdot \frac{L}{D} \cdot \frac{V^2}{2g} \tag{3.3}$$

where H_f = friction loss, ft of liquid
   L = pipe length, 100s of ft
   D = average inside diameter, ft
   f = friction factor

The friction factor f is usually derived from the Colebrook equation:

$$\frac{1}{\sqrt{f}} = -2 \log_{10}\left(\frac{\epsilon}{3.7D} + \frac{2.51}{R\sqrt{f}}\right) \tag{3.4}$$

where R = Reynolds number
  ϵ = Absolute roughness parameter (typically 0.00015 for steel pipe)

Williams and Hazen equation:

$$H_f = 0.002083 \cdot L \cdot \left(\frac{100}{C}\right)^{1.85} \cdot \frac{gal/min^{1.85}}{d^{4.8655}}$$ (3.5)

where L = pipe length, 100s of ft
  C = roughness factor
  d = inside pipe diameter, in

Reynolds number equation:

$$R = \frac{V \cdot D}{v}$$ (3.6)

where V = velocity, ft/s
  D = pipe diameter, ft
  v = kinematic viscosity of liquid, ft²/s

Friction loss for pipe fittings:

$$H_f = K \cdot \frac{V^2}{2g}$$ (3.7)

where K = friction coefficient determined by tests
  $V^2/2g$ = velocity head, ft

Special friction loss equations for cast iron tapered fittings:
  For tapered reducing fittings:

$$K = 0.8 \sin\frac{\theta}{2}\left(1 - \frac{d_1^2}{d_2^2}\right)$$ (3.8)

For tapered increasing fittings:

$$K = 2.6 \sin\frac{\theta}{2}\left(1 - \frac{d_1^2}{d_2^2}\right)^2$$ (3.9)

where θ = total angle of reduction or increase
  d_1 = smaller diameter
  d_2 = larger diameter

Friction loss for steel increaser or reducer fittings:

$$H_f = \frac{(V_1^2 - V_2^2)}{2g}$$ (3.10)

where V_1 = velocity in the smaller pipe
V_2 = velocity in the larger pipe

Chapter 4

Feet of head per pound of pressure:

$$\text{Feet of head per pound of pressure } H = \frac{144}{\gamma} \qquad (4.1)$$

Specific speed of a pump:

$$N_s = \frac{S \cdot \sqrt{Q}}{h^{3/4}} \qquad (4.2)$$

where S = pump speed
Q = pump flow
h = pump head, ft

Minimum flow for a pump:

$$\text{Minimum flow } Q_m = \frac{317.2 \cdot \text{pump bhp}}{\Delta T \, (°\text{F}) \cdot \gamma} \qquad (4.3)$$

Chapter 6

Affinity laws
  For a fixed-diameter impeller:

1. The pump capacity varies directly with the speed:

$$\frac{Q_1}{Q_2} = \frac{S_1}{S_2} \qquad (6.1)$$

2. The pump head varies as the square of the speed:

$$\frac{h_1}{h_2} = \frac{S_1{}^2}{S_2{}^2} \qquad (6.2)$$

3. The pump brake horsepower required varies as the cube of the speed:

$$\frac{\text{bhp}_1}{\text{bhp}_2} = \frac{S_1{}^3}{S_2{}^3} \qquad (6.3)$$

For a fixed pump speed:

1. The pump capacity varies directly with the impeller diameter:

$$\frac{Q_1}{Q_2} = \frac{d_1}{d_2}$$ (6.4)

2. The pump head varies as the square of the impeller diameter:

$$\frac{h_1}{h_2} = \frac{d_1{}^2}{d_2{}^2}$$ (6.5)

3. The pump brake horsepower required varies as the cube of the impeller diameter:

$$\frac{bhp_1}{bhp_2} = \frac{d_1{}^3}{d_2{}^3}$$ (6.6)

Net positive suction head:

$$NPSHA \geq NPSHR$$ (6.7)

where $NPSHA$ = net positive suction head available
$NPSHR$ = net positive suction head required

NPSHA formula
For water temperatures up to 85°F:

$$NPSHA = P_a + P_s - P_{vp} - P_f \text{ in feet of head}$$ (6.8)

where P_a = atmospheric pressure in feet at the installation altitude
P_s = static head of water above pump impeller (this is negative if the water level is below the pump impeller)
P_{vp} = vapor pressure of water in feet at operating temperature
P_f = friction of suction pipe, fittings, and valves in feet of head

For precise calculation of $NPSHA$ at any temperature and altitude:

$$P_a = \frac{144 \cdot P_e}{\gamma}$$ (6.9)

$$NPSHA = \frac{144 \cdot P_e}{\gamma} + P_s - P_{vp} - P_f$$ (6.10)

where P_e = atmospheric pressure, in lb/in^2
γ = specific weight of water at pumping temperature, in lb/ft^2

Water horsepower:

$$\text{whp} = \frac{Q \cdot h \cdot s}{3960} \tag{6.11}$$

where Q = flow, in gal/min
h = head, in feet of water
s = specific gravity of liquid

Pump horsepower:

$$\text{bhp} = \frac{\text{whp}}{P_\eta} \tag{6.12}$$

where P_η = pump efficiency as a decimal

Pump kilowatts:

$$\text{PkW} = \frac{\text{bhp} \cdot 0.746}{E_\eta} \tag{6.13}$$

where E_η = efficiency of the motor or the wire-to-shaft efficiency of the motor and the variable-speed drive as a decimal

or

$$\text{PkW} = \frac{Q \cdot H \cdot s}{5308 \cdot P_\eta \cdot E_\eta} \tag{6.14}$$

Chapter 7

Motor synchronous speed:

$$\text{rev/min} = \frac{60 \cdot 2 \cdot \text{frequency}}{\text{number of poles}} \tag{7.1}$$

Equivalent work:

$$1 \text{ hp} = 33,000 \text{ ft} \cdot \text{lb/min} \tag{7.2}$$

Power factor for a three-phase electric motor:
   Full load torque:

$$\text{lb-ft} = \frac{\text{bhp} \cdot 5250}{\text{max rev/min}} \tag{7.3}$$

$$PF = \frac{\text{watts applied}}{\sqrt{3} \cdot \text{volts} \cdot \text{amps}} \tag{7.4}$$

Electric motor efficiency:

$$E_\eta = \frac{\text{hp output} \cdot 746}{\text{watts input}} \tag{7.5}$$

Heat release from variable-frequency drive:

$$\text{Btu/h} = \text{max. kW of drive} \cdot 3412 \cdot (1-E_\eta) \tag{7.6}$$

where: 3,412 is the thermal equivalent of a KW in Btu per hour

E_η is the efficiency of the drive as a fraction at maximum speed.

Chapter 8

Efficient Energy for a Water System, Ke:

$$Ke = \frac{\text{useful frictions ft} \cdot \text{system flow-GPM} \cdot 0.746}{3,960}$$

or:

$$Ke = \frac{\text{useful frictions ft} \cdot \text{system flow-GPM}}{5,308} \tag{8.1}$$

Energy Consumed by a Water System, Ki:

$$Ki = \frac{\text{total system head ft} \cdot \text{system flow-GPM}}{5,308 \cdot P_\eta \cdot E_\eta} \tag{8.2}$$

or

$$Ki = PKW, \text{ the KW input to a pump driver}$$

System Efficiency:

$$WS_\gamma = \frac{K_e}{K_i} \cdot 100\% \tag{8.3}$$

Alternate equations for system energy consumption
In kW/ton for chilled water:

$$\text{kW/100 tons} = \frac{0.452 \cdot H}{P_\eta \cdot E_\eta \cdot \Delta T \, (°F)} \tag{8.4}$$

where H is the system head.
Or

$$kW/100 \text{ tons} = \frac{2400 \cdot \Sigma \text{pump kW}}{\text{gal/min} \cdot \Delta T \, (°F)} \qquad (8.5)$$

In kW/1000 mbh for hot water:

$$kW/1000 \text{ mbh} = \frac{23.48 \cdot h}{P_\eta \cdot E_\eta \cdot \Delta T \, (°F) \cdot \gamma} \qquad (8.6)$$

or

$$kW/1000 \text{ mbh} = \frac{124{,}700 \cdot \Sigma \text{pump kW}}{\text{gal/min} \cdot \Delta T \, (°F) \cdot \gamma} \qquad (8.7)$$

where gal/min = system flow
γ = specific weight of the water in lb/ft^3 at operating temperature

Energy loss for a system element such as a balance valve:

$$hp_v = \frac{Q_v \cdot H_v}{3960 \cdot P_\eta} \qquad (8.8)$$

$$kW_v = \frac{hp_v \cdot 0.746}{E_\eta} \qquad (8.9)$$

or

$$kW_v = \frac{Q_v \cdot H_v}{5308 \cdot P_\eta \cdot E_\eta} \qquad (8.10)$$

where kW_v = kilowatt loss through system element
Q_v = flow in gallons per minute through element
H_v = loss in feet through element

Approximate loss through any element of a water system:

$$kW_v = \frac{Q_v \cdot H_v}{3600} \qquad (8.11)$$

Assumes P_η = 75 percent and E_η = 90 percent
Annual energy lost through any element of a water system:

$$\text{Annual } kW_v = kW_v \cdot \text{annual hours of operation} \qquad (8.12)$$

Additional energy lost in a chiller annually:

$$\text{Added chiller kW} = \frac{\text{Annual kW}_v \cdot \text{average kW/ton}}{3.52} \qquad (8.13)$$

Energy credit to gas-fired hot water systems:
Equivalent fuel per kW, EF/kW:

$$EF/kW = \frac{3412/kW}{B_\eta \cdot 1000 \text{ Btu/ft}^3 \text{ ft}^3 \text{ gas/kW}} \qquad (8.14)$$

where B_η = boiler efficiency as a decimal, and the heating value of natural gas is assumed to be 1000 Btu ft^3

Heating effect in percent H_{ef}:

$$H_{ef} = \frac{EF/kW \cdot \text{fuel cost (¢/ft}^3) \cdot 100}{\text{overall power cost (¢/kW)}} \qquad (8.15)$$

$$= \% \text{ reduction}$$

Chapter 9

HVAC water system capacities in gallons per minute:

$$\text{Hot water system gal/min} = \frac{\text{system Btu per hour}}{500 \cdot \Delta T \, (°F)} \qquad (9.1)$$

$$\text{Chilled water system gal/min} = \frac{\text{system load (tons)} \cdot 24}{\Delta T \, (°F)} \qquad (9.2)$$

(*Note:* These are general calculations. See Chap. 2 for precise formula for calculation of system gallons per minute.)
Calculation of a system head curve for a water system:

$$H_a = H_2 + \left(\frac{Q_a}{Q_1}\right)^{1.90} \cdot (H_1 - H_2) \qquad (9.3)$$

where H_a = system head at any flow Q_a on the system head curve
H_1 = system head at maximum flow Q_1 in the water system
H_2 = constant head on the water system
Q_a = flow at any point on the system head curve

Static pressure in psig for a water system in a building:

$$\text{Static pressure} = \frac{Z_b - Z_p + Z_c}{144/\gamma} \text{ psig} \qquad (9.4)$$

where Z_b = elevation above sea level in feet at top of tallest building
Z_c = cushion in feet of head on top of remote buildings
Z_p = elevation above sea level in feet of operating floor of central energy plant
γ = specific weight of water at operating temperature

For distributed pumping:

$$\text{Static pressure} = \frac{Z_b - Z_p + H_d + Z_c}{144/\gamma} \qquad (9.5)$$

where Z_b = elevation above sea level of building that, combined with its system supply friction, determines the static head for the system
H_d = system supply friction in feet between the central plant and the remote buildings; the building that requires the highest static pressure, a combination of the system friction loss and the static pressure of that building, determines the static pressure for the system.

Chapter 10

Affinity laws for equivalent pump operating point:

$$Q_1 = \sqrt{\frac{Q_2{}^2}{h_2} \cdot h_1} \qquad (10.1)$$

where Q_2 and h_2 are the desired flow and head conditions required of the pump, while Q_1 and h_1 are the equivalent operating points on the known pump curve.

Wire-to-water efficiency *WWE:*

$$\text{WWE \%} = \frac{\text{water hp whp} \cdot 0.746 \cdot 100}{\text{input kW}}$$

$$= \frac{Q\ (\text{gal/min}) \cdot H\ (\text{ft})}{53.08 \cdot \text{input kW}} \qquad (10.2)$$

Chapter 11

Cooling tower flow:

$$\text{gal/min} = \frac{\text{total heat of rejection (Btu/h)}}{500 \cdot \Delta T\ (°F)} \qquad (11.1)$$

or

$$\text{gal/min} = \frac{\text{heat of rejection (Btu/ton)} \cdot \text{tons}}{500 \cdot \Delta T \, (\degree\text{F})} \tag{11.2}$$

Annual heat of rejection in Btu from cooling towers:

$$\text{Annual rejection} = \text{heat/ton} \cdot \text{tons} \cdot \text{hours} \tag{11.3}$$

Chapter 12

Process water flow in gallons per minute:

$$\text{gal/min} = \frac{\text{heat transferred (Btu/h)}}{8.02 \cdot \gamma \cdot \Delta T \, (\degree\text{F})} \tag{12.1}$$

where γ is the specific weight of water at cooling tower return water temperature.

Chapter 13

Pressure energy lost to open tank from pressurized system:

$$\text{kW lost} = \frac{\text{gal/min} \cdot \text{head} \cdot 0.746}{3960 \cdot P_\eta \cdot E_\eta}$$

$$= \frac{\text{gal/min} \cdot \text{head}}{5308 \cdot P_\eta \cdot E_\eta} \tag{13.1}$$

where head is the static pressure of the water system in feet.

Chapter 14

Coefficient of performance *COP*:

$$\text{COP} = \frac{\text{useful energy acquired}}{\text{energy applied}} \tag{14.1}$$

Energy efficiency ratio *EER*:

$$EER = \frac{\text{Btu/h net cooling effect}}{\text{Wh applied}} \tag{14.2}$$

or

$$EER = 3.412 \cdot \text{COP} \tag{14.3}$$

Total chilled water plant kilowatts per ton:

$$kW/ton = \frac{\Sigma plant\ consumptions\ in\ kW}{tons\ of\ cooling\ on\ system} \qquad (14.4)$$

Tons of cooling on a system:

$$tons = \frac{Q \cdot (T_2 - T_1)}{24} \qquad (14.5)$$

Chapter 17

Ton-hours stored:

$$Ton\text{-}hours = \frac{V_s \cdot (T_2 - T_1)}{1439} \qquad (17.1)$$

where V_s = volume of system, in gallons
T_2 = return water temperature, °F
T_1 = system supply temperature, °F

Chapter 19

Hot water system flow:

$$gal/min = \frac{125 \cdot system\ mbh}{\gamma \cdot \Delta T\ (°F)} \qquad (19.1)$$

where γ is the specific weight of water at operating temperature, in lb/ft^3.

Chapter 21

Water flow in medium- and high-temperature water systems:

$$lb/h = \frac{system\ Btu/h}{\Delta T\ (°F) \cdot c_p} \qquad (21.1)$$

where c_p is the specific heat of water at maximum operating temperature.

$$Pump\ gal/min = \frac{0.125 \cdot system\ Btu/h}{\gamma \cdot \Delta T\ (°F) \cdot C_p} \qquad (21.2)$$

Chapter 22

Discharge pressure for condensate return system
For pumping to gravity return system:

$$\text{Pressure} = \frac{h_f + \text{static lift}}{144/\gamma} \text{ psig} \qquad (22.1)$$

where h_f is the friction loss in feet of head for the piping between the condensate return system and the point of delivery of the condensate. For pumping to a boiler or deaerator:

$$\text{Pressure} = \frac{h_f + \text{static lift}}{144/\gamma} + I_p \text{ psig} \qquad (22.2)$$

where I_p is the internal pressure in psig for the boiler or deaerator.
Boiler feed pump capacity
Boiler rated in pounds of steam per hour:

$$\text{gal/min} = \frac{\text{boiler (lb steam/h)}}{7.22 \cdot \gamma} \qquad (22.3)$$

Boiler rated in boiler horsepower:

$$\text{gal/min} = \frac{4.78 \cdot \text{boiler capacity (hp)}}{\gamma} \qquad (22.4)$$

Boiler feed pump head:

$$h_b = P_f + P_d + \text{boiler psig} \cdot \frac{144}{\gamma} \qquad (22.5)$$

where P_f = suction piping friction loss in feet
P_b = discharge piping friction loss in feet

NPSH calculations for boiler feed pumps:

$$\text{NPSHA} = P_z - P_f \qquad (22.6)$$

where P_z is the static height difference in feet between the water level in the boiler feed system or deaerator and the suction level of the boiler feed pumps.

Conversion of English Units to SI

The following table lists the conversion factors for transferring physical values from the English units and standards to the International System (SI) of units. This conversion table has been provided because this *Handbook* has been written using English units.

An attempt was made in the HVAC industry to have the conversion of U.S. gallons per minute to liters per minute as a standard; this would have provided an approximate comparable number of 4 to 1 (3.786) for converting U.S. gallons to liters. However, the decision was made to use liters per second or cubic meters per hour as the acceptable SI values for pump capacity.

Pump head should be computed in meters, not kilopascals; if kilopascals are used, the specific gravity of the liquid must be included. This procedure is comparable with using feet of head instead of gauge pressure in psig in English units. Most pump companies provide head-capacity curves in either liters per second or cubic meters per hour and pump head in meters.

From English	Conversion factor	To SI
	Units of Length	
inches	2.54	centimeters
feet	0.3048	Meters
miles	1.609	Kilometers
	Units of Area	
square feet	9.29×10^{-2}	square meters
square inches	6.452	square centimeters
square yard	0.836	square meters

Units of Volume

cubic yards	0.7646	cubic meters
U.S. gallons	3.785	liters
U.S. gallons	3.785×10^{-3}	cubic meters
acre foot	1,233.482	cubic meters

Units of Capacity

gallons per minute (U.S. gal)	0.0631	liters/s
gallons per minute (U.S. gal)	0.2271	m^3/h
ft^3/s	28.32	liters/s
ft^3/s	101.9	m^3/h

Units of Mass (Weight)

pounds force	0.4536	kilograms
tons (short)	907.185	kilograms

Units of Pressure

pounds force per square inch (lb/in^2)	6.8948	kPa (kilopascals)
feet of water (39°F, 4°C)	2.989	kPa
atmosphere (sea level)	101.325	kPa
inches of mercury (32°F, 0°C)	3.386	kPa

Units of Energy, Work, or Heat

Btu	0.252	kilogram-calorie
Btu-heat	1,055.06	joule
footpounds	3.3239×10^{-4}	kilogram-calorie
footpounds	1.3558	joule
ton of cooling	3.517	kW

Units of Power or Rate of Doing Work

water horsepower (whp)	0.746	kilowatts
pump bhp	0.746	kilowatts
pump motor kW	1.0	kilowatts
footpounds per minute	22.59697	mW
boiler horsepower	9.81	kilowatts

Linear Velocity

feet per second	0.3048	meters per second
feet per minute	0.00508	meters per second
miles per hour	1.61	kilometers per hour

Linear Acceleration

feet per second squared	0.3048	m/s^2
universal constant g	9.805	m/s^2

Density or Mass/Unit Volume

pounds per cubic foot	16.02	kilograms per cubic meter

	Units of Torque	
pound-force, ft	1.356	newton-meter

	Specific Heat	
Btu/lb/°F	4.184	kilojoules

	Temperature	
Fahrenheit, °F	(°F-32) $\cdot \frac{5}{9}$	Celsius, °C

	Viscosity	
kinematic, in ft²/s	929.03	cm²/s
kinematic, in ft²/s	$\dfrac{\gamma}{6.7197 \times 10^{-4} \text{ cm}^2/\text{s}}$	

Index

ABOUT THE AUTHOR

James B. Rishel is chairman and founder of Systecon Inc. in Cincinnati, Ohio, a successful manufacturer of pumps and piping for the HVAC industry. One of the leading authorities on pumps and pumping applications in the world, he is a Fellow in ASHRAE, a member of many of ASHRAE's technical committees, and a lecturer for many ASHRAE functions.